Mathematical Modeling in Systems Biology

Mathematical Modeling in Systems Biology

Editor

Pavel Kraikivski

MDPI

Basel • Beijing • Wuhan • Barcelona • Belgrade • Novi Sad • Cluj • Manchester

Editor
Pavel Kraikivski
Academy of Integrated Science
Virginia Polytechnic Institute
and State University
Blacksburg
United States

Editorial Office
MDPI
St. Alban-Anlage 66
4052 Basel, Switzerland

This is a reprint of articles from the Special Issue published online in the open access journal *Entropy* (ISSN 1099-4300) (available at: www.mdpi.com/journal/entropy/special_issues/mathematical_systems_biology).

For citation purposes, cite each article independently as indicated on the article page online and as indicated below:

Lastname, A.A.; Lastname, B.B. Article Title. *Journal Name* **Year**, *Volume Number*, Page Range.

ISBN 978-3-0365-9037-0 (Hbk)
ISBN 978-3-0365-9036-3 (PDF)
doi.org/10.3390/books978-3-0365-9036-3

Contents

About the Editor

Pavel Kraikivski

Pavel Kraikivski is a Collegiate Associate Professor in the Academy of Integrated Science, Division of Systems Biology at Virginia Tech. Dr. Kraikivski has more than 15 years of experience in developing mathematical models to study the dynamic behavior of complex biological systems, especially to investigate molecular mechanisms regulating intracellular transport, cell growth, division, and cell death. He also teaches Integrated Science and Biology Systems courses at Virginia Tech. He teaches students to apply interdisciplinary approaches by giving them problem-oriented exercises that can be successfully attacked only with a high level of creativity and analytical thought.

entropy

MDPI

Editorial

Mathematical Modeling in Systems Biology

Pavel Kraikivski

Academy of Integrated Science, Division of Systems Biology, Virginia Polytechnic Institute and State University, Blacksburg, VA 24061, USA; pavelkr@vt.edu

Mathematical modeling is a key tool used in the field of systems biology to determine the mechanisms with which the elements of biological systems interact to produce complex dynamic behavior. It has become increasingly evident that the complex dynamic behavior of biological systems cannot be understood by intuitive reasoning alone. However, valuable insights with regard to the mechanisms governing the dynamic behavior of biological systems can be revealed through computational experiments by simulating these systems with mathematical models. The eleven contributions of this Special Issue demonstrate how computational and mathematical approaches can simultaneously be used to reveal important aspects of various biological problems.

Ute Deichmann presents a comprehensive analysis of the historical background of pattern formation models that are applied to describe morphogenesis and embryological structure development [1]. The analysis tracks the development of physical–chemical and genome-based pattern formation models and subsequently compares Alan Turing's 1952 reaction–diffusion-based models with more recent models that integrate gene regulatory networks with physical–chemical processes. The article concludes that Turing's models alone are not able to rigorously explain pattern generation in morphogenesis, but that mathematical models combining physical–chemical principles with gene regulatory networks, which govern embryological development, are the most successful in explaining pattern formation in organisms.

An information-based approach to quantify geometrical order in biological organizations using varying levels of information is introduced in the article by Juan Lopez-Sauceda et al. [2]. The approach employs Shannon entropy to measure the quantity of information in geometrical meshes of biological systems. The authors apply their approach to quantify spatial heterogeneity in thirty five biological and non-biological geometric aggregates and conclude that the differential entropy of geometrical organizations is an essential source of information in biological systems.

Steven Frank uses reservoir computing techniques to study how a biological system with an internal, randomly connected network receiving environmental inputs evolves to generate predictive responses [3]. The environmental inputs are generated using a mathematical model that exhibits chaotic dynamics. The biological system that interacts with the environment is represented by a random network reservoir that retains the memory of past inputs. The study quantifies the degree of effectiveness and accuracy with which the biological system predicts future input values, using the internal reservoir states as predictors.

The article by Yolocuauhtli Salazar et al. presents a mathematical model of biomass growth, glucose consumption, and ethanol production by *K. marxianus* yeast strains [4]. The model consists of three coupled, nonlinear first-order ordinary differential equations and ten parameters that can be well-constrained by experimental data. The model is successful in explaining the time-course data of alcoholic fermentation in batch culture for 17 different *K. marxianus* strains, and can be used to accurately predict the evolution of both biomass and ethanol in the system.

The review article by Madhumita Srinivasan, Robert Clarke, and Pavel Kraikivski presents mathematical models of different cell death execution mechanisms that have been

check for updates

Citation: Kraikivski, P. Mathematical Modeling in Systems Biology. *Entropy* **2023**, *25*, 1380. https://doi.org/10.3390/e25101380

Received: 12 September 2023
Accepted: 21 September 2023
Published: 25 September 2023

published over a period of twenty-two years [5]. The authors put forward a hypothesis that cell death can be controlled by a singular, highly integrated cell death decision network that enables cells to choose alternative cell death execution pathways within a single control network of cell death.

Francesco Cordoni's work presents a macroscopic deterministic approximation of microscopic systems, which is represented by a stochastic model for radiation-induced DNA damage kinetics and repair [6]. The approximation is used to compute the distribution of the number of DNA damages that result in cell death. It concludes that the distribution deviates from the Poisson law due to the clustering of the DNA damage.

The study conducted by Om Prakash et al. is related to DNA-based computing, which relies on error control coding techniques [7]. Coding theory is applied to construct a large set of DNA strings that satisfy certain combinatorial constraints. The authors study reversible DNA codes, as well as those of length n, and obtain new DNA codes with improved parameters.

Two articles in this Special Issue focus on the modeling of infection transmission dynamics. The article by Isa Abdullahi Baba et al. presents a fractional-order cholera model that is an extension of the Susceptible–Infected–Recovered epidemic model [8]. The model incorporates the saturated incidence rate to accurately represent the transmission dynamics of the disease. The article by Hosam Alhakami et al. uses a deterministic mathematical model of vector-borne viral plant disease dynamics to train a feed-forward neural network using Levenberg–Marquardt backpropagation algorithm [9]. The neural network is then used to study the implication of fluctuations on natural plant mortality and vector mortality rates.

Mathematical models of biological systems usually describe many interacting components and involve many parameters. Furthermore, it is common that only limited experimental data are available to calibrate the models. Therefore, reliable mathematical models of biological systems can only be developed with rigorous parameter estimation and model validation techniques. Samaneh Gholami and Silvana Ilie propose a parameter estimation method for stochastic discrete models of biochemical networks [10]. The method utilizes finite-difference approximations of the parameter sensitivities and the singular value decomposition of the sensitivity matrix. Several models of biochemical systems are used to demonstrate the advantages of the proposed method.

The article by Christopher Parker, Erik Nelson, and Tongli Zhang presents a computational framework named VeVaPy, which is designed to verify and validate mathematical models comprising many interacting components and parameters [11]. VeVaPy is a publicly available Python library that can help determine which model from the literature is the best for fitting new experimental data. The authors use several hypothalamic–pituitary–adrenal (HPA) axis models from the literature to demonstrate the way in which VeVaPy can help to verify and validate these models against new data: VeVaPy runs the differential evolution parameter optimization algorithm on each model against several novel datasets and ranks the models based on their average cost function value. In their demonstration, two out of five HPA models performed the best in elucidating the novel datasets. Overall, the model validation process is able to operate with significantly less effort when using VeVaPy.

Funding: This research received no external funding.

Conflicts of Interest: The author declares no conflict of interest.

References

1. Deichmann, U. Self-Organization and Genomic Causality in Models of Morphogenesis. *Entropy* **2023**, *25*, 873. [CrossRef] [PubMed]
2. Lopez-Sauceda, J.; von Bülow, P.; Ortega-Laurel, C.; Perez-Martinez, F.; Miranda-Perkins, K.; Carrillo-González, J.G. Entropy as a Geometrical Source of Information in Biological Organizations. *Entropy* **2022**, *24*, 1390. [CrossRef] [PubMed]
3. Frank, S.A. Precise Traits from Sloppy Components: Perception and the Origin of Phenotypic Response. *Entropy* **2023**, *25*, 1162. [CrossRef] [PubMed]

4. Salazar, Y.; Valle, P.A.; Rodríguez, E.; Soto-Cruz, N.O.; Páez-Lerma, J.B.; Reyes-Sánchez, F.J. Mechanistic Modelling of Biomass Growth, Glucose Consumption and Ethanol Production by Kluyveromyces marxianus in Batch Fermentation. *Entropy* **2023**, *25*, 497. [CrossRef] [PubMed]
5. Srinivasan, M.; Clarke, R.; Kraikivski, P. Mathematical Models of Death Signaling Networks. *Entropy* **2022**, *24*, 1402. [CrossRef] [PubMed]
6. Cordoni, F.G. On the Emergence of the Deviation from a Poisson Law in Stochastic Mathematical Models for Radiation-Induced DNA Damage: A System Size Expansion. *Entropy* **2023**, *25*, 1322. [CrossRef]
7. Prakash, O.; Singh, A.; Verma, R.K.; Solé, P.; Cheng, W. DNA Code from Cyclic and Skew Cyclic Codes over $F_4[v]/<v^3>$. *Entropy* **2023**, *25*, 239.
8. Baba, I.A.; Humphries, U.W.; Rihan, F.A. A Well-Posed Fractional Order Cholera Model with Saturated Incidence Rate. *Entropy* **2023**, *25*, 360. [CrossRef]
9. Alhakami, H.; Umar, M.; Sulaiman, M.; Alhakami, W.; Baz, A. A Numerical Study of the Dynamics of Vector-Born Viral Plant Disorders Using a Hybrid Artificial Neural Network Approach. *Entropy* **2022**, *24*, 1511. [CrossRef] [PubMed]
10. Gholami, S.; Ilie, S. Quantifying Parameter Interdependence in Stochastic Discrete Models of Biochemical Systems. *Entropy* **2023**, *25*, 1168. [CrossRef] [PubMed]
11. Parker, C.; Nelson, E.; Zhang, T. VeVaPy, a Python Platform for Efficient Verification and Validation of Systems Biology Models with Demonstrations Using Hypothalamic-Pituitary-Adrenal Axis Models. *Entropy* **2022**, *24*, 1747. [CrossRef] [PubMed]

entropy

MDPI

Review

Self-Organization and Genomic Causality in Models of Morphogenesis

Ute Deichmann

The Jacques Loeb Centre for the History and Philosophy of the Life Sciences, Ben-Gurion University of the Negev, Beer Sheva 84105, Israel; uted@post.bgu.ac.il

Abstract: The debate about what causes the generation of form and structure in embryological development goes back to antiquity. Most recently, it has focused on the divergent views as to whether the generation of patterns and form in development is a largely self-organized process or is mainly determined by the genome, in particular, complex developmental gene regulatory processes. This paper presents and analyzes pertinent models of pattern formation and form generation in a developing organism in the past and the present, with a special emphasis on Alan Turing's 1952 reaction–diffusion model. I first draw attention to the fact that Turing's paper remained, at first, without a noticeable impact on the community of biologists because purely physical–chemical models were unable to explain embryological development and often also simple repetitive patterns. I then show that from the year 2000 and onwards, Turing's 1952 paper was increasingly cited also by biologists. The model was updated to include gene products and now seemed able to account for the generation of biological patterns, though discrepancies between models and biological reality remained. I then point out Eric Davidson's successful theory of early embryogenesis based on gene-regulatory network analysis and its mathematical modeling that not only was able to provide a mechanistic and causal explanation for gene regulatory events controlling developmental cell fate specification but, unlike reaction–diffusion models, also addressed the effects of evolution and organisms' longstanding developmental and species stability. The paper concludes with an outlook on further developments of the gene regulatory network model.

Keywords: reaction–diffusion models in morphogenesis; pattern formation; developmental gene regulatory networks; Alan Turing; Eric Davidson

Citation: Deichmann, U. Self-Organization and Genomic Causality in Models of Morphogenesis. *Entropy* **2023**, *25*, 873. https://doi.org/10.3390/e25060873

Academic Editor: Pavel Kraikivski

Received: 10 May 2023
Revised: 25 May 2023
Accepted: 26 May 2023
Published: 30 May 2023

1. Introduction

Self-organization as the spontaneous emergence of spatio-temporal patterns through physical or chemical processes has been described in many different systems, for example, in non-living reaction–diffusion systems, such as the Belousov–Zhabotinsky reaction. It was used for an explanation of morphogenesis by Alan Turing in 1952 [1] More recently, it came to prominence in embryology with the use of stem cells and their in vitro differentiation into various tissues, and self-organization has become a fashionable topic in studies of the development of patterns and form.

The idea of self-organization—in various forms and terms—has a long history, and the question of the generation of shapes and structures in embryological development, in general, has occupied and fascinated philosophers and scientists for centuries, inducing them to adopt opposing views: the belief that the structures of adults were existent in miniature in the germ cells or programmed in the genes or genome, contrasted with the conviction that new forms and structures were newly created in the embryo.

This debate about what causes form and structure formation in the growing embryo goes back to antiquity. On the one hand, there was the idea of material continuity between generations that were expressed, for example, in the theories of pangenesis, according to which all organs of the body of a parent produce invisible "seeds" that were transmitted during sexual intercourse, or in the theory of preformation, according to which the structure of an adult organism was already preformed in the germ cells. On the other hand, development was understood as a process of increasing complexity from an unorganized egg that was brought about either by immaterial forces or by self-organizing matter. The former view originated in the School of Hippocrates, while the most prominent protagonist of the latter one (that in the 17th century was termed epigenesis) was Aristotle.

With the advent of experimental biology and particularly the enormous progress in cytology in the late 19th century, a new debate arose about self-organization in development. In the early 20th century, cytologists, such as Theodor Boveri in Germany and Edmund Wilson in the United States, provided ample experimental evidence for the central role of the cell nucleus, chromosomes, and genes in development [2]. However, the notion of a prominent role of the nucleus in development was strongly opposed by experimental embryologists, in particular the influential school of Hans Spemann in Germany. According to Spemann, the cytoplasm was the causal agent of development, not the nucleus; developmental steps were connected by a complicated web and determined by cytoplasmic factors as a kind of self-organized process.

In the 21st century, the debate continued between protagonists of the notion that regulation by genomic genes is the primary cause for the generation of form in embryonic development and those who believed that development is largely self-organized and that it is not genetically determined or regulated. The most prominent representative of the former view was Eric Davidson, who believed that the analysis of complex, hierarchical, multigene developmental gene regulatory networks offers an understanding of the precise spatial and temporal pattern of gene expression of an entire developmental process [3,4]. The latter view is held by embryologists and computational biologists who use modified reaction–diffusion models to simulate pattern formation in embryogenesis. An example is the group of Patrick Müller, according to which "embryonic development is a largely self-organizing process" and who have extended the reaction–diffusion theory to "realistic multi-component networks" [5].

In this paper, I present and examine (1) pertinent physical–chemical and genome-based models of pattern formation and morphogenesis in the past and present, with a special emphasis on Alan Turing's reaction–diffusion model and its reception in the community of biologists, and (2) recent attempts to combine physical–chemical models with models of gene regulation. By showing the insufficiency of purely physical–chemical models for the explanation of embryological development and often also of organisms' repetitive patterns, I claim the relevance to models of development of Brenner's dictum that "Biological systems are information-processing machines, and this must be an essential part of any theory we may construct" [6]. I point out Eric Davidson's successful model of early embryogenesis based on gene-regulatory network analysis and its further development by Ellen Rothenberg and James Briscoe, who also address some of the model's shortcomings, such as a lack of consideration of tissue mechanics and quantitation.

2. Prominent Models of Self-Organization in Morphogenesis and Their Critics

2.1. D'Arcy Thompson: Mathematical Modeling of Organisms' Growth and Form

While Mendel's mathematical modeling of hybridization in plants was one of the earliest and most fruitful models in the study of heredity and biology in general, British zoologist Wentworth D'Arcy Thompson was one of the early theoreticians of self-organization based on the laws of mathematics and physics. His major work, *On Growth and Form* (1942) [7], has often been commented on, and I review his major theses here only briefly because of their influence on Alan Turing. Like Mendel, Thompson perceived mathematics not only as a tool for representation and explanation but as an expression of biological reality. According to him, "the mathematical definition of a 'form' has a quality of precision which was quite lacking in our earlier stage of mere description"; this brings us "in touch with Galileo's aphorism that 'the Book of Nature is written in characters of Geometry'" [7].

Similarly, Plato's primacy of form over matter and Kant's dictum that the criterion of true science lay in its relation to mathematics played a major role in Thompson's reasoning.

In his widely read book, *On Growth and Form*, first published in 1917 [7], Thompson combined morphology with simple mathematics and Greek philosophy to find unifying principles in life's forms. According to him, the organic form was a diagram of forces predetermined by the physical organization of the system in which it developed. His "theory of transformations" aimed at showing how the differences between forms of related species, in particular fish, could be represented geometrically so that one form could be transformed into another one with the help of a simple equation. As an anti-materialist, he rejected theories that attributed specific properties to particles of the protoplasm, such as chromosomes. In his opinion, such an attribution would mean committing the "error of attributing to matter what is due to energy and is manifested in force: or more strictly speaking, of attributing to material particles individually what is due to the energy of their collocation." To him, August Weismann's term of a "hereditary substance" could only mean "that that particular portion of matter is the essential vehicle of a particular charge or distribution of energy, in which is involved the capability of producing motion, or of doing work" [7] (p. 288). Thompson also rejected Darwin's idea of gradual evolution through natural selection because, according to the Platonic idea of pure form (idea), mathematical shapes cannot be transformed through gradations, and organic forms are fashioned by the direct action of physical force, not by selection.

Thompson emphasized the importance of osmotic models of morphogenesis, for example, the work by physical chemist Stéphane Leduc, who claimed to have created artificial life by simulating phenomena such as karyokinesis and organisms' forms with the help of osmotic growth processes (ibid., pp. 324, 501). Leduc did not search for a causal explanation of these phenomena, and he called into question the validity of the generally accepted cell theory of Remak and Virchow of the 1850s, according to which cells arise only through the division of existing cells [8].

Thompson's view contradicted the convictions of prominent biologists at the time who had begun to examine the specificity of basic life processes and organisms' ability to regulate them. Examples are Jacques Loeb, according to whom the artificial creation of life was not only a physical process but had to involve the synthesis of specific molecules, in particular, self-replicating DNA (at the time referred to as nuclein) [9], and Hans Driesch, who held that these osmotic patterns and shapes lack the reproducible specificity of organic forms and the capacity to self-regulate [10].

Thompson's book has been widely admired and praised by a number of renowned scientists, but it had little direct scientific impact on research and never contributed to mainstream experimental biology at any time. However, there is a recently renewed appreciation for the mathematical and physical approaches of Thompson and his predecessors, such as Wilhelm His: morphologists have begun to combine the old, largely metaphoric approach of Thompson and others with insights from molecular biology, such as gene regulation and signaling molecules [11]. Molecular embryologists James Briscoe and Anna Kicheva [12] believe that Thompson's notion "that physical laws constrain biological systems has far reaching consequences". Thompson's book inspired many mathematicians and theoretical biologists to mathematically simulate pattern formation, and it pointed to the insufficiency of neo-Darwinian evolutionary theory, inspiring Stephen J. Gould's criticism of gradualism and adaptationism [13,14]. However, Briscoe's and Kicheva's reminder that mathematical constructions "do not in themselves provide a causal explanation for biological form. This requires molecular, genetic, or mechanical insight into the processes", is relevant not only to D'Arcy Thompson's models but also to all subsequent models of morphogenesis.

2.2. Alan Turing's Mathematical-Chemical Model of Self-Organization in Morphogenesis and Its Reception

The paper "The Chemical Basis of Morphogenesis" by mathematician and computer scientist Alan Turing [1] has played a central role in the discussion about self-organization in morphogenesis. This paper has been more frequently cited than the rest of his work taken together [15], though, interestingly, a citation analysis in the Web of Knowledge shows that a noteworthy increase in the number of citations per year only occurred in the early 2000s (see Section 3). It has recently inspired embryologists and computational biologists to generate models of pattern formation in development. As Peter Saunders [15] has pointed out, the title of the paper and the term "morphogen"—form producer—are not quite correct because the paper mostly dealt with the formation of patterns, not form.

In the introduction to his paper, Turing [1] suggested "that a system of chemical substances, called morphogens, reacting together and diffusing through a tissue, is adequate to account for the main phenomena of morphogenesis". His emphasis that "the theory does not make any new hypotheses; it merely suggests that certain well-known physical laws are sufficient to account for many of the facts", indicates the influence Turing received from D'Arcy Thompson.

Turing aimed to demonstrate that patterns can be created spontaneously in an originally homogeneous cell. To explain how this can happen, that is, how spatial patterns in an egg can form autonomously, he introduced reaction–diffusion equations into the modeling of development. He succeeded in showing mathematically that in a system of two or more diffusing reagents, a pattern of high and low concentrations can spontaneously emerge from an initially uniform distribution [1]. The idea was that two homogeneously distributed substances within a certain space, one "locally activated" and the other capable of "long-range inhibition", can produce novel shapes and gradients. The results of these substance interactions are dependent on just four variables per substance—the rate of production, the rate of degradation, the rate of diffusion, and the strength of their activating/inhibiting interactions.

Turing began to work on morphogenesis in the context of his work on the design of thinking machines, which raised his curiosity about the design of brain development. According to his biographer Alan Hodges [16] (pp. 541–542, cited in [17] p. 89), "There were two possibilities: either a brain learnt to think by dint of interaction with the world, or else it had something written in it at birth—which must be programmed, in a looser sense, by the genes. Brains were too complicated to consider at first. But how did anything know how to grow? There lay the question". Turing became fascinated with embryology, the taking shape of an animal from the sphere of a cell, and the fact that, as he believed, nobody had thought about what determined this growth (ibid.).

Another reason for Turing to become interested in biology was his desire to "defeat the argument from design" as proof of the existence of God. This argument was still widespread, although Darwin's materialistic theory of evolution was widely accepted at the time [15]. Therefore, Turing followed Thompson, who had urged biologists to attempt to explain forms in the same way physicists do, namely by reference to mechanical forces (ibid.), though, unlike Thompson, Turing focused on the generation of patterns, not of forms. It is surprising—and has been noticed by many commentators—that Turing approached the task he set himself mostly on his own, without consulting colleagues from biology or taking notice of what other modelers in biology did. There are only very few references in his paper; they include Thompson's *On Growth and Form*. According to Saunders [15], this reflected Turing's way of working, i.e., to determine what was important and not to be diverted from his view by what others did.

This attitude may, in part, explain the contradictions in the paper's premises and the grave shortcomings regarding the state of the art in biological research. Turing's "mathematical model of the growing embryo" was indeed, as intended, simple and elegant. However, to consider the embryo as a state function and eliminate growth is in direct contradiction to his stated goal. While diffusion and osmotic pressures are widely dealt with, the "chemical reactions" are not related to any particular type of molecules or their specificities, though at the time, biological specificity was largely related to proteins. Most importantly, the concepts of gene and cell were unclear, and the "genes as enzymes" theory that Turing advocated was long obsolete. In Turing's words: the genes may "be considered to be morphogens"; "it would be more accurate (...) to regard them as radicals of the giant molecules known as chromosomes; … the function of genes is presumed to be purely catalytic. They catalyze the production of other morphogens, which in turn may only be catalysts" [1].

The gene-as-enzyme hypothesis that was proposed by Richard Goldschmidt in 1927 soon proved to be untenable for various reasons [18]. Since the 1930s, several developmental geneticists have studied the cooperation of genes and their biochemical effects. In 1941, based on X-ray studies of mutants in the mold Neurospora, the American geneticists George Beadle and Edward Tatum found that each gene governed the production of one specific enzyme—the "one gene-one enzyme" concept. This means that the characteristic function of the gene was to supervise the formation of a particular enzyme. The authors determined that all biochemical reactions in an organism were controlled by specific genes, work for which they shared, with Joshua Lederberg, the 1958 Nobel Prize in Physiology and Medicine. Apart from overlooking this important advance in the chemistry of development, Turing disregarded the fact that the assumption of enzymes catalyzing the production of other enzymes (catalysts), etc., would lead to an infinite regress, an observation which, a few years later, led Francis Crick to conclude that the synthesis of enzymes must be radically different from the synthesis of other molecules and that the existence of a template seemed the only logical solution to this dilemma [19].

This disregard for biological knowledge and logic supports Evelyn Fox Keller's assessment that Turing was more interested in "mathematical fruitfulness and accessibility" than in the correspondence of his hypothetical reactions to real reactions in the cell [17]. Biologists, on the other hand, were not interested in whether the interactions could build patterns the way Turing suggested but whether they really do. Additionally, for a long period of time, there was no evidence of it. For this reason, the model was hardly cited by biologists for decades (a detailed historical analysis of Turing's model is in [17]. One of the few biologists interested in the model was Conrad Waddington, but he, too, in a letter to Turing in 1952, raised "several concerns about the applicability of Turing's reaction-diffusion model to biological developmental systems, questioning its limitation to reproduce some observed behaviors in embryonic development such as pattern scaling with tissue size or the generation of a spatial pattern of discrete cell types" [20]. Waddington believed that the model might apply to the formation of patterns such as spots and stripes but not to morphogenesis.

In the 1970s, Ilya Prigogine and his school of the irreversible thermodynamics of complex systems made the model popular for some time. The number of reaction–diffusion studies increased, particularly pattern formation in butterfly wings and animal coats. Scientists applied updated versions of Turing's model and other mathematical models to simulate pattern formation in a variety of different animal systems, such as the generation of periodic seashell patterns and body segmentation in Drosophila. These early studies of pattern formation, for example, the work by Hans Meinhardt and Alfred Gierer, have been described and analyzed in detail by Siegfried Roth [10].

However, for a variety of reasons, many of these simulation models did not reflect reality. One of the reasons is that, like Turing, their authors disregarded genes not only as causal factors for morphogenesis and development as a whole but also for many biochemical pathways for pattern formation. They disregarded the fact that the unfertilized egg, as was shown by Christiane Nüsslein-Volhard, was not a homogeneous sphere but rather a highly organized structure containing, among other things, a spatial pattern of information carrying mRNA and proteins [17] (p. 111). Additionally, many of these models also disregarded the difference between pattern formation and the complex processes of embryogenesis, a problem that, as Francis Crick remembered, may have been perceived even by Turing himself: At a meeting on mathematical models of development in 1972, Crick, one of the skeptics regarding the validity of Turing's model for development, quoted Turing's remark about the zebra: "Well, the stripes are easy but what about the horse part?" [21]. Pattern formation can be modeled elegantly and relatively simply, but morphogenesis and development would require modeling of the zebra itself, its body architecture, organs, etc., in a very complex way that would also have to take into consideration developmental constancy and evolution.

Evolutionary developmental biologist Michael Akam, who has studied the generation of the repeating stripes along the antero–posterior axis of Drosophila for many years, in 1989, wrote a widely discussed paper with the title "*Making Stripes Inelegantly*" [22], in which he discussed two possible ways of generating the exact periodicity of the stripes: An "elegant mechanism" that was favored by model builders such as Meinhardt [23] or Lacalli et al. [24], and that would "use an intrinsically periodic pattern-generating system, comprising the pair-rule genes [a class of segmentation genes] and their products." It only needed to be triggered by local stimuli from the gap genes that control the early cascade of the segmentation pathway. The alternative was that "unique instructions could be generated by the gap-gene proteins to define the position of each pair-rule stripe". His analysis of the interaction between the different kinds of genes (gap and pair-rule genes) showed that the less elegant "specific instruction" process was more likely to take place in the organism and that "the apparent simplicity of the repeating segment pattern" was deceptive. He concluded that spontaneous pattern-generating mechanisms might contribute to "sharpen the boundaries between stripes" but that they do not define periodicity.

Akam's paper [22] has been continuously cited since its publication, with a significantly higher average number of citations since 2005 (Figure 1). Most of the citing papers appeared in journals of developmental biology and computational biology and, more recently, also in physical and mathematical journals.

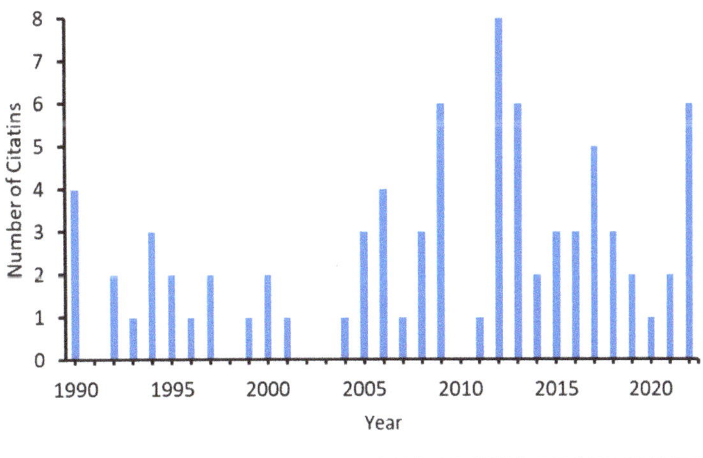

Figure 1. Number of citations of Akam (1989) in the Web of Science.

Among the citing authors are J. Sharpe and A. Economou (whose work is briefly presented in Section 3). More than 30 years later, Akam and his collaborators proposed a new mechanism for the segmentation of Drosophila and other arthropods, in which conserved gene regulatory networks play decisive roles, concluding that "over the past four decades, arthropod segmentation has contributed enormously to our understanding of developmental gene networks and their evolution" [25].

The strongest critic of mathematical simulation models that are not based on experimental perturbation was Eric Davidson. According to him, "one of the worst fallacies [in the field of modeling in biology] is the assumption that if you can make a model, which simulates a process, then the model must represent how it works. The great example is Meinhardt's explanation of Drosophila stripes in terms of reaction-diffusion equations. He explained it perfectly, except it doesn't happen to be how it works. [...] And what showed us how it works, of course, was taking the DNA out and experimentally finding out how it works" [26]. Davidson's successful attempt to causally explain the molecular events of early development in sea urchins with the help of his theory of gene regulatory networks and to generate a Boolean model for it is briefly discussed in Section 4.

3. The Recent Revival of Turing's Theory of Morphogenesis and Other Theories of Self-Organization in Biology; Merging with Genomic Models

Taken alone, methods based on Turing's model and updated reaction–diffusion models by others so far have been unable to explain the complex developmental program that is brought about by multiple genetic and molecular pathways; they even cannot account for many of the simpler patterns, such as stripes. However, according to Marcon and Sharpe [27], Turing-type reaction–diffusion models and other models of self-organization have recently started to be taken more seriously and applied to a variety of patterning processes by biologists and not only by mathematicians.

Citation analysis in the Web of Science shows that the number of citations of Turing's paper has drastically increased after 2000, especially since 2020 (Figure 2). Of the 214 citing papers since 2020, ca. 62 covered biological topics. Thirty-four of them dealt with topics of pattern formation and morphogenesis, several of them including gene regulatory networks in the title, and twenty-eight papers dealt with other biological topics, such as ecology, evolution, and neural networks.

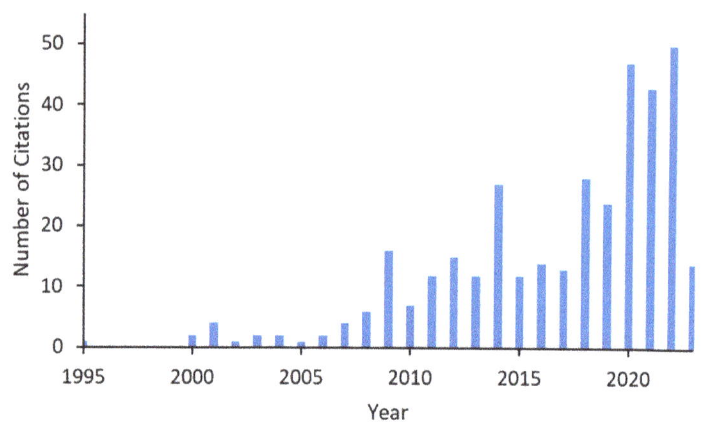

Figure 2. Number of citations of Turing's "The Chemical Basis of Morphogenesis" [1] in the Web of Knowledge, starting in 1995.

Multi-component Turing networks that do not require differential diffusivity have been proposed by Patrick Müller and his group; the authors believe that embryonic development is largely a self-organizing process [5]. Most of their simulation models have not been experimentally tested.

Self-organization has also been argued to be responsible for symmetry breaking (i.e., the acquisition of asymmetry along an axis) in the early mammalian embryo [28]. The authors hold that symmetry can be broken through stochastic variations in the cytoskeleton structure, but they perceive a difference between experiments conducted in vivo and in vitro. In in vivo studies, it is maternal and/or extraembryonic tissues that are instrumental in the establishment of an anterior–posterior axis through asymmetric signaling activity, whereas, in studies using in vitro cultures of blastocysts or stem cell aggregates, this does not seem to be necessary for symmetry breaking. According to Stas Shvartsman [29], genomic controls guide the self-organizing processes, selecting specific outcomes from many different possibilities.

Eric Karsenti [30] believes that the whole cell cycle in eukaryotes "can be seen as being based on the principle of self-organization by reaction-diffusion, both temporally and spatially. But he made it clear that though microtubule patterns appear self-organized, mutated cells (in the Ser-Thr protein kinase Orb6) have a different shape and microtubules cannot organize in long bundles. This means that genes are required for the self-organizing process. He also thinks it important to realize that none of these processes are true Turing patterns", because "the symmetry is not broken by spontaneous instabilities, but rather by deterministic effects", such as cyclin synthesis for the oscillator and stereospecific targeting of a small G-protein exchange factor to chromatin for nuclear and spindle assembly.

Tom Misteli [31] views the genome as a self-organizing system because this perception makes it possible to understand the conflicting aspects of genome organization, namely the stability of the transcriptional program of a given cell on the one hand and the dynamic and stochastic nature of gene expression on the other. According to Misteli, the available data support the notion that the major features of higher-order genome architecture are emergent properties in a self-organizing system that is driven by the functional status of the genome. He defines self-organization as "the inherent tendency for systems to form coherent patterns solely based on the dynamic interaction of its components," in the case of the genome, "the physical interactions of proteins with chromatin and of chromatin with chromatin" [32].

Misteli is of the opinion that "the architectural properties of the genome are driven by the sum of activities [such as gene activity] that occur along the genome", most of which are affected by DNA sequence. According to him, "the DNA sequence is a major contributor to determining these activities", although "most DNA-binding proteins bind far more promiscuously than we have previously thought". He believes that chromatin remodeling complexes promiscuously bind to chromatin and remodel it. The DNA sequence-specific proteins are important during the short period of time in which chromatin is open. All of this suggests that the term self-organization in Misteli's definition only applies to the genome after the activities along the genome have already been established, most of them by DNA sequence-specific events. From this, it can be concluded that: the genome is self-organized this self-organization is mainly based on previous DNA sequence-specific events.

Perspectives on the promiscuity of DNA binding proteins and remodeling events differ among different researchers. According to James Briscoe [33], promiscuity in the binding of chromatin remodeling complexes exists, but the modifying factors are guided by transcription factors and other regulatory factors that are specific to a DNA sequence. As an example, he mentioned that the starting point of polycomb group regulatory proteins is determined by DNA sequence [34].

Ellen Rothenberg thinks that the difference between sequence-specific transcription factors and chromatin remodeling factors is that the sequence-specific factors require interaction with some more-or-less specific DNA sequence in order to bind, whereas the chromatin remodelers and modifiers do not. She points to the highly complex nature of the binding of the factors: "For each transcription factor, there is still a range of variants of the preferred DNA sequence that are bound with different strengths. So, while all of these sequences ("motifs") are non-random and statistically far different from background DNA, they are not equivalently good targets for the transcription factor's binding and can be bound conditionally, for example, better if the chromatin is open than when it is closed" so that the binding is not 100% certain [35] (see also Section 4).

Some recent research has begun to study various cases of pattern formation in animals by combining reaction–diffusion models or other physical–chemical mechanisms with genome-based mechanisms. Examples are Sharpe et al.'s [36] work on the control of digit patterning by a Bmp-Sox9-Wnt Turing network modulated by morphogen gradients and Economou et al.'s [37] perturbation analysis of a Turing-like reaction–diffusion stripe patterning system. Sharpe et al. showed how digit patterning appears to be controlled by a Turing network implemented by gene products. The problem with this study is that though the findings are based on experimental data, some assumptions are speculative, derived from purely mathematical reasoning [38].

The perturbation analysis of a Turing-like reaction–diffusion stripe patterning system—of ridges in the mammalian palate—and the regulatory interactions involved in this process by Economou et al. is another attempt to integrate physical and genomic models. The study shows the cooperation of growth factor ligand proteins, their receptors, genes, and other factors, and it also reveals still existing discrepancies between the results of mathematical modeling and biological reality: the patterning of the palate uses five pathways in the organism, though only two would be required by mathematical modeling. Moreover, to my knowledge, none of these models or other mathematical models have addressed the questions of how models can account for the species specificity of the patterns and their stability in geological time.

Some authors attempt to establish a connection between the metaphor of "epigenetic landscape" that embryologist Conrad Waddington proposed in 1940 and physicists' notion of an "epigenetic state", a system-level stable state that arises from the interactions of genes as Waddington had envisioned. They regard Waddington's vision as a major contribution to the current convergence of molecular and physico-chemical approaches [39,40].

In their review of recent large-scale mathematical analyses of Turing patterns in biology that have attempted to narrow down potential design principles, Sean T. Vittadello et al. [41] showed that despite progress in many areas, the original problems related to the use of Turing models in the context of biology have not yet been fully resolved. One of them was the contradiction between "the beauty of mathematical models" and "the ugly truth of reality". They discuss an aspect of model development in biology that they consider essential for confronting this problem, namely "the extent to which the assumptions underlying our models are robust and in line with what we see in nature", describing the "caveats that need to be considered in designing a synthetic Turing-patterning mechanism that is viable in vivo".

4. Models Based on the Concept of Genomic Causality in Development

4.1. Eric Davidson's Model of a Complex Developmental Regulatory Gene Network (GRN)

The most successful model of the description and causal explanation of the early development of a complex organism, the sea urchin, is Eric Davidson's model of developmental gene regulatory networks (GRNs). Based on decades-long molecular biological research into how cell-type specific gene expression patterns appeared, Davidson adopted a systems approach that included almost all regulatory genes as soon as DNA sequencing was available [42]. However, "experimental perturbation and predictive challenge of the system" remained essential to reveal the underlying causal mechanisms [43].

Davidson created the concept of developmental GRNs in the early 2000s. Basic knowledge of genetic regulation in the development of higher organisms had already been obtained from Drosophila by Christiane Nüsslein-Volhard and Eric Wieschaus, who also demonstrated the hierarchy of maternal genes in the embryo that played an important role in Davidson's GRN. The early models of the temporal dynamics of already known gene networks in development included only a few genes [17] (pp. 250–251). Davidson was the first to achieve an almost complete model of a regulatory gene network for the development of a particular phenotype (endomesoderm specification) and to construct a mathematical model to account for observation in a complex biological object. Using big sequencing and expression data, he made a large quantitative step from a few regulatory genes to networks of hundreds of genes.

These developmental GRNs contain, as crucial elements, specific cis-regulatory modules (DNA regions binding the transcriptional machinery in the vicinity of the genes they regulate) that direct the expression of developmental transcription factors and signaling molecules. Cis-regulatory modules are causality-inferred regulatory regions of genes that are identified experimentally [44].

A world leader in molecular embryology, Davidson demonstrated that, at least in sea urchins, early development is entirely regulated by the genome. This was, to him, a logical necessity and requirement for evolution because without such a genomic regulatory program, it cannot be ensured that within each species, the outcome of development is extremely reproducible. Davidson's and his collaborators' attempts to explain gene regulation in development beyond the study of individual genes started in 1969 [45]. It not only led to the experimental construction of hierarchical GRNs for development and later to their mathematical modeling but also to the proposition that changes in the architecture of GRNs through changes in genomic sequences may be the engine of evolutionary changes in animal body architectures and other major characteristics.

Davidson was one of only a few scholars who, together with Douglas Erwin, not only proposed hypotheses for a causal mechanistic explanation of evolutionary change but also of evolutionary stasis based on the stability of developmental outcomes [46]. The extreme conservation of certain morphological features over immense geological periods gave rise to the question of how these parts of the GRN structure could be stabilized through deep time, questions that found a partial answer in the organizational hierarchy of these structures—the effect of changes differs fundamentally according to where in the network they occur. Small changes continuously occur at the periphery of networks, where effector genes code for proteins, while stasis of network patterning can be found in other parts.

Early on, Davidson joined forces with physicists and computer scientists to integrate computer-generated big data into a systems approach that was based on experiments and aimed at elucidating mechanisms and causal relationships. In 2006, he introduced the term "regulatory genome" for the interactions between regulatory genes and their products during development [3]. This concept was conceived and developed through decades-long, painstaking experimental research by Davidson and his collaborators. They systematically examined the cell-type-specific gene expression patterns before moving on from the "gene-by-gene characterization of the sea urchin embryo to full comprehensiveness" [42]. This systems approach was made possible when sequencing data of the whole sea urchin genome became available. Observations and descriptions were crucial as a starting point. A perturbation analysis was essential because "only by deliberate experimental perturbation and predictive challenge of the system can the mechanisms by which it operates be revealed" [43].

Davidson's first mathematical models in the 1990s not only contained logic functions (AND, OR, NOT) between the input of different regulatory proteins (transcription factors) but also assumed that they were quantitatively modulated; thus, he and his collaborators created systems of differential equations with continuously variable inputs and outputs [47] (see the overview by Ellen Rothenberg [42]). However, because the key rate constants and concentrations needed to render these models predictive did not exist, Davidson envisaged that a Boolean model, in which the status of each gene is assumed to be either "on" or "off", might be sufficient as a predictive systems model of development. Together with Isabelle Peter, Davidson converted the whole GRN system into a Boolean model, a "grueling effort lasting many months of concentrated work" [4] (p. 309). This model contained all current data for the logic of transcriptional inputs at each gene (cis-regulatory) system and for the location of each cell at each time point of development [48].

The first results of the Boolean modeling showed that there were only a few inconsistencies between model predictions and measured in vivo gene expression. This meant not only that the key regulatory elements of the GRN and their interactions were almost complete but also that other factors, such as changes in chromatin structure, did not appear to be relevant at this early developmental stage. The model was based on experiments that proposed linkages based on perturbations, not only correlations. It provided a means for experimentally testing the relevance and consistency of the GRN concept, thus fundamentally contrasting with models that merely simulated phenomenological features without analyzing their mechanisms and causes.

Davidson's GRN model underlines the relevance of three principles that are relevant for the explanation of complex biological systems, namely, informational hierarchy, genomic causality, and biological specificity (see [49]).

4.2. Assessments and Further Developments of Davidson's Developmental GRN Model

Davidson's developmental GRN model has been successfully used to explain aspects of development in a wide range of different organisms. Like any other model, it has been challenged by new research and is being transformed accordingly. Thus, the original proposal of deep evolutionary conservation of network kernels does not seem to be maintainable—as Douglas Erwin [50] has shown, there is an extensive rewiring of GRN sub-circuits. Despite these changes in the original network concept, the ideas of the central role of GRNs for embryological development and developmental constancy, their hierarchical organization, at least concerning genetic information, and the causal role of genes, have been confirmed in numerous different studies and have remained fruitful.

Ellen Rothenberg emphasized the general importance that Davidson's and his collaborators' work on developmental GRNs has had for many different systems. According to her, it is now widely recognized that GRN analysis is a major step to advancing from a descriptive to a mechanistic understanding of biological systems [51], and she showed how much researchers in her own field, hematopoiesis, have benefitted from the pioneering work of Davidson on network control in sea urchins. At the same time, she points to differences between the network models for nonvertebrate embryogenesis and for hematopoietic systems in mammals, particularly regarding dose dependence and timing: in contrast to a rapidly unfolding cascade of transcriptional change guided by transcription factors (TFs) in embryological development, the development of lymphocytes from stem cells in mammals is slow, cell fate choices have a strong stochastic component, and timing is highly variable [52]. Another difference is that though TFs that read regulatory sequences in the genome to initiate changes in the expression of specific genes in development as well as in physiological processes, their actions in the latter are constrained by slow-changing chromatin states and by interactions with other TFs. She uses the development of T lymphocytes to show how binding specificity and dynamics, TF cooperativity, and chromatin state changes impact the regulatory functions of key TFs (ibid). An example is Runx transcription factors. They are always binding specific Runx motifs but choose a different ~10,000 sites out of the possible ~1 million to bind in early T cells than the ones they choose in B cells, stem cells, or even in later T cells [53]. This shows that though this binding is not promiscuous, it is also not predictively deterministic. The action of the Runx factors depends on the different contexts at different DNA sites, such as other regulatory proteins in different pathways [54]. There are also cell type-specific factors that affect the modification of the site choice within the constraints of the sequence motif-specific binding, an area that is currently being explored.

Rothenberg also broadened the view on the role of TFs in developmental processes. They not only bind to regulatory sequences, but certain TFs in the T cell specification network model also play an important role in opening chromatin, displacing nucleosomes, and initiating activating histone modifications [55]. She thus echoes the view that was already brought forward by Gary Felsenfeld [56] that histone modification is preceded by a DNA sequence-specific event. Closely examining the collaboration of TFs in their system, Shin and Rothenberg [57] show how "transcription factors collaborate to initiate, stabilize, synergize, oppose, or silence gene expression programs".

James Briscoe considers the theory of GRN by Davidson and colleagues to be a logical and formal framework in which to describe the transcriptional programs that have to be activated at the right time and place during development, programs that are encoded in the genome. Because the functional output of a developmental GRN is the "organized expression of genes", the "analysis of the architecture and dynamics of these networks offers an understanding and a rationale for the precise spatial and temporal pattern of expression of the thousands of genes necessary for tissue patterning" [33]. Referring to the example of Davidson's "rigorous and comprehensive dissection of sea urchin endomesoderm development", he concluded that this work illustrates the "potential of the GRN approach to provide a mechanistic and causal explanation to a complex set of gene regulatory events controlling development cell fate specification" (ibid.). He also points to the importance of GRNs that have been reconstructed from other species and tissues.

Briscoe also demonstrated some limitations of the GRN approach, in particular, its lack of quantitation and its emphasis on structure and topology, that is, connections between genes and transcription factors. According to him, this "underplays the dynamics and quantitative aspects of a system, which is crucial when feedback and nonlinearity are involved" [33]. Briscoe made it clear that Davidson, in his last years, became increasingly interested in getting these quantitative data, and he is of the opinion that advances in experimental techniques now make it possible to collect and analyze more high-resolution and quantitative data [34]. Briscoe believes that the combination of GRN analysis and dynamic systems approaches also serves to overcome some of the limitations of the GRN approach. According to him, dynamic systems and complexity theory help explain and predict behavior that is not easy to understand otherwise, such as sudden changes in behavior in a deterministic dynamical system, known as bifurcation. Briscoe uses a framework based on Catastrophe theory in order to generate quantitative geometric models of cell differentiation [58].Dynamic systems theory is used as a framework to describe specific developmental mechanisms, define simplifying abstractions, and explain principles. Thus, the interactions within the GRN serve as an example of multilevel behavior that explains how tissue patterns of gene expression arise from the molecular interactions of transcription factors in individual cells [33]. However, many questions still remain open, such as how stochastic fluctuations that are inherent to gene regulation affect performance and are propagated through a gene regulatory network or the role of cell and tissue mechanics in pattern formation.

5. Conclusions

Until now, updated Turing models by themselves do not appear to be able to explain either robust morphogenesis or pattern generation in development. A combination of theoretical and experimental approaches and the integration of gene products into models of self-organization or other mathematical models appears most promising for a causal explanation of morphogenesis and pattern formation in organisms. In many cases, however, empirical confirmation is still missing, and the old problem of the discrepancy between model prediction and biological reality has not been solved in most cases. The models usually disregard the effect of evolution and do not address longstanding developmental and species stability. The GRN approach based on a complex set of gene regulatory events has proved most successful in controlling developmental cell fate specification and in providing an explanation for developmental constancy and evolutionary change and stability. However, it also has some limitations, such as its focus on gene regulation and disregard of cell tissue mechanics in morphogenesis [33].

The metaphor of "genetic program" as a molecular genetic concept was introduced by Francois Jacob and Jacques Monod in the context of their work on gene regulation in bacteria [59]. Despite the fact that the analogy to a computer program, which this term may suggest, has been strongly criticized, in particular by historians and philosophers of science, the metaphor has remained influential in biology as a succession of steps, not the equivalent of a computer program [60]. Some years later, Jacob [61] used the metaphor to suggest a modern vision of development that combines the ancient idea of preformation, which for him was the genetic program, and epigenesis that he understood, for example, as feedback regulation of enzyme activities. We may extend this view of epigenesis by including other events that are not directly controlled by the genome, such as the mechanics of cells, adhesion processes between molecules, and geometric constraints of development.

In 2016, Eric Davidson took a similar philosophical stand when he distinguished between two types of experimentally supported causal explanations in his field, animal developmental biology and also in the evolutionary biology of the animal body plan. He described them as "rooted" and "unrooted" explanations. In his words, "rooted causal explanation provides logical links to and from the genomic regulatory code, extending right into the genomic sequences that control regulatory gene expression"; "Unrooted explanations" are those "in which the only causality is to be located within a process considered, for example within a synthesis pathway (without reference to why the enzymes are expressed where they are in the first place), or within a signaling event (without reference to why the signal is expressed in the sending cells, or what it does to gene activation in the receiving cells)" [43]. The many cases of the spontaneous emergence of patterns or forms in biology that seem to be driven by physical or chemical forces and that are therefore labeled self-organized are good candidates for other examples of Jacob's "epigenesis" and Davidson's "unrooted explanations". The concept of self-organization in development as modern epigenesis seems to be most fruitful when it is included in the frame of genomic causality, and models of genomic causality have to integrate physical–chemical models to be complete.

Funding: This research received no external funding.

Conflicts of Interest: The author declares no conflict of interest.

References

1. Turing, A.M. The chemical basis of morphogenesis. *Philos. Trans. R. Soc. Lond. B Biol. Sci.* **1952**, *237*, 37–72.
2. Wilson, E.B. *The Cell in Development and Heredity*, 3rd ed.; Macmillan: New York, NY, USA, 1928.
3. Davidson, E.H. *The Regulatory Genome. Gene Regulatory Networks in Development and Evolution*, 1st ed.; Elsevier: San Diego, CA, USA, 2006.
4. Peter, I.; Davidson, E.H. *Genomic Control Process: Development and Evolution*; Academic Press: Amsterdam, The Netherlands, 2015.
5. Landge, A.N.; Jordan, B.M.; Diego, X.; Müller, P. Pattern formation mechanisms of self-organizing reaction-diffusion systems. *Dev. Biol.* **2020**, *460*, 2–11. [CrossRef] [PubMed]
6. Brenner, S. Theoretical biology in the third millennium. *Philos. Trans. R. Soc. London. Ser. B* **1999**, *354*, 1963–1965. [CrossRef]
7. Thompson, D. *On Growth and Form*, 2nd ed.; Cambridge University Press: Cambridge, UK, 1942.
8. Deichmann, U. Crystals, colloids or molecules? Early controversies about the origin of life and synthetic life. *Perspect. Biol. Med.* **2012**, *55*, 521–542. [CrossRef]
9. Loeb, J. Experimental study of the influence of environment on animals. In *The Mechanistic Conception of Life*; Fleming, D., Ed.; Harvard University Press: Cambridge, MA, USA, 1964; pp. 178–210.
10. Roth, S. Mathematics and biology: A Kantian view on the history of pattern formation theory. *Dev. Genes Evol.* **2011**, *221*, 255–279. [CrossRef] [PubMed]
11. Savin, T.; Kurpios, N.A.; Shyer, A.E.; Florescu, P.; Liang, H.; Mahadevan, L.; Tabin, C.J. On the growth and form of the gut. *Nature* **2011**, *476*, 57–62. [CrossRef]
12. Briscoe, J.; Kicheva, A. The physics of development 100 years after D'Arcy Thompson's "On Growth and Form". *Mech. Dev.* **2017**, *145*, 26–31. [CrossRef]
13. Eldredge, N.; Gould, S.J. Punctuated equilibria: An alternative to phyletic gradualism. In *Models in Paleobiology*; Schopf, T.J.M., Ed.; Freeman Cooper: San Francisco, CA, USA, 1972; pp. 82–115.
14. Gould, S.J.; Lewontin, R. The spandrels of San Marco and the Panglossian paradigm: A critique of the adaptationist programme. *Proc. R. Soc. Lond. B Biol. Sci.* **1979**, *205*, 581–598.
15. Saunders, P. Alan Turing and biology. *IEEE Ann. Hist. Comput.* **1993**, *15*, 33–36. [CrossRef]
16. Hodges, A. *Alan Turing: The Enigma*; Princeton Univ. Press: Princeton, NJ, USA, 1983.
17. Fox Keller, E. *Making Sense of Life: Explaining Biological Development with Models, Metaphors and Machines*; Harvard Univ. Press: Cambridge, MA, USA, 2003.
18. Deichmann, U. The Concept of the Causal Role of Chromosomes and Genes in Heredity and Development: Opponents from Darwin to Lysenko. *Perspect. Biol. Med.* **2014**, *57*, 57–77. [CrossRef]
19. Crick, F. On protein synthesis. *Symp. Soc. Exp. Biol.* **1958**, *12*, 138–163.
20. Hallou, A. *When Turing Meets Waddington: Theory of Mechanochemical Patterning in Biphasic Biological Tissues*; Lecture at Gurdon Institute & Cavendish Laboratory: Cambridge, UK, 4 November 2021. Available online: https://talks.cam.ac.uk/talk/index/165172 (accessed on 1 December 2022.).
21. Hogeweg, P. The roots of bioinformatics in theoretical biology. *PLoS Comput. Biol.* **2011**, *7*, e1002021. [CrossRef] [PubMed]
22. Akam, M. Making stripes inelegantly. *Nature* **1989**, *341*, 282–283. [CrossRef] [PubMed]
23. Meinhardt, H. Hierarchical inductions of cell states: A model for segmentation in Drosophila. *J. Cell Sci. Suppl.* **1986**, *4*, 357–381. [CrossRef] [PubMed]

24. Lacalli, T.C.; Wilkinson, D.A.; Harrison, L.G. Theoretical aspects of stripe formation in relation to Drosophila segmentation. *Development* **1988**, *4*, 105–113. [CrossRef] [PubMed]
25. Clark, E.; Peel, A.D.; Akam, M. Arthropod segmentation. *Development* **2019**, *146*, dev170480. [CrossRef]
26. Davidson, E.H. Interview by Ute Deichmann. *Dev. Biol.* **2016**, *212*, S20–S29.
27. Marcon, L.; Sharpe, J. Turing patterns in development: What about the horse part? *Curr. Opin. Genet. Dev.* **2012**, *22*, 578–584. [CrossRef]
28. Zhang, H.T.; Hiiragi, T. Symmetry breaking in the mammalian embryo. *Annu. Rev. Cell Dev. Biol.* **2018**, *34*, 405–426. [CrossRef]
29. Shvartsman, S.; (Princeton University, Princeton, NJ, USA). Personal communication, 2022.
30. Karsenti, E. Self-organization in cell biology: A brief history. *Nat. Rev. Mol. Cell Biol.* **2008**, *9*, 255–262. [CrossRef]
31. Misteli, T. The self-organizing genome: Principles of genome architecture and function. *Cell* **2020**, *183*, 28–45. [CrossRef]
32. Misteli, T.; (Centre for Cancer Research, Bethesda, MD, USA). Personal communication, 2023.
33. Briscoe, J. Understanding pattern formation in embryos: Experiment, theory, and simulation. *J. Comput. Biol.* **2019**, *26*, 696–702. [CrossRef] [PubMed]
34. Briscoe, J.; (Francis Crick Institute, London, UK). Personal communication, 2023.
35. Rothenberg, E.; (California Institute of Technology, Pasadena, CA, USA). Personal communication, 2023.
36. Sharpe, J.; Raspopovic, L.; Marcon, L.; Russo, J. Digit patterning is controlled by a Bmp-Sox9-Wnt Turing network modulated by morphogen gradients. *Science* **2014**, *345*, 566–570.
37. Economou, A.D.; Monk, N.A.M.; Jeremy, B.A.; Green, J.B.A. Perturbation analysis of a multi-morphogen Turing reaction-diffusion stripe patterning system reveals key regulatory interactions. *Development* **2020**, *147*, dev190553. [CrossRef] [PubMed]
38. Deichmann, U. The idea of constancy in development and evolution—Scientific and philosophical perspectives. *Biosystems* **2022**, *221*, 104773. [CrossRef] [PubMed]
39. Huang, S.; Ernberg, I.; Kauffman, S. Cancer attractors: A systems view of tumors from a gene network dynamics and developmental perspective. *Semin. Cell Dev. Biol.* **2009**, *20*, 869–876. [CrossRef] [PubMed]
40. Huang, S. The molecular and mathematical basis of Waddington's epigenetic landscape: A framework for post-Darwinian biology? *Bioessays* **2012**, *34*, 149–157. [CrossRef]
41. Vittadello, S.T.; Leyshon, T.; Schnoerr, D.; Stumpf, M.P.H. Turing pattern design principles and their robustness. *Philos. Trans. A Math. Phys. Eng. Sci.* **2021**, *379*, 20200272. [CrossRef]
42. Rothenberg, E.V. Eric Davidson: Steps to a gene regulatory network for development. *Dev. Biol.* **2016**, *412*, S7–S19. [CrossRef]
43. Davidson, E.H. Genomics, "discovery science", systems biology, and causal explanation. What really works? *Perspect. Biol. Med.* **2016**, *58*, 165–181. [CrossRef]
44. Istrail, S. Eric Davidson's regulatory genome for computer science: Causality, logic, and proof principles of the genomic cis-regulatory code. *Comput. Biol.* **2019**, *26*, 653–684. [CrossRef]
45. Britten, R.J.; Davidson, E.H. Gene regulation for higher cells: A theory. *Science* **1969**, *165*, 349–357. [CrossRef] [PubMed]
46. Davidson, E.H.; Erwin, D. Gene regulatory networks and the evolution of animal body plans. *Science* **2006**, *311*, 796–800. [CrossRef] [PubMed]
47. Bolouri, H.; Davidson, E. Modeling transcriptional regulatory networks. *Bioassays* **2002**, *24*, 1118–1129. [CrossRef] [PubMed]
48. Peter, I.S.; Faure, E.; Davidson, E.H. Predictive computation of genomic logic processing functions in embryonic development. *Proc. Natl. Acad. Sci. USA* **2012**, *109*, 16434–16442. [CrossRef]
49. Deichmann, U. Hierarchy, determinism, and specificity in theories of development and evolution. *Hist. Philos. Life Sci.* **2017**, *39*, 33. [CrossRef]
50. Erwin, D. Body Plans and Evolvability: How Does the Structure of the Regulatory Genome Deliver Novelty While Ensuring Continuity? In Proceedings of the international workshop Constancy and Plasticity in Development and Evolution, Jacques Loeb Centre, Ben-Gurion University of the Negev, Beer Sheva, Israel, 9–10 June 2022.
51. Rothenberg, E.V.; Goettgens, B. How haematopoiesis research became a fertile ground for regulatory network biology as pioneered by Eric Davidson. *Curr. Opini. Hematol.* **2021**, *28*, 1–10. [CrossRef]
52. Rothenberg, E.V. Logic and lineage impacts on functional transcription factor deployment for T-cell fate commitment. *Biophys. J.* **2021**, *120*, 4162–4181. [CrossRef]
53. Shin, B.; Hosokawa, H.; Romero-Wolf, M.; Zhou, W.; Masuhara, K.; Tobin, V.R.; Levanon, D.; Groner, Y.; Rothenberg, E.V. Runx1 and Runx3 drive progenitor to T-lineage transcriptome conversion in mouse T cell commitment via dynamic genomic site switching. *Proc. Natl. Acad. Sci. USA* **2021**, *118*, e2019655118. [CrossRef]
54. Rothenberg, E.V. Modular Organization and Mechanisms of the T-cell Specification Gene Regulatory Network. In Proceedings of the Conference Gene Regulatory Networks in Development and Disease, Oxford University, Oxford, UK, 13 April 2023.
55. Rothenberg, E.V. Dynamic control of the T-cell specification gene regulatory network. *Curr. Opin. Syst. Biol.* **2019**, *18*, 62–76. [CrossRef]
56. Felsenfeld, G. The evolution of epigenetics. *Perspect. Biol. Med.* **2014**, *57*, 130–146. [CrossRef] [PubMed]
57. Shin, B.; Rothenberg, E.V. Multi-modular structure of the gene regulatory network for specification and commitment of murine T cells. *Front. Immunol.* **2023**, *14*, 1108368. [CrossRef] [PubMed]
58. Sáez, M.; Briscoe, J.; Rand, D.A. Dynamical landscapes of cell fate decisions. *Interface Focus* **2022**, *12*, 20220002. [CrossRef] [PubMed]

59. Jacob, F.; Monod, J. Genetic regulatory mechanisms in the synthesis of proteins. *J. Mol. Biol.* **1961**, *3*, 318–356. [CrossRef]
60. Morange, M. The death of molecular biology? *Hist. Philos. Life Sci.* **2008**, *30*, 31–42. [PubMed]
61. Jacob, F. *The Logic of Life: A History of Heredity*; Princeton Univ. Press: Princeton, NJ, USA, 1973.

entropy

MDPI

Article

Entropy as a Geometrical Source of Information in Biological Organizations

Juan Lopez-Sauceda [1,2], Philipp von Bülow [2], Carlos Ortega-Laurel [3], Francisco Perez-Martinez [3], Kalina Miranda-Perkins [4] and José Gerardo Carrillo-González [1,3,*]

1 Consejo Nacional de Ciencia y Tecnología (CONACYT), Avenida Insurgentes Sur 1582, Colonia Crédito Constructor, Alcaldía Benito Juárez, Mexico City 03940, Mexico
2 Departamento de Procesos Productivos, Universidad Autónoma Metropolitana, Avenida de las Garzas No. 10, Colonia El Panteón, Lerma de Villada 52005, Mexico
3 Departamento de Sistemas de Información y Comunicaciones, Universidad Autónoma Metropolitana, Avenida de las Garzas No. 10, Colonia El Panteón, Lerma de Villada 52005, Mexico
4 Secretaría de Medio Ambiente y Recursos Naturales (SEMARNAT), Avenida Ejército Nacional 223, Colonia Anáhuac, Alcaldía Miguel Hidalgo, Mexico City 11320, Mexico
* Correspondence: j.carrillo@correo.ler.uam.mx

Abstract: Considering both biological and non-biological polygonal shape organizations, in this paper we introduce a quantitative method which is able to determine informational entropy as spatial differences between heterogeneity of internal areas from simulation and experimental samples. According to these data (i.e., heterogeneity), we are able to establish levels of informational entropy using statistical insights of spatial orders using discrete and continuous values. Given a particular state of entropy, we establish levels of information as a novel approach which can unveil general principles of biological organization. Thirty-five geometric aggregates are tested (biological, non-biological, and polygonal simulations) in order to obtain the theoretical and experimental results of their spatial heterogeneity. Geometrical aggregates (meshes) include a spectrum of organizations ranging from cell meshes to ecological patterns. Experimental results for discrete entropy using a bin width of 0.5 show that a particular range of informational entropy (0.08 to 0.27 bits) is intrinsically associated with low rates of heterogeneity, which indicates a high degree of uncertainty in finding non-homogeneous configurations. In contrast, differential entropy (continuous) results reflect negative entropy within a particular range (−0.4 to −0.9) for all bin widths. We conclude that the differential entropy of geometrical organizations is an important source of neglected information in biological systems.

Keywords: differential entropy; discrete entropy; geometrical information; heterogeneity; information theory

Citation: Lopez-Sauceda, J.; von Bülow, P.; Ortega-Laurel, C.; Perez-Martinez, F.; Miranda-Perkins, K.; Carrillo-González, J.G. Entropy as a Geometrical Source of Information in Biological Organizations. *Entropy* **2022**, *24*, 1390. https://doi.org/10.3390/e24101390

Academic Editor: Pavel Kraikivski

Received: 6 August 2022
Accepted: 22 September 2022
Published: 29 September 2022

1. Introduction

In the context of shapes and forms in biology, there has been an historical effort to find the source of some patterns and the fundamental nature of their seemingly steady basic arrangement. Traditionally, the bottom-up logic of biological developed structures as dynamical time-space expression processes has been extensively approached either by Neo-Darwinism (e.g., genetic blueprint or functional viewpoints) or by biological structuralism (e.g., fractal biologic patterns from chaos theory). In fact, these approaches are supported by epistemological conceptions defining traditions of research work, such as positioning whole organisms as being made of atomic and separate parts (i.e., systematics) or the holistic dynamical system approach of the structuralist point of view (e.g., Turing patterns). In contrast, our perspective employs Shannon entropy to understand biological organizations as a geometric whole whose configurations defining their steady state result from an inherent and specific level of information. One important example of steady states is derived from a prevailing and well stereotyped distribution of cellular polygons in metazoans tissues (epithelium). The question concerning whether patterns and shapes are an active source of geometrical information, stability, and variability during developmental processes and evolution represents an intriguing issue that requires further study. Although that "geometrical information" sounds very similar to the interdisciplinary field of Information geometry is important to say that they are different approaches with some important convergences that will be treated briefly at discussion.

In this work, the geometric properties of individual discrete elements in forms are not simple intrinsic features biologically exposed as outcomes. Instead, we understand them as both independent spaces in a larger whole and as units defining interacting properties inside of a larger whole of geometric information. In this line of reasoning, our main methodological question about shapes arises: Is there a way to quantify geometrical order in biological organizations using levels of information? This question has been tackled from other perspectives. There is an important amount of work related to quantifying information at different complexity levels in biological networks [1–4], ecosystems [5–9], molecular entropy [10], and cellular entropy [11], to name a few approaches. Furthermore, the characterization of ecological landscape heterogeneity (e.g., urban, sociological, and economical properties at multiple scales associated with them) have been approached [12–14] using spatial entropy and complexity tools. However, in the context of pure biology, the underlying informational order behind the geometry of general biological organizations is still not quite clear. However, there are some intuitions regarding quantitative values for biology and architecture [15]. Nevertheless, the limits defining quantitative parameters of order according to entropy, as a generic value for biological organizations, is still an issue which has yet to be solved. We maintain that an important aspect of our research is the selection of a set of biological forms to work on.

The core idea of measuring spatial heterogeneity to determine the geometrical entropy of a particular biological form is derived from a previous analysis of geometrical constrictions in five-fold morphologies (polygonal random disc organizations) [16]. In that work, it was found that spatial organization of five-fold morphologies is statistically lower than all other planar disc organizations (three to ten-fold morphologies) in terms of spatial heterogeneity (unequal distribution of space inside polygons). In fact, the authors found a statistical basis corresponding to the most frequent morphologies in biological disc organizations (three, four, five, and six disk partitions are typically found in flowers, fruits, and other biological organizations) [16]. While we found a statistical value to approach the key idea of low heterogeneity for related morphologies in nature, the authors were unable to capture quantitatively the geometrical limits of biological organizations within a formal framework of reference.

Much work has been carried out regarding the larger geometrical context of cells and the physical causalities of interactions into cell aggregates using meshes [17–23], which notably enforces our background. This work derives from physical parameters and describes geometrical properties while not strictly determining levels of geometric information. The characterization of 'self-assembled 2D patterns with Voronoi entropy' represents a certain approach for achieving geometry as a source of organization, employing levels of spatial heterogeneity at different scales [24,25]. However, the main results of this kind of work provide some insight about entropy in the context of matter organization and ecological dynamics, even stirring research on material sciences and cellular aspects (including topics such as the informational limits of generic order in biology). Living systems show an important reduction of entropy, reaching very low values along self-organization as an ostensibly consistent rule [1,3,6,10,13,16,24–28]. Such behavior has been associated with biological and physical constraints [29,30], with some proposals linking it to pure geometry [26–28,31,32]. The hypothesis we will try to verify in this work is as follows: the more self-organized a system, the less entropic is its behavior. Hence, we expect that this information is related with the ordering of geometric parts throughout biological structures.

Accordingly, Shannon entropy shall indicate the amount of information considered (besides being a proxy of geometrical heterogeneity). In addition, Shannon entropy is the average of a variable's uncertainty that reflects how much information is associated with the probability of a given event. In this paper, we propose that its range [i.e., heterogeneity, non-heterogeneity] can be translated into bits of information between 0 and 1.

The set of organizations that we choose is based on looking for strictly biological samples made of polygons at two size levels (cellular and ecological), simulations of biological samples, and experimental controls (random simulations and poisson tessellations). The main idea was to generate a proper collection of biological samples to detect particular levels of informational entropy using the unique simplicity of polygons as a general feature for a data source. Those polygons have levels of heterogeneity which will be our source of data used to establish levels of entropy in order to identify biological particularities.

To develop this idea, this paper is organized as follows. First, there is an exposition of the collecting method and features and categorization of biological images samples and non-biological mesh simulations in Section 2. These data will define the material to work on besides random polygons with different numbers of sides (Appendix A). The mathematical framework and the statistical motivation to work on these polygons and the main background used to define heterogeneity in spatial organization of polygonal shapes and meshes are given in the 'Methods' section. The procedure used to measure the quantity of information in geometrical meshes of biological and non-biological systems using Shannon entropy and the associated statistical distributions of internal partitioning in shapes is shown in Section 3. Finally, Sections 4 and 5 correspond to discussion and conclusions, respectively.

2. Materials and Methods

2.1. Materials

The outline of an area or figure is a shape that can be a determined configuration of discrete elements, which sometimes can be understood as a population of geometric parts which serve as constitutive elements. Our approach here is to determine levels of geometric information using Shannon entropy as the main theoretical framework. Therefore, informational entropy would allow for the quantification of order and disorder levels from discrete and continuous geometric variables. Continuous approaches which are able to characterize chemical, physical, and biological patterns, based on the continuous measure of symmetry, were introduced [33–38]. Suitably, the first focus of our research is on extract basic discrete and continuous geometric principles of polygons immersed into larger whole organizations (called polygonally shaped patterns, or PSP) in order to standardize levels of biological information given several amounts of heterogeneity (i.e., unequal distribution of space inside a given area). Rather than just looking at polygons and their aggregates as mathematical outcomes derived from computing simulations, in this paper we developed a statistical process to detect levels of information from them. Our method points out to spatial heterogeneity of polygons as a free scale informational substrate that can be approached on a wide range of biological size scales (which also can be easily translated into an informational entropy metric description). Hence, the material of our work will be the areas (polygons) and their associated sub-areas (triangles) defining levels of heterogeneity. Our procedure satisfies the fact that we may work with sets of polygonal shapes as an informational substrate upon we can discern levels of geometrical heterogeneity getting a width spectrum of numerical data. This metric was tested into meshes (biological, non-biological, and random polygonal arrangements) and simulated random discs with different number of sides. Finally, we will retrieve the informational limits of biological structures whose geometry would potentially be biologically representative in terms of their closeness with nature images due to the informational entropy associated.

The first step was the establishment of a collection of biological images and sample data to work on. Several biological cell organizations have been used as models to define geometric parameters. In that sense, an important number of studies have analyzed the topological properties of many cell organizations [18–22,26–28,32,39–44]. Also, a lot of epithelium models have extensively used anatomical parts, developmental stages, and tissue variations images. In fact, there is a prevailing and well stereotyped distribution of cellular polygons (SDCP) conserved in proliferating metazoans tissues with a polygonal frequency of 29% of 5-sided polygons, 49% of 6-sided polygons, and 20% of 7-sided polygons [20,40,41]. In that context, some other images of biological cell organizations are available online, such as histological samples derived from different human tissues [20,42]. Currently, it is widely accepted that although variation in those organizations exists, there is just a narrow range of variations of cellular polygonal distributions [20,40]. In this regard, samples of polygonal meshes are directly comparable even if some of them are from different origin or scale due to all of them are PSP [45,46], including biological natural images, biological simulations, non-biological simulations (such as random meshes and Poisson-Voronoi tessellations), and random polygons. Therefore, levels of Shannon entropy in polygonal meshes and sets of random polygons turn into a window of universal and comparable information if we approach them from a pure geometric perspective.

Collecting Samples

We collected samples of images (online) looking for a broad and representative set of biological organizations in order to support our main hypothesis (i.e., that geometric information defined by the Shannon entropy of spatial polygonal heterogeneity is a proper parameter able to define the limits of a generic biological organizational value using PSP). Thus, the establishment of a measure of spatial organization able to determine the geometrical entropy of order for biological forms must be analyzed measuring biological and non-biological organizations (Figure 1). At the tissue level, we used images from proliferating drosophila prepupal wing discs (dWP) [20,41,43], middle third instar wing discs (dWL) [41,43], normal human biceps (BCA) [20], muscular dystrophy from skeletal muscles (MD) [42], and pseudo stratified drosophila wing disk epithelium (PSD) [40]. Also, at the ecological level polygonal meshes derived from Namibia fairy circles (ecological patterns associated with SDCP convergences) images were integrated into the analysis (NFC) and ecological oak patterns (EOP) [45–48]. The global tag to encompass MD, dWP, dWL, BCA, PSD, and NFC is called BIO. The non-biological meshes were different diagrams resulting from different vertex model simulations. Those simulations were based on quantified distances from SDCP, which is traditionally used as reference in epithelial studies [49–51]. The closeness with SDCP can be defined through an optimal paths approach using iterations of Lloyd's algorithm and other cellular biophysical conditions in order to investigate the effects of cell divisions on topology [20]. In contrast, other work reached equilibrium states by seeking minimal potential energy [50]. Given this, there were epithelium simulations which we defined as control simulations (CS) [20,32], simulation out of equilibrium (SOE) [20,32], simulation at equilibrium (SAE) [20,32], atrophy simulation (AS) [20], and Poisson–Voronoi tessellation (PT) [20]. We consider CS, SOE, SAE, AS, and PT altogether as non-biological meshes (non BIO), since they were derived from algorithms and not from actual biological samples. In addition, in order to have a reference to contrast numerical values of nature typical arrangements we also include planar discrete areas (PDA; Section 2.2.1). Finally, we incorporated an algorithmic routine [45] to develop random arrangements (RA) into the global analysis as a control. Therefore, the analysis will include three PSP mesh categories, BIO, non BIO, RA, and data from PDA (Table 1; summary of category, abbreviation, name and number of samples).

2.2. Methods

2.2.1. Mathematical Description of Shapes Γ and Heterogeneity of Spatial Organization

The establishment of a measure of heterogeneity able to determine the geometrical entropy of biological organizations is derived from a previous analysis of spatial constrictions in five-fold morphologies [16]. The algorithm to simulate partitions and shapes Γ-PDA (planar discrete areas inside a disc; Box 1) is extensively supported in Appendix A. Here, our main methodology goes beyond, focused on statistical measurements of geometrical heterogeneity onto biological and non-biological PSP, associating levels of entropy to them using fundamentals features of shapes Γ.

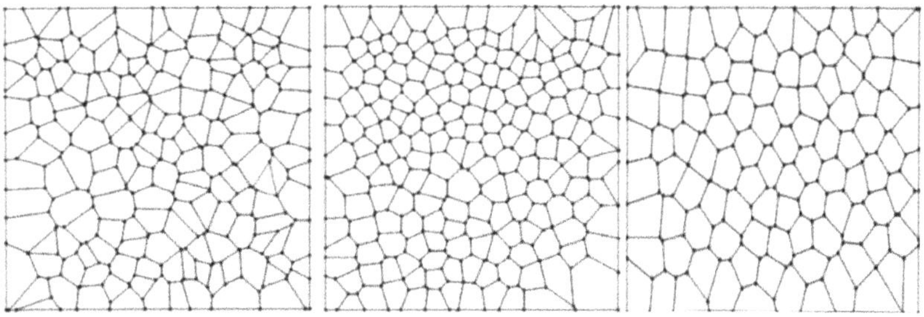

Figure 1. Three general types of schematic mosaics were analyzed. Left: random arrangements (RA). Center: natural images (BIO, schemes of aggregates of cells and ecological polygonal meshes) extracted from the web; muscular dystrophy (MD), drosophila prepupal wing discs (dWP), middle third instar wing discs (dWL), normal human biceps (BCA), pseudo-stratified drosophila wing disk epithelium (PSD), and ecological patterns (NFC and EOP). Right: processed non biological images (non BIO) extracted from the web which we named, control simulation (CS), simulation at equilibrium (SAE), atrophy simulation (AS), simulation out of equilibrium (SOE), and Poisson–Voronoi tessellation (PT).

Table 1. Summary of category, abbreviation, particular name, and number of samples.

Mesh Categories	Abbreviation	Name and Number of Samples
-	PSP	Polygonal shape pattern (total number of samples 38)
-	Γ-PDA	Planar discrete areas (8)
Bio	dWP	Drosophila prepupal wing discs (3)
Bio	dWL	Middle third instar wing discs (4)
Bio	BCA	Normal human biceps (2)
Bio	MD	Muscular dystrophy from skeletal muscles (1)
Bio	PSD	Pseudo stratified Drosophila disk epithelium (4)
Bio	NFC	Namibia fairy circles (2)
Bio	EOP	Ecological Oak Patterns (3)
Non-Bio	CS	Control simulations (5)
Non-Bio	SOE	Simulation out of equilibrium (1)
Non-Bio	SAE	Simulation at equilibrium (2)
Non-Bio	AS	Atrophy simulation (2)
Non-Bio	PT	Poisson–Voronoi tessellation (1)
RA	RA	Random arrangements (50)

A former statistical analysis is derived from the study of partitions (areas) and their sub-localities (sub-areas) arising from computational constructions named Γ shapes. Generically, a shape Γ is a set of numerical values able to be analyzed statistically which is composed of sub-localities which are areas inside a partition P_i (Box 1). Therefore, there are two particular cases of Γ shapes. Tthe first particular case of shape Γ can be a set of sub-areas derived from a partition P_i being a disc simulation with a given number N_i of sub-localities (Γ-PDA). The second one is a regular or irregular polygon with any number of sides. In that sense, each shape Γ can be achieved as a set of numerical sub-areas that can be subject to be statistically analyzed. The main idea used to establish the generic name of shape Γ is that it is useful to name either geometric objects (e.g., irregular and regular polygons or PDA) or areas (numeric values inside discs simulations or Γ-PDA) associated with either discs or any 2D simulated or not simulated polygonal shape derived from meshes.

Box 1. Partition number.

Figure a–c shows the process of partitioning using, as an example, five sub-localities. The concentric scheme at figure d shows three levels of variability (shadow zones limited by 1, 4, and 8) according to the scale given by the first circle radius. These shadow restricted zones are areas whose random points define sub-localities according to a particular partition number (figure a–c). This methodology is applied to partition number P_i using discs with 3, 4, 5, 6, 7, 8, 9, and 10 sub-localities. The second concentric circle limits the variation of area once that Voronoi algorithm is running in order to limit as much as possible the area variability.

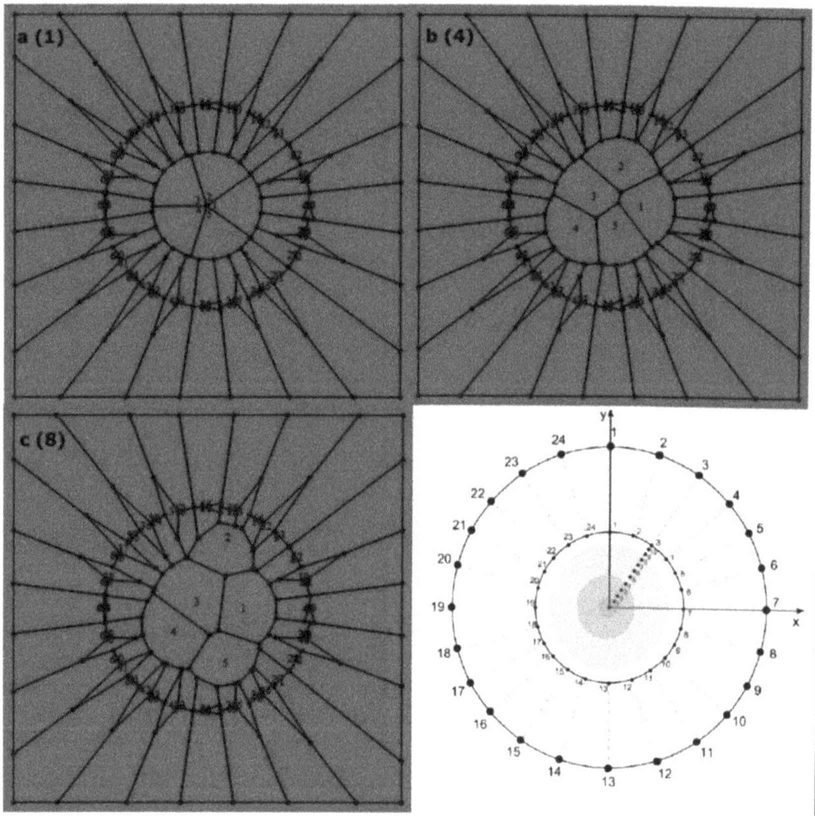

Partitions P_i are sets of areas where each partition is constituted by a subset of a given number N_i of sub-localities, $S_{i1}, S_{i2}, \ldots, S_{iN_i}$ such that $P_i = \cup_{j=1}^{N_i} S_{ij}$, where P_i is a spatial region which could be either a set of areas as numerical values or any convex polygon in $\mathbb{R}^2$. In order to start with a statistical description, let A_{ij} be the area of each sub-locality. If $A_{ij} = A_{ik} \forall\, j, k$, then we said that P_i is non heterogeneous. In contrast, if exists some $j \neq k$ such that $A_{ij} \neq A_{ik}$ then we say that P_i is heterogeneous. Therefore, let $A_i = \sum_{j=1}^{N_i} A_{ij}$ be the sum of all of the associated areas of a partition; this set determines a shape $\Gamma = \{A_i\}$. We consider a polygon as a first example of a particular shape Γ with a set of sub-areas which are considered in numerical terms. In that example, the numerical values correspond to triangle areas from a hexagon representing a particular case of a shape Γ (Figure 2).

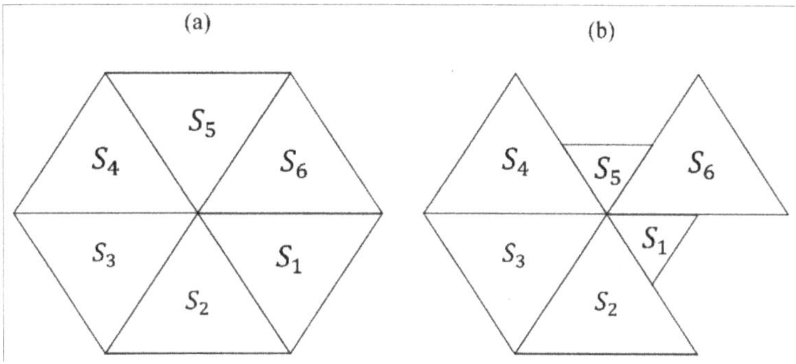

Figure 2. Schematic properties of two different examples of a particular shape Γ. (**a**) A regular hexagon is a partition associated with six sub-areas from six sub-localities $S_1, S_2, \ldots, S_6$ which are all equal. Then it is non heterogeneous. (**b**) A shape Γ with a six-fold heterogeneous partition such that the areas defined by sub-localities S_1 and S_5 are smaller than those of S_2, S_3, S_4, and S_6, then this is heterogeneous.

Therefore, the area average of a partition P_i is:

$$\overline{A}_i = \frac{1}{N_i} \sum_{j=1}^{N_i} A_{ij} \tag{1}$$

and

$$\sigma_i = \sqrt{\frac{1}{N_i - 1} \sum_{j=1}^{N_i} \left(A_{ij} - \overline{A}_i\right)^2} \tag{2}$$

is the standard deviation of each partition. Notice that if $\sigma_i = 0 \Rightarrow A_{ij} = A_{ik} \; \forall \, j,k$. Therefore, that partition is non heterogeneous.

Equation (3) reflects the amount of heterogeneity in a given shape Γ and is inside a scale from 0 to ~1. Generalization for equations defining heterogeneity of polygons is written x_i, where sub-index i reflects the number of sides of each polygon.

$$x_i - \sigma_i / \overline{A}_i \tag{3}$$

The main objective of our research here is the establishment of a measure of spatial organization which is able to determine the geometrical entropy for biological and non-biological organizations. Therefore, we must relate (3) with a proper collection of shapes Γ reflecting numerical data of spatial heterogeneity in PSP, quantifying indexes of heterogeneity in all of our samples (those from Section 2.2.1). Before defining entropy in mosaics of cells we have to develop a proper methodology to get the coordinates of individual polygons. As mentioned in Section 2.2.1, we used biological (natural) and non-biological processed images (from web sites and references) to define the coordinates of polygons using the centroid of each polygon as the origin of polygonal coordinates. The heterogeneity of each polygon in mosaics was derived with (1), (2), and (3), see Figure 3. With this data and the statistical description of polygons as Γ shapes, we get frequency distributions of heterogeneities for each mesh.

We relate (3) with a proper collection of data reflecting rates of spatial heterogeneity, then quantifying indexes of heterogeneity in all of our theoretical samples. Consequently, our results will be the standard deviation of heterogeneity derived from levels of variability in a collection of shapes Γ from a set of PSP samples. In order to define the standard deviation of heterogeneity we have to determine first the average of PSP heterogeneity for all samples with (4),

$$\overline{x} = \frac{1}{N_s} \sum_{s=1}^{N_s} x_{is} \tag{4}$$

now, the first subscript i of x_{is} correspond to the number of sub-areas, s is the index of a shape, and N_s is the total of shapes Γ in a mesh of polygons. The standard deviation can be obtained with (5),

$$\sigma = \sqrt{\frac{1}{N_s - 1} \sum_{s=1}^{N_s} (x_{is} - \overline{x})^2} \tag{5}$$

Equation (5) reflects a global statistical value aiming to determine area variability and the informational entropy.

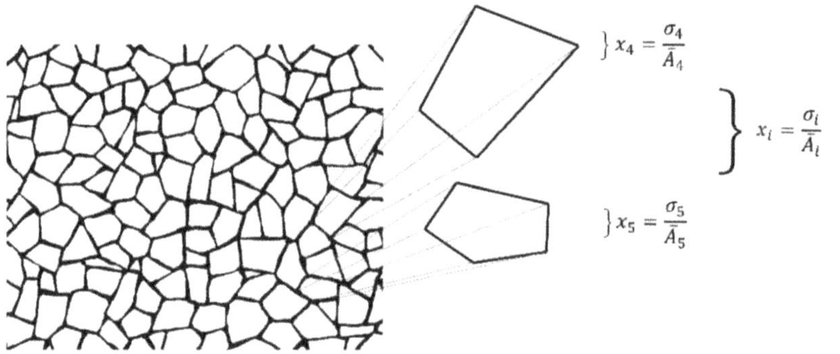

Figure 3. Symbology of equations for individual polygons extracted from a mesh. The expression given by (3) is used in order to obtain individual polygon heterogeneity. Also, the distribution of spatial heterogeneity derived from frequencies levels of heterogeneity in meshes of polygons of BIO, Non-BIO, and RA was defined using the values given by their heterogeneity.

3. Results

3.1. Continuous Distribution of Heterogeneity for Shapes Γ-PDA

An important question is whether the variable side number of polygonal shapes in the context of heterogeneity might lead to a continual progression in terms of informational amount or not (that is, entropy as a function of polygonal side number). Since sampled meshes (PSP) are sets of mixed polygons with different number of sides this question must be approached using frequency distributions of heterogeneity in the first case of shapes Γ using planar discrete areas inside a disc (Γ-PDA) with a fixed number of sides as the independent variable (algorithm and methodology are provided in Appendix A). The main aim for all of these data is whether statistical variations of spatial distributions in polygons have particular attributes to obtain some clues of biological configurations. Initially, we have discrete distributions for heterogeneity data extracted from shapes Γ-PDA, which will be transformed into continuous ones applying the probability density function algorithm (Wolfram Mathematica 9.0; Champaign, IL 61820-7237, USA. Figure 4).

In order to start with a continuous approach to infer levels of entropy, we decided to use a Kolmogorov-Smirnov test between normal distribution of a Gaussian random variable of heterogeneity and the remaining ones to detect distribution differences. For continuous distributions, the maximum entropy corresponds to normal distribution, since a Gaussian random variable has the largest entropy amongst all random variables [52,53]. Therefore, we consider that Kolmogorov–Smirnov test will give us a good proxy of closeness with normal distribution as a first hint of high entropy. According to Figure 5, the Log base 10 of p-values of a Kolmogorov–Smirnov test is applied in Wolfram Mathematica 9.0, resulting in a proper comparative of entropy in continuous terms. Frequency distributions of three, four, five, and six Γ-PDA are the four lowest values which is an indicative of low entropy and dissimilarity with normal distribution. In that sense, p-value is positively related to entropy, low p-values indicate low entropy, high p-values indicate high entropy and normality. Kolmogorov–Smirnov test performs the Kolmogorov–Smirnov goodness-of-fit test with null hypothesis H_0 that data was drawn from a population with a normal distribution and alternative hypothesis H_a that it was not (Wolfram Mathematica software 9.0). Also, as an initial experiment one BIO sample (PSD) [40] and one random sample are included, showing that BIO sample has an important contrast with random sample in terms of Kolmogorov–Smirnov test results. The BIO sample reach a value of −38.54 while the random value is −1.23. The two local minima (four-side and BIO samples) are indicative of small p-values which reflect that both configurations come from samples with low entropy that is hypothetically frequent in biological arrangements [1].

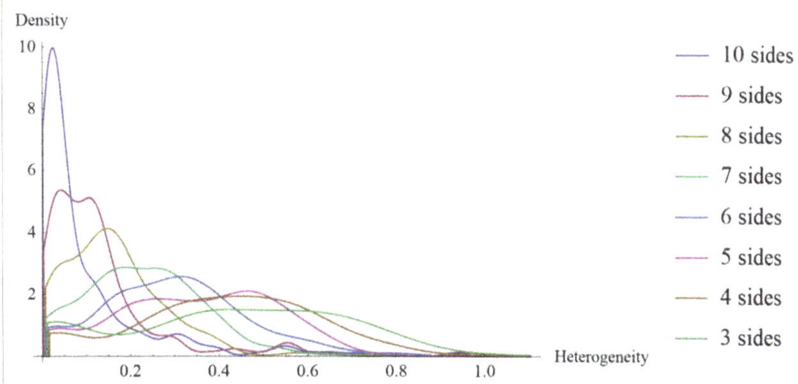

Figure 4. Probability density function for distributions of Γ-PDA applied to transform histograms with discrete values (modified from [16]) into continuous graphics. The horizontal axis shows heterogeneity levels derived from Equation (3).

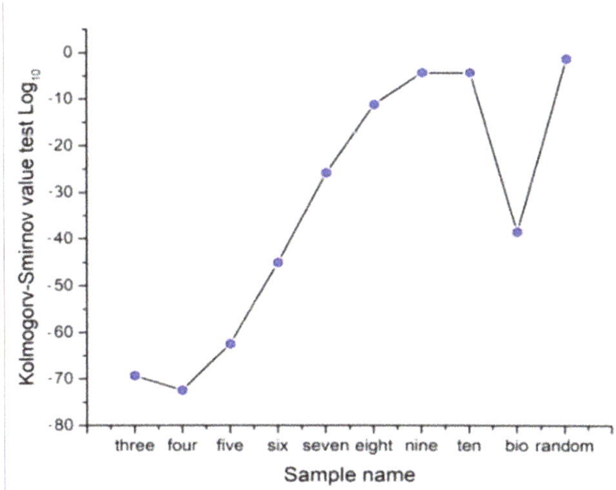

Figure 5. Contrasting continuous distributions of frequencies using heterogeneity data. In order to start with a continuous approach to detect levels of entropy we use Kolmogorov–Smirnov test as a parameter to detect distribution differences between normal distributions and the remaining ones. Heterogeneity values of random sample have the closest value to normal distribution. According to the Log base 10 Kolmogorov–Smirnov test values, partition number four has the lowest values of entropy in continuous terms.

3.2. Bin Categorizations for Measuring Discrete and Continuous Entropy Using Polygons

The Shannon entropy is a parameter indicating a degree of information approaching a resolution of uncertainty. Our description model satisfies the fact that we may work either with frequencies of numerical variables which are sub-areas of polygonal shapes in the context of PSP or with areas using Γ-PDA. Shannon elucidates the convenience of the use of a logarithmic function in the definition of entropy, mainly due to the fact that it is more suitable mathematically since many operations in terms of the logarithm are simpler than in terms of the statistical behavior (the number of possibilities or frequency). In fact, one of our main source ideas is the finding of a practical procedure to retrieve PSP given a geometric informational entropy value. The average surprise of a variable X, which has a distribution $p(X)$, is called the entropy of $p(X)$ and is represented as $H(X)$. For convenience, we often speak of the entropy of the variable X even though (strictly speaking) entropy refers to the distribution $p(X)$ of X [53]. Thus, the entropy of the heterogeneity variable x_i from (3) can be derived from the general formula for discrete values:

$$H(X) \approx \frac{1}{N_j} \sum_{j=1}^{N_j} log \frac{1}{p(x_i)_j} \tag{6}$$

where the subscript j represents the variable number.

The choice of a logarithmic base regards for a proper election of a unit for measuring information. In consonance with this last idea, we consider frequency values of heterogeneity in a range of bin width. The entropy values using different bin widths (0.1, 0.2, 0.25, 0.33, and 0.5) show that this variable gives an important difference in terms of the decrease of values (Figure 6a from black to green) even in terms of a statistical correlation with raw heterogeneity data (Table 1; for discrete and differential entropy). Regarding differential entropy as a continuous technique, we can consider a formal approximation using:

$$H_{dif}\left(X^{\Delta}\right) \approx \left[\sum_i P_i log \frac{1}{P_i}\right] - \log \frac{1}{\Delta x} \tag{7}$$

where i is a subscript referring for the ith bin and Δx is the bin width. The count of the ith bin is n_i whose area is $a_i = n_i \times \Delta x$. The total area is $A = \sum_i a_i$, and the proportion $P_i = a_i / A$. Equation (7) derives from:

$$H_{dif}(X) = \int_{x=-\infty}^{\infty} p(x) log \frac{1}{p(x)} dx \tag{8}$$

which is a measure of entropy called differential entropy (continuous entropy) of a variable. Equation (8) can be suited to ignore infinity, so (7) can be derived from (8). For completeness, we measure differential entropy with the data of each sub-areas number considering the five bin width values for discrete Γ-PDA datasets, see Figure 6b (from black to green). Even though each value of a continuous variable can, in principle, convey infinite information, the amount of information it conveys in practice depends on the accuracy of our measurements. In effect, measurement noise divides up the range of a continuous variable into a finite number of discrete intervals; the number of intervals increases as the measurement noise decreases. Therefore, bin width 0.5 represents the noisiest interval of our samples and bin width 0.1 the least of all [53]. Finally, the total standard deviation for discrete entropy values is 0.109905 and for differential entropy is 0.083602. In addition, the total range for discrete entropy goes from 0 to 3, in contrast with differential entropy which reach −1.2 from 0.

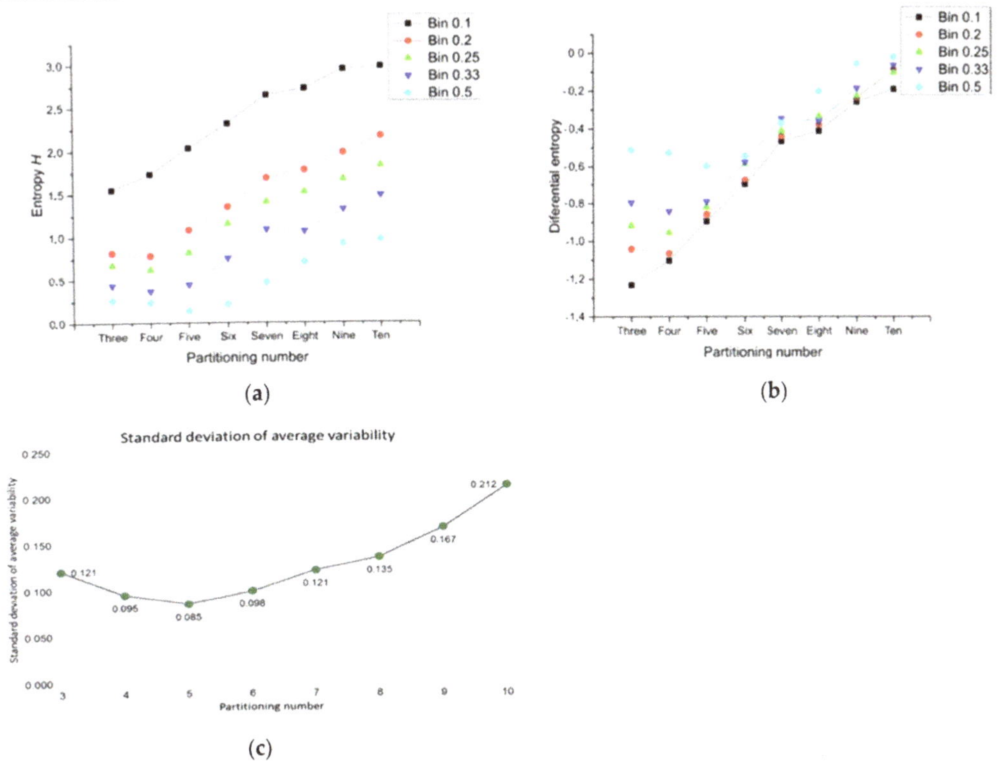

Figure 6. Entropy for Γ-PDA datasets. (**a**) Partitioning number and their associated entropy can be derived from different bin statistical discrete categorization. Graphic shows five bin widths and their associated entropy. Bin width 0.5 has the lowest values of entropy for every partitioning number, meanwhile bin 0.1 statistical categorization has an approximately linear incremental behavior in contrast with the remaining categorizations. In addition, this graphic also shows that there is a similar pattern between discrete and standard deviation of variability (**c**) in terms of the distance from zero using Bin 0.5. (**b**) The associated differential entropy of a partitioning number was derived from Equation (8). Differential entropy datasets show that negative entropy goes from −0.0181 to −1.2309. (**c**) The graphic shows the standard deviation of raw heterogeneity for Γ -PDA using the logarithm base 10, using Equations (4) and (5).

3.3. Statistical Frequency Distributions of Internal Partition in Γ-PDA and Binary Localities in Bio, Non-Bio, and RA Samples

In order to start with a proper analysis of entropy, we must consider the correlation values of Table 2. Higher correlation values imply a first hint for bin correlation. In spite to seem a weak statistical argument to detect both, the closest bin category and the right mathematical variable to use (either discrete or continuous); this correlation remains variable considering all bin categories. According to partition number the statistical frequency distribution of heterogeneity of Γ-PDA is showed in Figure 7. The bar area determines levels of heterogeneity conforming the binary categorization x_{il} and x_{ih} as a first pragmatic approach. The green area shows levels of high heterogeneity where $0.5 \leq x_i < 1 \Rightarrow X_i = x_{ih}$. On the other side, low levels of heterogeneity correspond to the grey area where $0 \leq x_i < 0.5 \Rightarrow X_i = x_{il}$, with x_{il} values derived from Equation (3).

Table 2. Correlation values between discrete and differential entropy with standard deviation of heterogeneity raw data.

Bin Width	r between Dis_E and STD_HRD	r between Dif_E and STD_HRD
0.1	0.7215	0.7405
0.2	0.8129	0.8191
0.25	0.8161	0.8221
0.333	0.8642	0.8667
0.5	0.9311	0.9308

Dif_E = differential entropy; Dis_E = discrete entropy; r = correlation; STD_HRD = standard deviation of heterogeneity raw data.

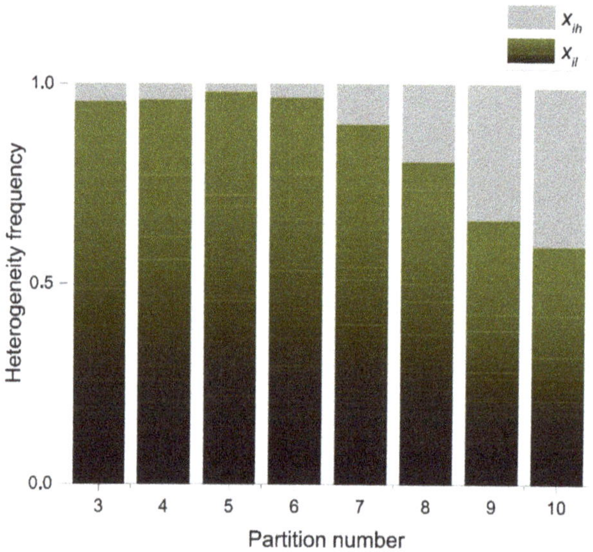

Figure 7. Levels of heterogeneity according to the binary categorization x_{il} and x_{ih}. Grey zones are frequency values associated to x_{il} and the green ones are associated with x_{ih}. The highest level of homogeneity is for partition number five (grey area), even though, three, four, and six have similar levels. The highest level of heterogeneity is for partitioning number ten.

In order to link entropy and raw geometrical information, our evidence suggests that five-folding organization depicts a sort of spatial organization with low values of information (besides three, four, and six folding organizations, which are also frequent in nature). In fact, this sort of arrangement shows the highest correlation with raw low spatial heterogeneity data for both, differential and discrete entropy (Table 2). As we were pointing out before we consider that binarity must give us some clues in order to understand heterogeneity and discrete entropy (Figure 7) from a simple perspective. The fact that five-fold partitions reflect the most equal distribution of internal space in contrast with other partitions, it is a non-trivial result since this is not a function derived from the polygonal number of sides (Figure 6a,b). According to Figure 8, such as five-fold arrangement, biological organizations of cell aggregates that were derived directly from natural images, or even biological simulations, have a constant high proportion of low heterogeneity in terms of spatial distribution. That implies a clear high degree of homogeneity lying on that proportion of low heterogeneity that is found in all samples derived from biological approximations.

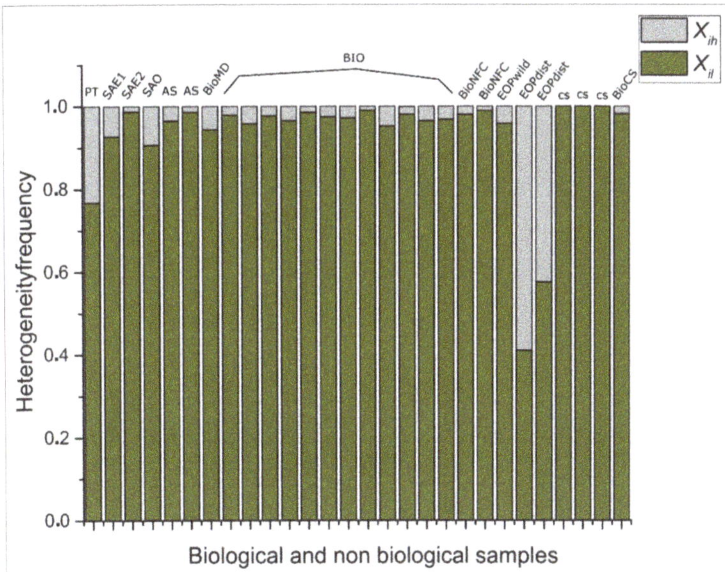

Figure 8. Twenty-eight samples of biological and non-biological simulations of organizations of cells aggregates have a constant high proportion of homogeneity in terms of spatial distribution of inner areas (for name samples and nomenclature of BIO and non-BIO see Table 1). Data from columns BioNFC (Namibia fairy circles), EOP wild (non-disturbed ecological oak pattern) and EOPdist (disturbed ecological oak pattern) shows that at ecological level a wild zone has less heterogeneity polygons that a disturbed zone. The last four samples are biological simulations [20]. The first three simulations result with an entropy of 0. All of these samples result from a dynamical configuration derived from a fine tuning of biophysical parameter variation (line tension and tension values). Even this is happening just when the impairment of the cell division when tension value threshold reaches a 40 percentage with cell proliferation and heterogeneous reduction of line tension among the tissue cells the informational entropy increases up to 0.132065 (BIO CS sample). The first column represents a Poisson–Voronoi tessellation which was used as control.

This last result shows the simplicity of approaching the geometry of biological organizations focusing on the binarization data in order to (may) see the main facts of the organizational nature of biological geometries that are often found. In addition, Figure 8 indicates that binarizations works well as an indicator to realize heterogeneity levels in complex meshes of polygonal arrangements since we can see the statistical behavior of data, a high degree of low heterogeneity (homogeneity) and a very low degree of heterogeneity. Random samples are used as a control experiment to visualize contrast in terms of those distributions in Figure 9.

3.4. Discrete Entropy for Shapes Γ from Bio, Non-Bio, and RA Samples Using Binarization

In terms of discrete entropy, there are an important number of simulations extracted from the open access figures that are excluded from the BIO zone despite of being considered as simulations of biological samples (Figure 10). All of those collected simulations were based on quantified distances from SDCP conserved in proliferating metazoans tissues with a polygonal frequency of 49% six-sided polygons, 29% five-sided polygons, and 20% seven-sided polygons, which is traditionally used as reference in epithelial studies. Control simulation (CS), simulation at equilibrium (SAE), atrophy simulation (AS), and simulation out of equilibrium (SOE) were computational simulations of cells aggregates assuming variations as metric distances from Lewis's Law values or holders of another kind of biological or physical properties. The Shannon entropy associated with biological simulations, such as CS reaches levels of zero entropy, which implies null information which is not the case for either Γ-PDA associated with hypothetical biological morphologies or samples constricted to the BIO zone. However, there was a sample that increase their entropy according to some simulation conditions (Figure 10). Certainly, Poisson–Voronoi tessellation (PT) was used as control since we assumed that its arrangement would be far away from the order zone. Hence, the geometry between BIO and non-BIO arrangements of internal space are underlying important differences whose consequences and effects would define particular behavior in actual biological organizations.

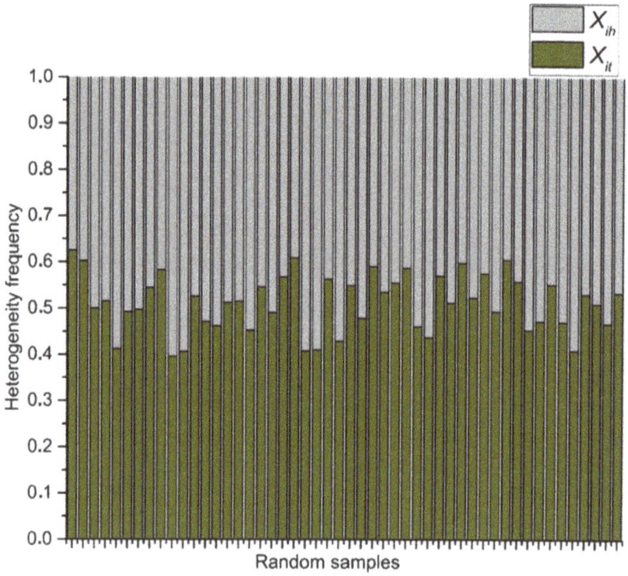

Figure 9. Random arrangements of cells and their heterogeneity frequency. Data shows that random aggregates have an average of an almost half proportion of low heterogeneity (blue) of spatial distribution on internal areas in polygons, and a half of spatial high heterogeneity (grey).

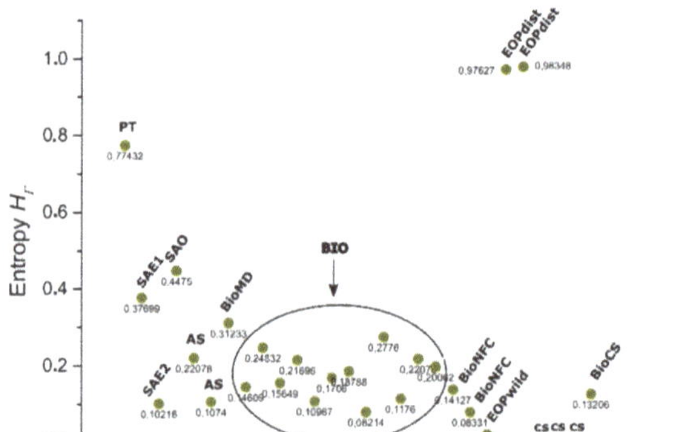

Figure 10. The entropy of cell aggregates groups: biological collected images (BIO; descriptions in Table 1) and processed images which we named non BIO extracted directly from online open access figures; control simulation (CS), simulation at equilibrium (1 and 2) following four interactions of

Lloyd's algorithm (SAE), atrophy simulation (AS), simulation out of equilibrium (SOE), muscular dystrophy (BioMD), and Poisson–Voronoi tessellation (PT). The most abundant area (ellipse) includes BIO data, which is close in terms of entropy with AS, SAE, and SOE. BioNFC (Namibia fairy circles) and EOPwild (ecological oak pattern wild) are also defined by a low degree of entropy. That is not the case for EOPdist (ecological oak pattern disturbed). Control simulation of biological organizations reaches a 0 entropy value. That value can change when biophysical manipulation of parameters is included [20].

As we can see at Figure 10 the BIO zone includes images that are not being simulated, such as dWP, dWL, and BCA (Table 1). However, simulations that have some kind of manipulation can increase their heterogeneity resulting in highest entropy than control simulations without parameter variation. One interesting point comes from the ecological oak patterns which are disturbed and non disturbed oak zones [48]. The level of entropy increases whether the zone is a perturbed ecosystem or not.

The Shannon entropy associated with RA frequency distribution (Figure 9) is an expected result, whose discrete entropy values are around 1 (Figure 11).

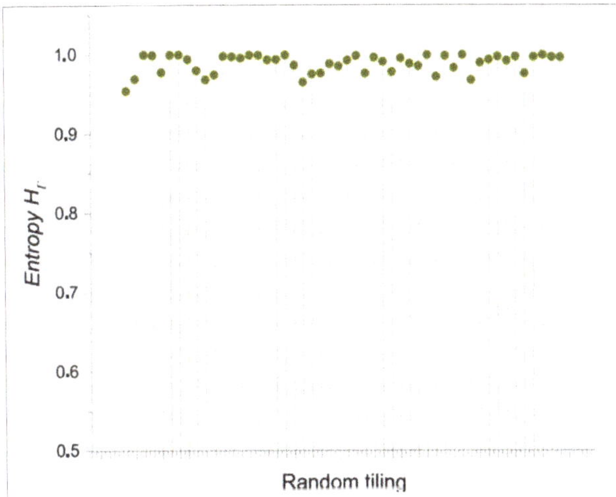

Figure 11. The discrete entropy of random samples (dataset derived from Figure 9). The entropy values are almost constantly in line with maximum entropy.

3.5. Continuous Entropy for Shapes Γ from Bio, Non-Bio, and RA Samples

To estimate the entropy of any variable, it is necessary to know the probability associated with each of its possible values [53]. As we point out (Section 3.1) probability density function is a well-accepted starting reference to estimate a continuous distribution from discrete possible values. In fact, it has been an important mathematical trouble that has been solved arriving to Equation (8). In addition, Equation (8) can be useful even with discrete values using bin areas a_i. In order to establish a panoramic view of continuous entropy values we consider getting the values from Bio, non-Bio (SOE, AS, and SAE) and RA samples. According to Figure 12 there are two negative intervals for all bin categories.

Despite being working with the same data we have an important gap among discrete entropy and differential entropy (Figure 6a,b) given that the first results are positive and the second negative. Regarding this last point, we will develop some more hypotheses at a later time.

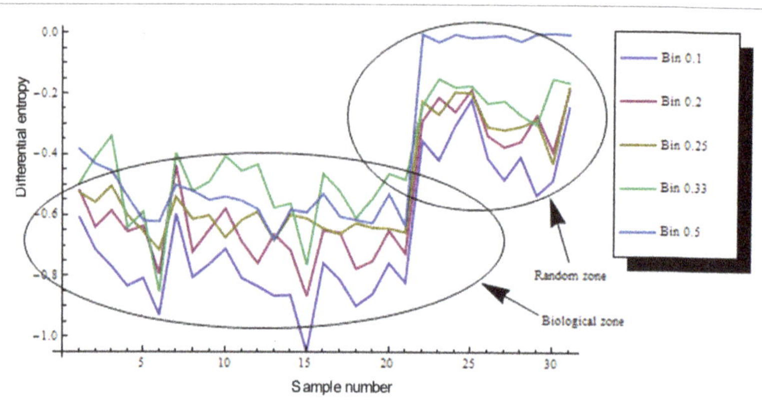

Figure 12. Differential entropy for total datasets. The continuous approach reflects that entropy values are negative in 21 Bio samples and the results are related Bio, non-Bio (SOE, SAE, and AS) and RA samples.

4. Discussion

We may see that three, four, five, and six-fold Γ shapes as planar discrete areas (Γ-PDA) behave as almost egalitarians in terms of raw spatial inner heterogeneity (Figure 6c) which we consider as a first reference of geometrical constraint in biological organizations. Beside this last fact, we consider as an important issue to be the differential entropy derived from the geometry of polygonal shape patterns (PSP) samples whose values remain close to those of Γ-PDA (Figure 6b). In addition, both partitioning number of shapes Γ-PDA and their associated entropy and the differential entropy derived from the geometry of PSP can be derived from different bin statistical discrete categorization. Figure 6a,b show five bin categorizations and their associated entropy (bin width 0.1, 0.2, 0.25, 0.33, and 0.5) for discrete and continuous values. Given that bin width 0.5 has the highest correlation with raw spatial heterogeneity for both values of entropy (Table 1), we decided to use it as the main dataset to observe discrete entropy at Section 3.3. On the other side, bin width 0.1 statistical categorization has a linear incremental behavior in contrast with the remaining categorizations. According to the elected binary system, where low heterogeneity is in the range $0 < = X_{il} < 0.5$ and high heterogeneity is in the range of $0.5 < = X_{ih} < = 1$, biotypical arrangements distributes internal space in a very egalitarian statistical way. For discrete values an interval of entropy values emerges, clustering arrangements from biological samples (around 0.08 and 0.27 bits of entropy; Figure 10). Section 3.3 shows the methodology to analyze discrete entropy using three types of mosaics (PSP): Random arrangements (RA), natural images extracted from the web (BIO), and processed images also extracted from the web (Non-BIO): which we named control simulation (CS), simulation at equilibrium (SAE), atrophy simulation (AS), simulation out of equilibrium (SOE), and Poisson-Voronoi tessellation (PT). Spatial heterogeneity in mosaics of polygons was derived using (3) for each polygon and discrete entropy using (6). Random arrangements of cells and their heterogeneity frequency shows that random polygonal aggregates representing cell aggregates have an average of an almost half proportion of heterogeneity of spatial distribution on internal areas in polygons with a nearby equal half of spatial homogeneity (Figure 9). In fact, this result explains by itself how is that highly heterogeneous partitions gives a highly entropic result.

Biological simulations (which we have included both as part of non BIO samples) of organizations of cells aggregates have a constant high proportion of homogeneity in terms of spatial distribution of inner areas. Some other approaches have found similar results, such as that analyzing avian photoreceptor patterns representing a disordered hyperuniform solution to a multiscale packing problem [54]. In fact, the penultimate three samples (CS) areas in biological simulations assuming Lewis's Law have a 100% degree of homogeneity (Figure 8). Then, a high degree of homogeneity in a computational simulation following some algorithmic instructions could derive in a beautiful representation following the SDCP of a real biological sample but a considerable lack of substantive geometric information. Thus, levels of intrinsic disorder (heterogeneity) emerging from the actual biological forms are necessary to have a proper simulation. A typical statistical approach using just statistical differences between different polygonal organizations shall not integrate this last key issue. Despite found statistical variations between BIO and Non-BIO organizations for PSP in terms of discrete entropy, differential entropy shows a better resolution (with an $\overline{\sigma}$ of 0.115982 in contrast with $\overline{\sigma}$ of 0.187632 for discrete values) resulting in an interesting gap for all bin categorizations ($\overline{X} = -0.61872$). To finish with the discussion about the continuous subject, we shall remark that this research is not inside the interdisciplinary field of information geometry. Despite this, there are some interesting methodological convergences that can be visited at [55,56]. In addition, we considered that the main convergence lay on a very interesting epistemological subject, geometry as a source of information. On the other hand, regarding discrete entropy, BIO group is between 0.08 and 0.27 bits which is a range for entropy values including three, four, five, and six folding partitions which are very common in nature. Also, in Figures 8 and 10 the first value represents a Poisson–Voronoi tessellation (PT) which was used as a control since this mesh is derived from a well know non-ordered organization of points. Even this sort of organization is not biological it seems not be inside the gap of random organizations for discrete entropy. The most abundant grey area of Figure 10 is considered as the BIO zone, which also include AS (that is a non-Bio sample). Hence, the atrophy of some simulations increases their heterogeneity degree which finally derives in a biological-like outcome. Regarding the differential entropy the Bio zone is a clear interval showed at Figure 12 which remains with a notable distance from random differential entropy. In that sense, considering the continuous approach where the inclusion of Non-Bio into BIO group seems clear is not an unexpected result since computational simulations representing algorithmic instructions are perturbed in a way that could easily derive in a biological entropy position. It does not happen with control simulations since heterogeneity does not appear at all. Hence, the algorithmic constructions showed on this paper are following hidden mathematical prescriptions reveling high levels of homogeneity beside another fundamental nature of the BIO group, a lightly bias disruption of order. In fact, five control simulation group whose main feature has been the closeness with SDCP (CS right side) have values of zero entropy (Figures 8 and 10).

On the other hand, MD seems to be a close object to BIO realm. However, it is not inside the limits. We consider that it is an important find since our parametric measure of geometric information can give us some clues about pathological routes in a very simple way, that important finding agrees with [43]. At the level of ecological scales, we include just two image samples that were very representative. Namibia fairy circles are one of the most interesting results since we confirmed some previous hypothesis about the potential of free scale approaches to understand biological organizations [46].

5. Conclusions

The main goal of this research lies on the intriguing question whether geometry is an actual source of information defining biological arrangements. The Shannon information of an outcome is also called surprisal since it reflects the amount of surprise when that outcome is observed [53]. In the context of information theory, the fact of being surprised requires knowing which outcomes are more surprising and which are less surprising. According to this last idea, we have specific statistical distribution of spatial heterogeneity frequencies for Bio, Non-Bio, RA, and Γ PDA using collections of individual polygons and disc simulations. All of these outcome frequencies are treated as outcome probabilities that are giving us particular levels of discrete and differential entropy for biological organizations using pure geometry. High levels of heterogeneity imply an intrinsic amount of surprise in contrast with a high degree of heterogeneity using the binarization approach. Therefore, our results reflect that there is a potential informational limit for biological organizations in terms of discrete and differential entropy. Despite of the value of this result there is still a broad distance to conclude that the differential entropy interval represents a unique range since it is not the same for discrete entropy. A deep mathematical and computational research is still lacking in order to define the limits of biological geometric information of polygonal aggregates. However, biological organizations are complex spatial systems which should be constrained into a narrow window of variability depending on levels of heterogeneity that can be translated into informational entropy. Paradoxically, we can see a myriad of morphological variations in nature. We conclude that the statistical properties of biological architectures can be manifested into an overwhelming number of morphologies since all of them are singular possibilities in a realm of pure organization with particular geometrical attributes (such as heterogeneity). In that sense, shape is a constant dynamical composition of arrangements and an opening infinite possibility of configurations with spatial confined attributes as a consequence of its essential organization which depends on their own informational limits. According to our results, we consider that homogeneity with very low levels of heterogeneity in biological systems is a fundamental factor for biological organizations (e.g., network theory calls it sparsity). Hypothetically, in the context of complex adaptive systems spatial heterogeneity could be associated with a source of variation (or noise) and degrees of freedom, which is notably a different perspective from the pure blueprint genetic approach, whose information lies exclusively onto molecular and ontogenetical basis. With this in mind, we consider that the value and limits of informational entropy for geometrical systems in biology is a novelty approach with a potentially width domain of impact.

Author Contributions: Conceptualization, J.L.-S. and J.G.C.G.; data curation, J.L.-S. and J.G.C.G.; formal analysis, J.L.-S. and J.G.C.G.; funding acquisition, J.L.-S., P.v.B., C.O.-L., F.P.-M., K.M.-P. and J.G.C.G.; investigation, J.L.-S., C.O.-L., F.P.-M. and J.G.C.G.; methodology, J.L.-S., P.v.B., C.O.-L., F.P.-M., K.M.-P. and J.G.C.G.; project administration, J.L.-S.; resources, J.L.-S., P.v.B., C.O.-L., F.P.-M., K.M.-P. and J.G.C.G.; software, J.L.-S. and J.G.C.G.; supervision, J.L.-S.; validation, J.L.-S., C.O.-L., F.P.-M., K.M.-P. and J.G.C.G.; visualization, J.L.-S., C.O.-L., F.P.-M. and K.M.-P.; writing—original draft, J.L.-S. and J.G.C.G.; writing—review and editing, J.L.-S., P.v.B., C.O.-L., F.P.-M., K.M.-P. and J.G.C.G. All authors have read and agreed to the published version of the manuscript.

Funding: This research received no external funding.

Institutional Review Board Statement: Not applicable.

Informed Consent Statement: Not applicable.

Data Availability Statement: The sources of the data used in this study are mentioned throughout the paper.

Conflicts of Interest: The authors declare no conflict of interest.

Appendix A. A Numerical Approach Using Partitions of Shapes Γ-PDA (Planar Discrete Areas)

A complete view of a wide spectrum of planar discrete areas (PDA) is obtained if we design a numerical model. Our geometrical design has as a first condition, namely the fact that shapes Γ-PDA with different number of sub-localities remains with a constant area during the experiment in preparation for obtaining normalized data. In order to establish variability inside a constant area, we consider two conditions for shapes Γ-PDA: (a) they must remain with an almost constant area during the experiment where partition P_i range from 3 to 10 sub-localities (eight categories); and (b) also each partition P_i must include 10 levels of variability. Therefore, each partition P_i with a particular constant area has 10 levels of variability during the experiment. We must be aware that shapes Γ-PDA is a particular case of a partition P_i.

For this purpose, we use Voronoi diagrams to model space of shapes Γ-PDA with different number of parts (from 3 to 10) where two variables are studied, namely partitioning number (pn) and partition variability (pv), which are defined as follows:

a. The partitioning number (pn) defines the number of partitions inside a disc (ranging from 3 to 10): Each partition P_i is constituted by a subset of a given number N_i of sub-localities, $S_{i1}, S_{i2}, \ldots, S_{iN_i}$ such that $P_i = \cup_{j=1}^{N_i} S_{ij}$, where P_i is a spatial region which could be any Γ -PDA in $\mathbb{R}^2$.
b. Partition variability (pv) determines multiple levels of variability (10) inside each pn by using random points, which in turn will define the Voronoi diagrams.

The algorithm to build pn and pv is described in the next seven steps as follows:

1. Features of the external disc: the boundaries of the external limit are defined by 24 fixed points generated as follows: The radius of the external disk is set to $r = 1$ and consecutive points are separated by an angle $\theta/24$ (where θ corresponds to 2π). Point 1 is aligned with axis y (Figure A1).
2. Features of the internal disc: the boundaries of the internal limit are defined by 24 fixed points generated as follows: The radius of the internal disk is initially set to $r = 0.53 \pm 0.4$ with 24 points consecutively separated by an angle $\theta/24$. Point 1 is aligned with axis y. (Figure A1).
3. Partitioning number (pn): once the number of partitions is defined, say n (where $3 \leq n \leq 10$ and $n \in \mathbb{Z}$), points are located in the disk at angles $2\pi/n \pm 0.069$ radians but at different radius. These radius values will define the pv, as described in the next item.
4. Partition variability (pv). For each angular region defined above, 10 points are located at radius (between r = 0 and r = 10) at different positions to define different degrees of variability (diagonal points of internal disc at Figure A1). The first point (first level of variability) is at r = 1. After the second point, all of them are located at random radius between 1 to 10. Hence, each level of variability (10) is given by radii ranges except 1 which is fixed at 1 (diagonal points of internal disc); (a) 0 to 1, (b) 0 to 2, (c) 0 to 3, (d) 0 to 4, (e) 0 to 5, (f) 0 to 6, (g) 0 to 7, (h) 0 to 8, (i) 0 to 9 and (j) 0 to 10.
5. Voronoi tessellations: the partition variability will define the broad spectrum of possibilities for area distribution inside discs without losing partitioning number using Voronoi tessellations.
6. Area average: according to Equation (1), the average of areas requires a summation of sub-localities areas (A_{ij}) which were derived from pn with a changing variability pv.
7. Data mining: once the partition areas (A_{ij}) inside discs were obtained and (1) was solved, (2) is used to obtain standard deviations (σ_i) of variability for each disc. In order to normalize the level of variability for each pn, an index dividing the standard deviation of partitions and the particular area average of each partition was obtained (variability average; Figure A2). There are eight particular area averages of partitions since we have a sample of 8 discs with different pn (from 3 to 10). These particular area averages are derived from a value $n/(\approx 108.5 \pm 1.5)$ which are n values obtained from the first level of variability (pv) at r = 1. It is important to say that the radius of the external disc (1) and the radius of the internal disc ($r = 0.53 \pm 0.4$) was modified in order to get the particular area averages. However, despite the modification, the index between external discs and the internal ones remains constant. A sample of 20 discs to get 20 standard deviations was generated for each pn, and for each level of pv (10) giving a sample of 200 discs for each pn. An average of standard deviations ($\overline{\sigma}_i$; variability average) was derived for each level of variability.
8. Standard deviation. Finally, a standard deviation of all variability averages is obtained for each pn.

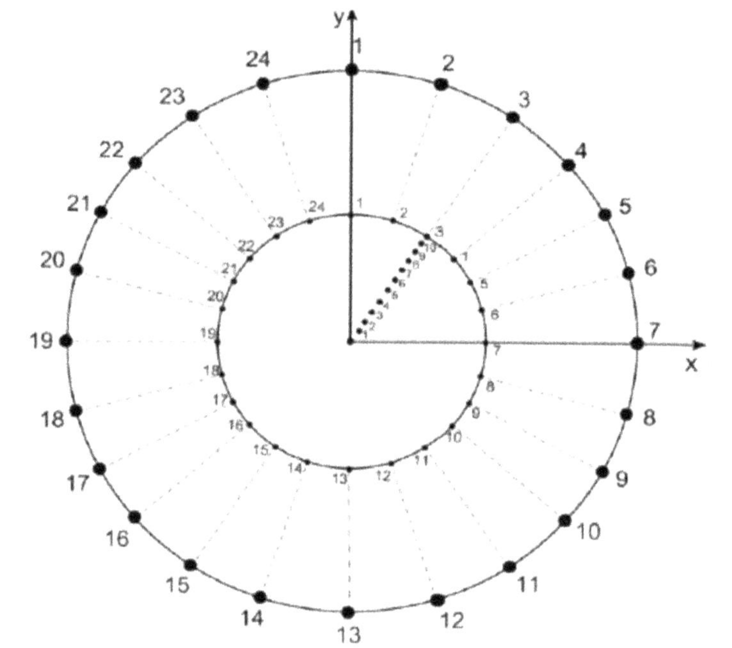

Figure A1. Defining partitioning number and partition variability. A disc is constructed to get Voronoi diagrams with constant area despite variability. The magnitude of the radius defines ten levels of partition variability: (a) 1, (b) 1-2, (c) 1-3, (d) 1-4, (e) 1-5, (f) 1-6, (g) 1-7, (h) 1-8, (i) 1-9, and (j) 1-10. Each level of variability is given by radii ranges except (a) which is fixed at 1.

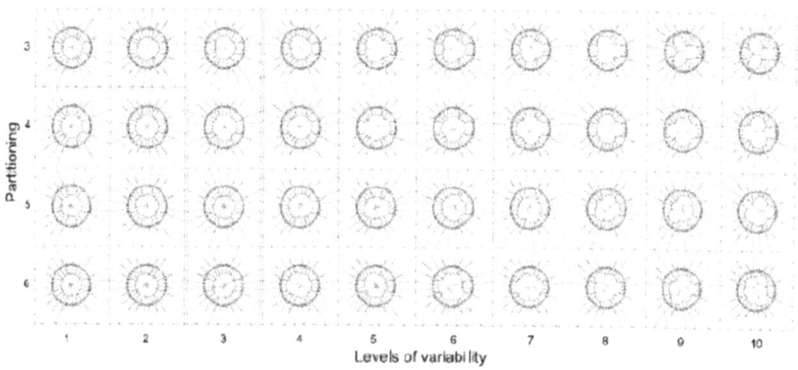

Figure A2. Partitioning number and partition variation of planar discs. A sample of 40 planar discs shows how partitioning number (vertical left side) determines segmentation of an almost constant area (≈108.5 ± 1.5) into a particular number of sub-localities. Partition variability (bottom horizontal numbers) installs levels of variability giving 10 constant and subtle increases of area to generate random segmentations using Voronoi tessellations.

Table A1 shows the area at internal disc, and the area average, for particular partition numbers.

Table A1. Level of variability and area average according to the partition number.

Partition Number	Area at Internal Disc (Level of Variability Pv1)	Particular Area Average
3	107.2	35.7354
4	108.7	27.1963
5	109.5	21.9155
6	109.9	18.3248
7	110.1	15.74
8	110.32	13.7959
9	110.51	12.2794
10	110.605	11.0605

References

1. Busiello, D.M.; Suweis, S.; Hidalgo, J.; Maritan, A. Explorability and the Origin of Network Sparsity in Living Systems. *Sci. Rep.* **2017**, *7*, 12323. [CrossRef] [PubMed]
2. Demongeot, J.; Jelassi, M.; Hazgui, H.; Ben Miled, S.; Bellamine Ben Saoud, N.; Taramasco, C. Biological Networks Entropies: Examples in Neural Memory Networks, Genetic Regulation Networks and Social Epidemic Networks. *Entropy* **2018**, *20*, 36. [CrossRef] [PubMed]
3. Bianconi, G. The Entropy of Randomized Network Ensembles. *Eur. Lett.* **2007**, *81*, 28005. [CrossRef]
4. Demetrius, L.; Manke, T. Robustness and Network Evolution—an Entropic Principle. *Phys. A Stat. Mech. Appl.* **2005**, *346*, 682–696. [CrossRef]
5. Cushman, S.A. Thermodynamics in Landscape Ecology: The Importance of Integrating Measurement and Modeling of Landscape Entropy. *Landsc. Ecol.* **2015**, *30*, 7–10. [CrossRef]
6. Vranken, I.; Baudry, J.; Aubinet, M.; Visser, M.; Bogaert, J. A Review on the Use of Entropy in Landscape Ecology: Heterogeneity, Unpredictability, Scale Dependence and Their Links with Thermodynamics. *Landsc. Ecol.* **2015**, *30*, 51–65. [CrossRef]
7. Parrott, L. Measuring Ecological Complexity. *Ecol. Indic.* **2010**, *10*, 1069–1076. [CrossRef]
8. Proulx, R.; Parrott, L. Measures of Structural Complexity in Digital Images for Monitoring the Ecological Signature of an Old-Growth Forest Ecosystem. *Ecol. Indic.* **2008**, *8*, 270–284. [CrossRef]
9. Frost, N.J.; Burrows, M.T.; Johnson, M.P.; Hanley, M.E.; Hawkins, S.J. Measuring Surface Complexity in Ecological Studies. *Limnol. Oceanogr. Methods* **2005**, *3*, 203–210. [CrossRef]
10. Davies, P.C.W.; Rieper, E.; Tuszynski, J.A. Self-Organization and Entropy Reduction in a Living Cell. *Biosystems* **2013**, *111*, 1–10. [CrossRef]
11. Buskermolen, A.B.C.; Suresh, H.; Shishvan, S.S.; Vigliotti, A.; DeSimone, A.; Kurniawan, N.A.; Bouten, C.V.C.; Deshpande, V.S. Entropic Forces Drive Cellular Contact Guidance. *Biophys. J.* **2019**, *116*, 1994–2008. [CrossRef] [PubMed]
12. Cabral, P.; Augusto, G.; Tewolde, M.; Araya, Y. Entropy in Urban Systems. *Entropy* **2013**, *15*, 5223–5236. [CrossRef]
13. Gershenson, C.; Fernández, N. Complexity and Information: Measuring Emergence, Self-Organization, and Homeostasis at Multiple Scales. *Complexity* **2012**, *18*, 29–44. [CrossRef]
14. Martínez-Berumen, H.A.; López-Torres, G.C.; Romo-Rojas, L. Developing a Method to Evaluate Entropy in Organizational Systems. *Procedia Comput. Sci.* **2014**, *28*, 389–397. [CrossRef]
15. Alexander, C. *The Nature of Order: An Essay on the Art of Building and the Nature of the Universe. Book 3, A Vision of a Living World*; Center for Environmental Structure: Berkeley, CA, USA, 2005.
16. López-Sauceda, J.; López-Ortega, J.; Laguna Sánchez, G.A.; Sandoval Gutiérrez, J.; Rojas Meza, A.P.; Aragón, J.L. Spatial Organization of Five-Fold Morphology as a Source of Geometrical Constraint in Biology. *Entropy* **2018**, *20*, 705. [CrossRef]
17. Gómez-Gálvez, P.; Vicente-Munuera, P.; Tagua, A.; Forja, C.; Castro, A.M.; Letrán, M.; Valencia-Expósito, A.; Grima, C.; Bermúdez-Gallardo, M.; Serrano-Pérez-Higueras, Ó.; et al. Scutoids are a Geometrical Solution to Three-Dimensional Packing of Epithelia. *Nat. Commun.* **2018**, *9*, 2960. [CrossRef]
18. Klatt, M.A.; Lovrić, J.; Chen, D.; Kapfer, S.C.; Schaller, F.M.; Schönhöfer, P.W.A.; Gardiner, B.S.; Smith, A.-S.; Schröder-Turk, G.E.; Torquato, S. Universal Hidden Order in Amorphous Cellular Geometries. *Nat. Commun.* **2019**, *10*, 811. [CrossRef]
19. Rejniak, K.A.; Wang, S.E.; Bryce, N.S.; Chang, H.; Parvin, B.; Jourquin, J.; Estrada, L.; Gray, J.W.; Arteaga, C.L.; Weaver, A.M.; et al. Linking Changes in Epithelial Morphogenesis to Cancer Mutations Using Computational Modeling. *PLoS Comput. Biol.* **2010**, *6*, e1000900. [CrossRef]
20. Sánchez-Gutiérrez, D.; Tozluoglu, M.; Barry, J.D.; Pascual, A.; Mao, Y.; Escudero, L.M. Fundamental Physical Cellular Constraints Drive Self-Organization of Tissues. *EMBO J.* **2016**, *35*, 77–88. [CrossRef]
21. Sandersius, S.A.; Chuai, M.; Weijer, C.J.; Newman, T.J. Correlating Cell Behavior with Tissue Topology in Embryonic Epithelia. *PLoS ONE* **2011**, *6*, e18081. [CrossRef]
22. Stooke-Vaughan, G.A.; Campàs, O. Physical Control of Tissue Morphogenesis across Scales. *Curr. Opin. Genet. Dev.* **2018**, *51*, 111–119. [CrossRef] [PubMed]

23. Yan, L.; Bi, D. Multicellular Rosettes Drive Fluid-Solid Transition in Epithelial Tissues. *Phys. Rev. X* **2019**, *9*, 11029. [CrossRef]
24. Bormashenko, E.; Frenkel, M.; Vilk, A.; Legchenkova, I.; Fedorets, A.A.; Aktaev, N.E.; Dombrovsky, L.A.; Nosonovsky, M. Characterization of Self-Assembled 2D Patterns with Voronoi Entropy. *Entropy* **2018**, *20*, 956. [CrossRef] [PubMed]
25. Wang, C.; Zhao, H. Spatial Heterogeneity Analysis: Introducing a New Form of Spatial Entropy. *Entropy* **2018**, *20*, 398. [CrossRef]
26. Van Anders, G.; Klotsa, D.; Ahmed, N.K.; Engel, M.; Glotzer, S.C. Understanding Shape Entropy through Local Dense Packing. *Proc. Natl. Acad. Sci. USA* **2014**, *111*, E4812-21. [CrossRef]
27. Tsuboi, A.; Ohsawa, S.; Umetsu, D.; Sando, Y.; Kuranaga, E.; Igaki, T.; Fujimoto, K. Competition for Space Is Controlled by Apoptosis-Induced Change of Local Epithelial Topology. *Curr. Biol.* **2018**, *28*, 2115–2128.e5. [CrossRef]
28. Boghaert, E.; Gleghorn, J.P.; Lee, K.; Gjorevski, N.; Radisky, D.C.; Nelson, C.M. Host Epithelial Geometry Regulates Breast Cancer Cell Invasiveness. *Proc. Natl. Acad. Sci. USA* **2012**, *109*, 19632–19637. [CrossRef]
29. Nicolis, G.; Prigogine, I. *Self-Organization in Nonequilibrium Systems: From Dissipative Structures to Order Through Fluctuations*; Wiley: Hoboken, NJ, USA, 1977; pp. 339–426.
30. Klimontovich, Y.L. Turbulent Motion. The Structure of Chaos. In *Turbulent Motion and the Structure of Chaos*; Springer: Berlin/Heidelberg, Germany, 1991; Fundamental Theories of Physics; Volume 42, pp. 329–371. [CrossRef]
31. González Valerio, M.A. Agenciamientos Materiales y Formales: Variaciones Sobre Morfologías. *Agenciamientos Mater. Y Form. Var. Sobre Morfol.* **2017**, *19*, 63–89. [CrossRef]
32. Drag, M.I. Epithelium: The Lightweight, Customizable Epithelial Tissue Simulator. Master's Thesis, The Ohio State University, Columbus, OH, USA, 2015.
33. Zabrodsky, H.; Peleg, S.; Avnir, D. Continuous Symmetry Measures. *J. Am. Chem. Soc.* **1992**, *114*, 7843–7851. [CrossRef]
34. Alemany, P.; Casanova, D.; Alvarez, S.; Dryzun, C.; Avnir, D. Continuous Symmetry Measures: A New Tool in Quantum Chemistry. *Rev. Comput. Chem.* **2017**, *30*, 289–352.
35. Zabrodsky, H.; Avnir, D. Continuous Symmetry Measures. 4. Chirality. *J. Am. Chem. Soc.* **1995**, *117*, 462–473. [CrossRef]
36. Zabrodsky, H.; Peleg, S.; Avnir, D. Symmetry as a Continuous Feature. *IEEE Trans. Pattern Anal. Mach. Intell.* **1995**, *17*, 1154–1166. [CrossRef]
37. Bonjack, M.; Avnir, D. The Near-Symmetry of Protein Oligomers: NMR-Derived Structures. *Sci. Rep.* **2020**, *10*, 8367. [CrossRef] [PubMed]
38. Frenkel, M.; Fedorets, A.A.; Dombrovsky, L.A.; Nosonovsky, M.; Legchenkova, I.; Bormashenko, E. Continuous Symmetry Measure vs Voronoi Entropy of Droplet Clusters. *J. Phys. Chem. C* **2021**, *125*, 2431–2436. [CrossRef]
39. Atia, L.; Bi, D.; Sharma, Y.; Mitchel, J.A.; Gweon, B.; Koehler, A.S.; DeCamp, S.J.; Lan, B.; Kim, J.H.; Hirsch, R.; et al. Geometric Constraints during Epithelial Jamming. *Nat. Phys.* **2018**, *14*, 613–620. [CrossRef]
40. Gibson, W.T.; Gibson, M.C. Cell Topology, Geometry, and Morphogenesis in Proliferating Epithelia. *Curr. Top. Dev. Biol.* **2009**, *89*, 87–114. [CrossRef] [PubMed]
41. Sánchez-Gutiérrez, D.; Sáez, A.; Pascual, A.; Escudero, L.M. Topological Progression in Proliferating Epithelia Is Driven by a Unique Variation in Polygon Distribution. *PLoS ONE* **2013**, *8*, e79227. [CrossRef] [PubMed]
42. Sáez, A.; Rivas, E.; Montero-Sánchez, A.; Paradas, C.; Acha, B.; Pascual, A.; Serrano, C.; Escudero, L.M. Quantifiable Diagnosis of Muscular Dystrophies and Neurogenic Atrophies through Network Analysis. *BMC Med.* **2013**, *11*, 77. [CrossRef]
43. Escudero, L.M.; Costa, L.D.F.; Kicheva, A.; Briscoe, J.; Freeman, M.; Babu, M.M. Epithelial Organisation Revealed by a Network of Cellular Contacts. *Nat. Commun.* **2011**, *2*, 526. [CrossRef]
44. Pilot, F.; Lecuit, T. Compartmentalized Morphogenesis in Epithelia: From Cell to Tissue Shape. *Dev. Dyn. Off. Publ. Am. Assoc. Anat.* **2005**, *232*, 685–694. [CrossRef]
45. López-Sauceda, J.; Rueda-Contreras, M.D. A Method to Categorize 2-Dimensional Patterns Using Statistics of Spatial Organization. *Evol. Bioinforma.* **2017**, *13*, 1176934317697978. [CrossRef] [PubMed]
46. Zhang, H.; Sinclair, R. Namibian Fairy Circles and Epithelial Cells Share Emergent Geometric Order. *Ecol. Complex.* **2015**, *22*, 32–35. [CrossRef]
47. Getzin, S.; Wiegand, K.; Wiegand, T.; Yizhaq, H.; Hardenberg, J.; Meron, E. Adopting a Spatially Explicit Perspective to Study the Mysterious Fairy Circles of Namibia. *Ecography* **2015**, *38*, 1–11. [CrossRef]
48. Contreras-Figueroa, G.; Hernandez-Sandoval, L.; Aragon-Vera, J.L. A measure of regularity for polygonal mosaics in biological systems. *Theor. Biol. Med. Model.* **2015**, *12*, 27. [CrossRef] [PubMed]
49. Gibson, M.C.; Patel, A.B.; Nagpal, R.; Perrimon, N. The Emergence of Geometric Order in Proliferating Metazoan Epithelia. *Nature* **2006**, *442*, 1038–1041. [CrossRef]
50. Nagpal, R.; Patel, A.; Gibson, M.C. Epithelial Topology. *Bioessays* **2008**, *30*, 260–266. [CrossRef]
51. Patel, A.B.; Gibson, W.T.; Gibson, M.C.; Nagpal, R. Modeling and Inferring Cleavage Patterns in Proliferating Epithelia. *PLOS Comput. Biol.* **2009**, *5*, e1000412. [CrossRef]
52. Stone, J.V. *Information Theory: A Tutorial Introduction*; Sebtel Press: LaVergne, TN, USA, 2015.
53. Shannon, C.E. A Mathematical Theory of Communication. *Bell Syst. Tech. J.* **1948**, *27*, 379–423. [CrossRef]
54. Jiao, Y.; Lau, T.; Hatzikirou, H.; Meyer-Hermann, M.; Corbo, J.C.; Torquato, S. Avian Photoreceptor Patterns Represent a Disordered Hyperuniform Solution to a Multiscale Packing Problem. *Phys. Rev. E* **2014**, *89*, 22721. [CrossRef]

55. Cafaro, C.; Ali, S.A. Information Geometric Measures of Complexity with Applications to Classical and Quantum Physical Settings. *Foundations* **2021**, *1*, 45–62. [CrossRef]
56. Summers, R.L. An Action Principle for Biological Systems. In Proceedings of the 10th International Conference on Mathematical Modeling in Physical Sciences (IC-MSQUARE 2021), Journal of Physics: Conference Series, Virtual, 6–9 September 2021.

entropy

MDPI

Article

Precise Traits from Sloppy Components: Perception and the Origin of Phenotypic Response

Steven A. Frank

Department of Ecology and Evolutionary Biology, University of California, Irvine, CA 92697-2525, USA; safrank@uci.edu

Abstract: Organisms perceive their environment and respond. The origin of perception–response traits presents a puzzle. Perception provides no value without response. Response requires perception. Recent advances in machine learning may provide a solution. A randomly connected network creates a reservoir of perceptive information about the recent history of environmental states. In each time step, a relatively small number of inputs drives the dynamics of the relatively large network. Over time, the internal network states retain a memory of past inputs. To achieve a functional response to past states or to predict future states, a system must learn only how to match states of the reservoir to the target response. In the same way, a random biochemical or neural network of an organism can provide an initial perceptive basis. With a solution for one side of the two-step perception–response challenge, evolving an adaptive response may not be so difficult. Two broader themes emerge. First, organisms may often achieve precise traits from sloppy components. Second, evolutionary puzzles often follow the same outlines as the challenges of machine learning. In each case, the basic problem is how to learn, either by artificial computational methods or by natural selection.

Keywords: evolutionary origins; critical learning period; machine learning; liquid state machine; reservoir computing; echo state network

Citation: Frank, S.A. Precise Traits from Sloppy Components: Perception and the Origin of Phenotypic Response. *Entropy* **2023**, 25, 1162. https://doi.org/10.3390/e25081162

Academic Editors: Antonio M. Scarfone and Pavel Kraikivski

Received: 3 April 2023
Revised: 12 June 2023
Accepted: 1 August 2023
Published: 3 August 2023

1. Introduction

Response to an environmental signal requires two steps. First, the signal must be perceived. Second, a response must follow. The evolutionary origin of two-part traits presents a puzzle. Perception without response provides no benefit. Response without perception cannot happen.

Pre-existing perceptions or responses may be modified. With a partial step on one side, an evolutionary path opens to solve the new challenge. The modification of prior adaptive traits may be a common pathway.

This article poses an alternative solution. In essence, a purely random pre-existing biochemical or neural network within the organism can provide the initial perceptive basis for the evolution of precise responsiveness. If so, then we gain an understanding of how organisms may acquire truly novel responsiveness.

In addition, we may begin to understand one of the great puzzles in life. How do organisms acquire a wide array of relatively precise traits given that biological components are inherently stochastic and often unreliable? How does precision arise from sloppiness?

Consider perception. We require that external signals induce an internal change in state. To analyze how random systems can acquire and store information, the computational literature has recently built on the idea of liquid state machines.

Think of the smooth surface of a liquid in a container. Drop a pebble on the surface. Waves move across the surface. Drop another pebble, and then another. At any point in time, the pattern of surface waves contains a reservoir of information about the temporal history.

Randomly connected networks act similarly. External inputs enter via sensor nodes. Those signals propagate through the network based on the random patterns of internal connectivity and rules for updating. At any point in time, the network contains information about the temporal history of inputs. The network functions as a dimensional expansion reservoir, transforming time into extent.

A random biochemical or neural network may act as a perceptive internal reservoir. The two-step challenge of perception and response reduces to the much easier problem of evolving an internal response to the perceptive reservoir. It may be possible to achieve an adaptively responsive trait arising from sloppy underlying components.

The remainder of this article provides details. The next subsection gives additional background and references to the computational and biological literature. The following analysis develops a model to illustrate how random networks store information about environmental inputs, creating the basis to predict future environmental states and respond accordingly.

A following subsection speculates that critical learning periods allow individuals to adjust their responses to their unique internal wiring and pattern of reservoir information. The Conclusions consider some possible tests of the ideas and some future directions.

Background and Literature

Maass et al. [1] introduced the liquid state machine. The concept, outlined in the introduction, describes a general way in which large dynamical systems retain a memory of their past inputs. At any point in time, that memory encoded in the current state of the system can be used to compute responses. The responses may achieve particular goals or predict future inputs.

Computationally, liquid state machines have a recurrent architecture. Roughly speaking, recurrence means feedback loops between internal states [2]. For example, a recurrent computational neural network updates internal states sequentially. External inputs modify the first layer of the network. The first layer then modifies the second layer, which may then modify the third layer, and so on. Recurrent connections flow updates backwards, from a later layer to an earlier layer. Recurrence greatly enhances the computational power of neural networks, in part by storing an internal memory of past inputs.

Recurrent neural networks led to many of the great recent advances in artificial intelligence. However, it can be very difficult to tune the particular connections and dynamic update rules in a network to achieve a particular function.

To solve the tuning problem, one may separate the accumulation of environmental information and memory from the computation of a response to that information. In the simplest application, one can use a randomly connected dynamic system as a reservoir of information and memory about inputs. One can then use a relatively simple computational learning or optimization method to match the current internal state of the reservoir to the desired goal. Often, basic regression methods such as ridge regression are sufficient.

This two-step solution has led to many developments in the computational literature, typically under the topics of reservoir computing or echo state networks [3–5]. Reservoir computing has also grown into a common approach in neuroscience modeling [6], with additional applications using biochemical networks as reservoirs [7,8]. In both computational and neuroscience models, reservoir connectivity patterns other than purely random patterns often arise [5,9–11]. For nonrandom reservoirs, the idea is that particular kinds of information may be better retained by particular architectures. Typically, the architectures are not optimized for each application. Instead, a few broad architectural varieties are explored in relation to a particular challenge.

Two articles have noted the potential of reservoirs to help in the understanding of various evolutionary problems [12,13]. My own focus is also evolutionary but limited here to two particular questions. First, can random reservoirs be a potential solution to the puzzle of jointly evolving perception and response? Second, can we place the perception response problem within the broader frame of precise traits from sloppy components?

2. Materials and Methods

2.1. Perception and Response

The joint evolution of perception and response may be easier if an initially random reservoir can solve the perception side of the puzzle. If random reservoirs provide information that can be the basis for perception, then the evolutionary path to a perception–response system may not be so difficult. In essence, a random system provides sufficient perception to get started and so, initially, only the single response trait must improve evolutionarily to make a workable system. The origin of a workable system provides the opportunity for further evolutionary refinement.

In this article, I limit the analysis to illustrating how random reservoirs provide the capacity for perception and the basis for developing a predictive response. The model brings the key ideas into the evolutionary literature within the context of a simple but important evolutionary puzzle.

The model has three parts. First, environmental inputs come from a chaotic dynamical system. A single parameter of the chaotic system describes the difficulty of predicting future input values. Second, the chaotic environmental inputs feed into a random network that acts as the reservoir. Third, an optimized regression model predicts future input values by using the internal reservoir states as predictors. The quality of the predictions is measured by evaluating additional input data and reservoir dynamics not used in the regression fitting procedure.

2.2. Chaotic Dynamics

I use the classic Lorenz–96 model for chaotic dynamics [14–16], which is

$$\frac{\mathrm{d}x_i}{\mathrm{d}t} = (x_{i+1} - x_{i-2})x_{i-1} - x_i + F \tag{1}$$

for $i = 1, \ldots, N$, with $x_{-k} = x_{N-k}$ and $x_{N+k} = x_k$, and F as the single parameter that describes a constant forcing input. The symmetry of the model means that the long-run trajectories for each dimension have similar properties. I use $N = 5$ for all analyses in this article.

The system tends to be more chaotic as F rises above 8 (Figure 1). Chaos means that a small perturbation at a particular time causes the future system trajectory to diverge from the trajectory of an unperturbed system. The greater the rate of divergence, the less predictable the system.

Typically, one quantifies the rate of divergence by the dominant Lyapunov exponent, λ. Similarly, the system predictability can be quantified by the doubling time of the distance between divergent trajectories, which is $dbl = \log 2/\lambda$, with dbl denoting a variable. A faster doubling time means that future values of the trajectory are harder to predict. I calculated the dynamics of Equation (1) and the Lyapunov exponent with the Julia package DynamicalSystems [17]. The system becomes increasingly chaotic as F rises above 8, which means that λ increases and dbl (predictability) declines.

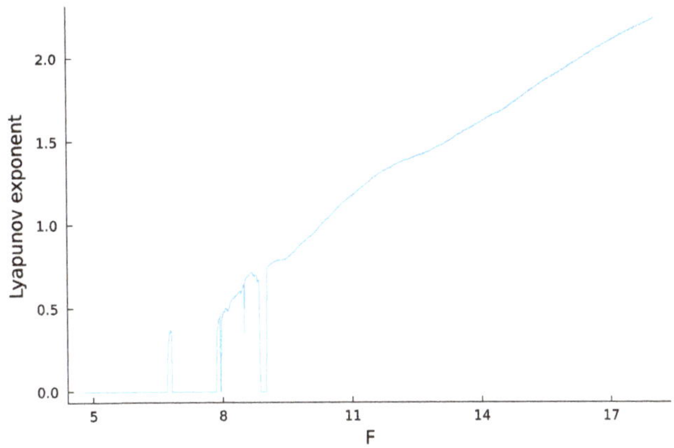

Figure 1. Estimate for the relative speed of chaotic divergence in the dynamics of the Lorenz–96 equations given in Equation (1), with $N = 5$. Here, the Lyapunov exponent, λ, estimates the relative divergence rate. The analysis in this article focuses on the doubling time for divergence, $dbl = \log 2/\lambda$, in which a lower doubling time means that future values of the trajectory are harder to predict. For a few limited regions of smaller F values, the estimated Lyapunov exponent drops below the trend. Those deviations may arise from numerical limitations or a complex pattern of nearly stable periodicity. Sufficiently complex periodicity poses a significant challenge for prediction. The analyses in this article avoid those erratic regions.

2.3. Random Reservoir

I computed the random reservoir state using the Julia package ReservoirComputing [18]. The reservoir takes the N inputs from Equation (1) and updates its *size* internal states. The cited documentation gives the details of the reservoir dynamics architecture and calculations. The outcome arises from the common principles of liquid state machines.

A particular run starts with random initial conditions for the input dynamics and a randomly structured reservoir. Then, over the T time units of a run, the inputs are fed into the reservoir every 0.01 time units, which triggers an update to the reservoir states. For each of the $T/0.01$ time steps, the reservoir has *size* different state values. Those state values can be used to predict future values of the inputs.

3. Results

3.1. Predicting Future Inputs

Briefly, a random reservoir provides sufficient information for the system to predict future inputs of the chaotic environmental dynamics. The more strongly chaotic the system, the shorter the divergence doubling time, *dbl*, and the shorter the time forward for successful predictions. Larger random reservoirs improve the system's ability to predict future input values. Supporting details follow.

I first calculated the external inputs from Equation (1) at each of the $T/0.01$ time steps, with T = 20,000 for all analyses. I then split the time periods into a training set for the first $0.7T$ = 14,000 of the time units and a test set for the remaining $0.3T$ = 6000 time units. Time units are arbitrary. Predictions provide value if the time extent of predictive success corresponds to a biologically valuable foresight.

Figure 2a shows an example run of the model predictions. The blue curve is the external input value for the first dimension of the Lorenz–96 system, x_1, in Equation (1). The plotted value is rescaled so that the range over the training set is $[-1, 1]$. The plot shows the final 20 time units of the test set, the time period 19,980–20,000.

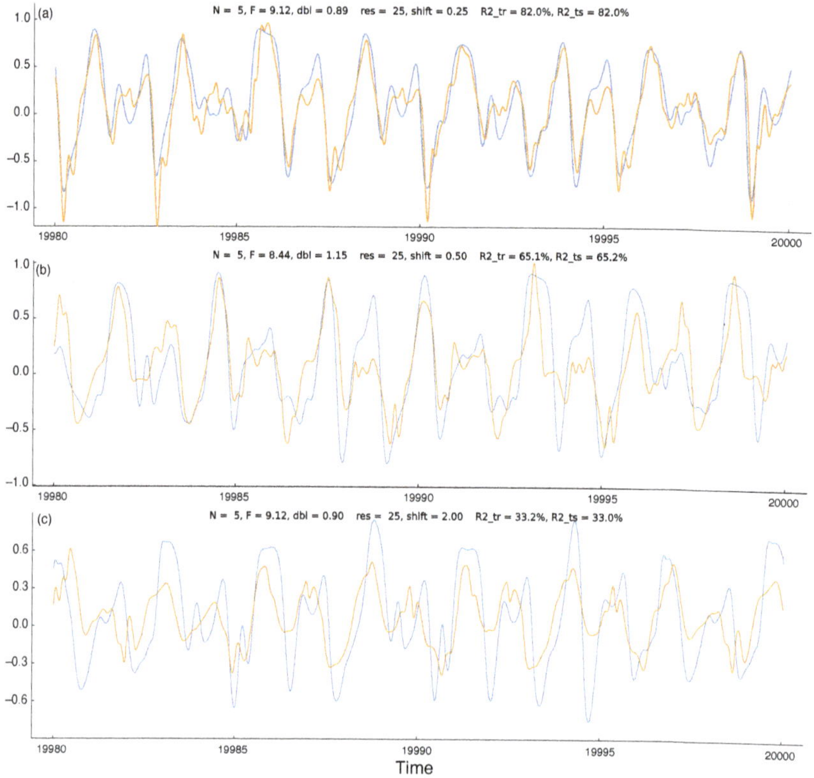

Figure 2. Temporal dynamics of environmental state (blue) and system prediction for the environmental state (gold). At each time point, the internal system uses the information in its reservoir to predict the environmental state *shift* time units into the future. The gold prediction curve is shifted to the right by *shift* time units, so that the closeness of the match between the two curves describes the quality of the predictions. Above each panel, the parameters N and F describe the environmental dynamics in Equation (1); *dbl* gives the doubling time for the deviation distance of a small perturbation to the dynamics; *res* is the reservoir *size*; and *R2_tr* and *R2_ts* are the R-squared values that describe the percentage of the variation in the blue dynamics curve captured by the gold prediction curve for the training and test periods, respectively, as described in the text. The panels (**a**–**c**) have corresponding labels on the curves in Figure 3a. Time units are nondimensional and can be chosen to match the scaling of the environmental process under study. Here, the plots show the 20 time units at the end of the test period of the machine learning procedure used to generate the curves. The abbreviations *res*, *shift*, *size*, *dbl*, *R2_tr*, and *R2_ts* denote variables. Execution times for the parameters in (**b**) with reservoir sizes (*res*) of 25, 50, and 100 are approximately 58 s, 118 s, and 253 s. Timing was carried out on Apple Mac Studio M1 Ultra with Julia 1.9.1, source code git commit a7f74f1. The code was not optimized for execution speed.

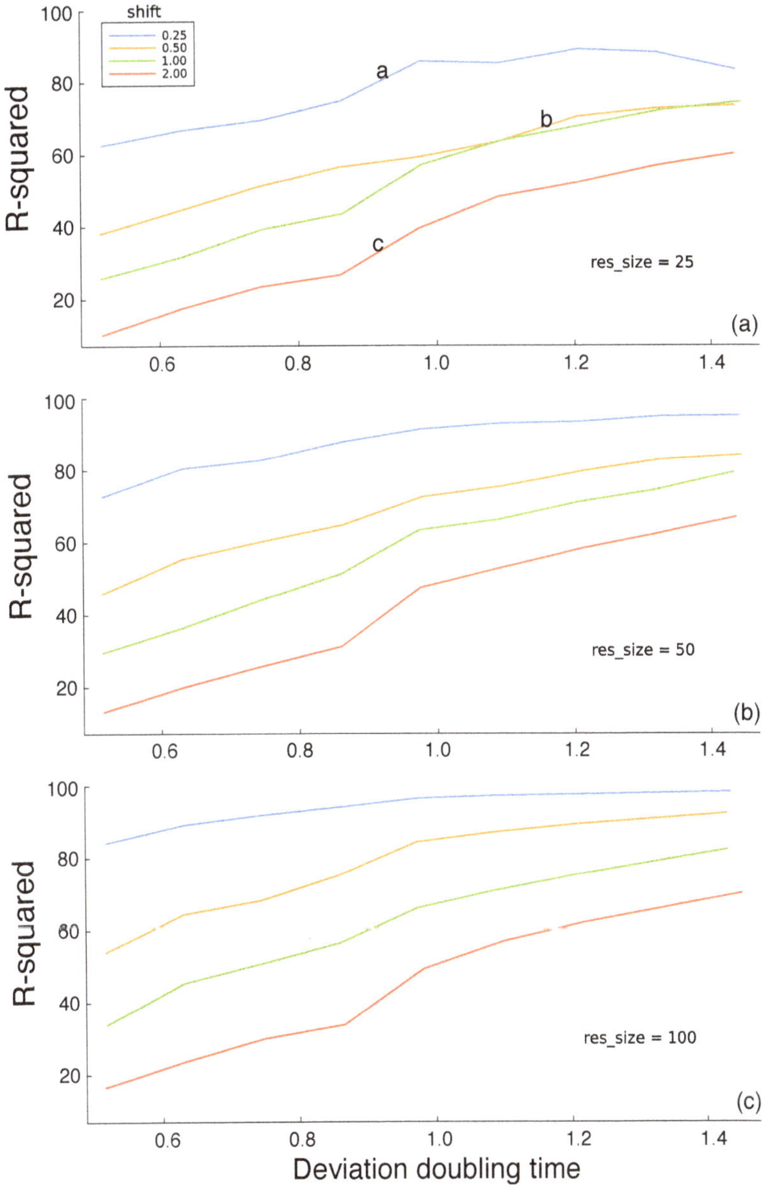

Figure 3. Prediction of future environmental state based on the information in a random reservoir network. Figure 2 shows the environmental dynamics and the prediction challenge. In this figure, the y-axis measures the percentage of the total variance (R-squared) in the environmental state explained by the predictions generated from the internal reservoir, reflecting the potential for adaptive response. The x-axis shows the intrinsic predictability of the environment, measured by the time required to double a small initial perturbation to the dynamic trajectory. The different colored lines describe the time shift into the future at which predictions are compared to actual future dynamics. The *res_size* parameter in each panel gives the *size* of the random reservoir. The a, b, and c labels in panel (**a**) match the corresponding panels in Figure 2. Each line connects the outcomes at the following 11 approximate doubling times: 0.52, 0.54, 0.58, 0.64, 0.70, 0.77, 0.86, 0.90, 0.99, 1.15, and 1.43. Panels (**a**–**c**) show three different reservoir sizes denoted by the *res_size* parameter labels.

The gold curve shows the system's prediction for future values of the external chaotic input, x_1. For a time point, t, the system predicts x_1 at time $t + shift$. To compare the predicted input value to the actual input value, I shifted the gold curve by *shift* time units to the right. Thus, each time point on the plot shows the system's observed and predicted value for time t.

I calculated the predicted values by fitting a Bayesian ridge regression model to the training set of observed x_1 values based on the *size* predictors from the internal reservoir states. In Figure 2, $size = 25$ for all three panels. I obtained the fitted model by the BayesRidge function of the Python scikit-learn 1.2.0 package [19]. I accessed the Python code via the Julia machine learning package MLJ [20].

In Figure 2, I show the actual input values and predicted input values over the test set of observations. Those test data were not used during the fitting of the ridge regression model and so describe how well the model predictions fit additional observations from the chaotic inputs. I measured the quality of the predictions by the R-squared value, which is the fraction of the variance in the actual input values of the blue curves explained by the predicted input values of the gold curves. For example, the R-squared value for Figure 2a is 82%, a close fit.

To avoid overfitting the ridge regression model, I used MLJ's TunedModel function to optimize the BayesRidge hyperparameters for the training period data. That procedure shuffled the data provided for fitting in a way that minimized overfitting. To test for overfitting on the training data, above each panel in Figure 2 I show the R-squared values for the training period (*R2_tr*) and the test period (*R2_ts*). The close match of those values demonstrates that the model was not overfitted to the training data.

In Figure 3a, the different colored curves show the quality of the predictions for different *shift* time values into the future. The prediction quality on the y-axis is given by the R-squared values of the test period. Shorter time shifts into the future provide better predictions, as expected. The x-axis shows the doubling time, *dbl*, for trajectory divergence. Greater doubling times correspond to weaker chaotic dynamics and greater predictability. The a, b, and c labels on the curves in Figure 3a match the three panels of Figure 2. The different panels of Figure 3 show that increasing the reservoir size leads to better predictions.

I calculated the test R-squared value *R2_ts* for each parameter combination from one replicate. In Figure 3, the consistency of the trends across different doubling times and reservoir sizes implies that the variability within a parameter combination is low. If that were not true, then the trends would be much noisier than observed.

To check the actual variability among replicates for a parameter combination, I calculated *R2_ts* for a sample of 20 independent runs for each reservoir size of 25, 50, and 100, using for the other parameters $N = 5$, as in all reported results; $F = 8.75$, corresponding to a doubling time of about 1.0; and a shift value of 1.0.

For any given reservoir size, the variation among samples is small. Reporting results as (minimum, median, maximum) for each set of 20 replicates, the results for reservoir size 25 are (56.0, 57.8, 58.9), for size 50 are (60.0, 61.4, 62.7), and for size 100 are (66.0, 67.1, 68.3).

Figure 4 shows that increasing the reservoir size improves the prediction of the future environmental state.

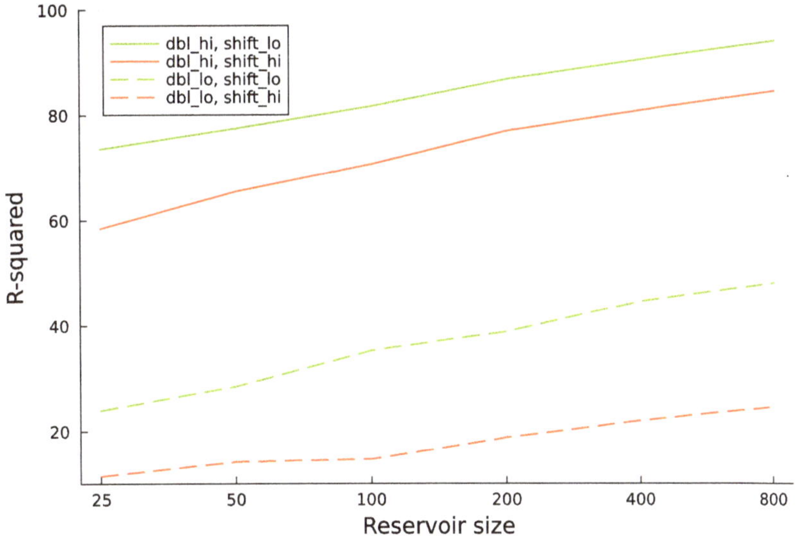

Figure 4. Increasing reservoir size provides better predictions for future environmental state. The analysis follows the methods used in Figure 3. Here, dbl_lo denotes a doubling time of approximately 0.52, and dbl_hi denotes a doubling time of approximately 1.42. The value of shift_lo denotes a prediction into the future over 1.0 time units, and shift_hi denotes a prediction into the future over 2.0 time units. The six different reservoir sizes used in the computer runs are shown as labels for the tick marks along the x-axis.

3.2. Critical Learning Period

The wiring of internal reservoirs may be fixed. For example, the parameters of a simple biochemical network within a cell may be determined primarily by DNA sequence. The network may be random in the sense that it was not shaped by natural selection to capture specific information. But such a random network may be relatively consistent from one individual to another. If so, then the readout of the network to achieve a function may also be fixed among individuals.

Simple neural networks may also be relatively consistent from one individual to another. However, larger networks likely have some stochasticity in wiring. Stochasticity means that random reservoirs of perceptual information may vary from one individual to another. If so, then the way in which individuals read their reservoirs to achieve a function may have to be partially learned.

The demand for such learning may impose the need for a primitive kind of critical learning period in which individuals associate their particular internal reservoir state with successful actions. Such learning periods would be simpler than the kinds of learning that are sometimes observed in the advanced neural systems of vertebrates. Although speculative, the logic for such kinds of critical learning seems compelling.

3.3. Other Ideas for Future Study

Comments arising in the review process for this manuscript raised three interesting ideas for future study. First, heritable variation in network size and wiring architecture may provide the opportunity for selection to improve environmental perception. The computational literature on reservoir computing provides insight into how different reservoir networks perform with respect to different kinds of environmental challenges [5,18].

Second, environmental change often requires organisms to modify some aspect of their perception or response. In the reservoir model, a change in response means a modification of the readout from the perceptional information stored in the reservoir. This sort of tuning may happen relatively quickly within an individual's lifetime, as in the critical learning period. Alternatively, the readout may be altered over evolutionary time by typical variation and selection processes. The puzzles concern how this happens physiologically and genetically.

Third, larger networks improve performance. Larger networks also tend to have greater redundancy with regard to storing information about the environment. Redundancy enhances robustness, provides opportunity for greater complexity, and alters evolutionary dynamics in many interesting ways [21]. This perspective raises many interesting questions about the origin and evolution of perception.

4. Conclusions

Random perceptual networks may solve the puzzle of how two-step perception–response traits evolve. If a response can build on a random perceptual reservoir, then the initial evolutionary path requires adaptation only on the response side. Subsequent refinement may modify the perceptual side, changing random aspects of the initial network into more highly structured forms.

Studying the origin of traits can be difficult because we rarely observe such origins directly. Synthetic biology may provide a way to gain some insight and to test specific hypotheses. If technology advances sufficiently, it may be possible to create various types of biochemical networks that have random properties with respect to specific adaptive functions [22]. One could then use experimental evolution to analyze the conditions under which cells can improve their ability to read the information in the random biochemical reservoir to achieve those specific functions.

Comparative biology could provide insight into the historical pathways and modifications of perception–response pairs. But it is not clear how easily one could find traces of evolutionary historical sequence among extant organisms. The great variety of single-cell microbial life is both promising and challenging.

Funding: This research was funded by The Donald Bren Foundation, National Science Foundation grant DEB-1939423, and DoD grant W911NF2010227.

Institutional Review Board Statement: Not applicable.

Informed Consent Statement: Not applicable.

Data Availability Statement: The Julia source code for the analysis and figures is available on GitHub via Zenodo at https://doi.org/10.5281/zenodo.8030653 accessed on 12 June 2023.

Conflicts of Interest: The author declares no conflict of interest.

References

1. Maass, W.; Natschläger, T.; Markram, H. Real-time computing without stable states: A new framework for neural computation based on perturbations. *Neural Comput.* **2002**, *14*, 2531–2560. [CrossRef] [PubMed]
2. Goodfellow, I.; Bengio, Y.; Courville, A. *Deep Learning*; MIT Press: Cambridge, MA, USA, 2016.
3. Jaeger, H. Echo state network. *Scholarpedia* **2007**, *2*, 2330. [CrossRef]
4. Gauthier, D.J.; Bollt, E.; Griffith, A.; Barbosa, W.A. Next generation reservoir computing. *Nat. Commun.* **2021**, *12*, 5564. [PubMed]
5. Cucchi, M.; Abreu, S.; Ciccone, G.; Brunner, D.; Kleemann, H. Hands-on reservoir computing: A tutorial for practical implementation. *Neuromorphic Comput. Eng.* **2022**, *2*, 032002. [CrossRef]
6. Damicelli, F.; Hilgetag, C.C.; Goulas, A. Brain connectivity meets reservoir computing. *PLoS Comput. Biol.* **2022**, *18*, e1010639. [CrossRef] [PubMed]
7. Goudarzi, A.; Lakin, M.R.; Stefanovic, D. DNA reservoir computing: A novel molecular computing approach. In *International Workshop on DNA-Based Computers*; Springer: Berlin/Heidelberg, Germany, 2013; pp. 76–89.
8. Yahiro, W.; Aubert-Kato, N.; Hagiya, M. A reservoir computing approach for molecular computing. In Proceedings of the ALIFE 2018: The 2018 Conference on Artificial Life, Tokyo, Japan, 23–27 July 2018; pp. 31–38.
9. Loeffler, A.; Zhu, R.; Hochstetter, J.; Li, M.; Fu, K.; Diaz-Alvarez, A.; Nakayama, T.; Shine, J.M.; Kuncic, Z. Topological properties of neuromorphic nanowire networks. *Front. Neurosci.* **2020**, *14*, 184. [CrossRef] [PubMed]
10. Loeffler, A.; Zhu, R.; Hochstetter, J.; Diaz-Alvarez, A.; Nakayama, T.; Shine, J.M.; Kuncic, Z. Modularity and multitasking in neuro-memristive reservoir networks. *Neuromorphic Comput. Eng.* **2021**, *1*, 014003. [CrossRef]

11. Lukoševičius, M. A practical guide to applying echo state networks. In *Neural Networks: Tricks of the Trade*; Montavon, G., Orr, G.B., Müller, K.R., Eds.; Springer: Berlin/Heidelberg, Germany, 2012; pp. 659–686.
12. Seoane, L.F. Evolutionary aspects of reservoir computing. *Philos. Trans. R. Soc. B* **2019**, *374*, 20180377. [CrossRef] [PubMed]
13. Solé, R.; Seoane, L.F. Evolution of brains and computers: The roads not taken. *Entropy* **2022**, *24*, 665. [CrossRef] [PubMed]
14. Lorenz, E.N. Predictability—A problem partly solved. In *Predictability of Weather and Climate*; Palmer, T., Hagedorn, R., Eds.; Cambridge Univerity Press: Cambridge, UK, 2006; pp. 40–58.
15. Karimi, A.; Paul, M.R. Extensive chaos in the Lorenz–96 model. *Chaos* **2010**, *20*, 043105. [CrossRef]
16. Bedrossian, J.; Blumenthal, A.; Punshon-Smith, S. A regularity method for lower bounds on the Lyapunov exponent for stochastic differential equations. *Invent. Math.* **2022**, *227*, 429–516. [CrossRef]
17. Datseris, G. DynamicalSystems.jl: A Julia software library for chaos and nonlinear dynamics. *J. Open Source Softw.* **2018**, *3*, 598. [CrossRef]
18. Martinuzzi, F.; Rackauckas, C.; Abdelrehim, A.; Mahecha, M.D.; Mora, K. ReservoirComputing.jl: An efficient and modular library for reservoir computing models. *J. Mach. Learn. Res.* **2022**, *23*, 13093–13100. [CrossRef]
19. Pedregosa, F.; Varoquaux, G.; Gramfort, A.; Michel, V.; Thirion, B.; Grisel, O.; Blondel, M.; Prettenhofer, P.; Weiss, R.; Dubourg, V.; et al. Scikit-learn: Machine learning in Python. *J. Mach. Learn. Res.* **2011**, *12*, 2825–2830.
20. Blaom, A.D.; Kiraly, F.; Lienart, T.; Simillides, Y.; Arenas, D.; Vollmer, S.J. MLJ: A Julia package for composable machine learning. *J. Open Source Softw.* **2020**, *5*, 2704. [CrossRef]
21. Frank, S.A. Robustness and complexity. *arXiv* **2023**. [CrossRef]
22. Levin, M. Life, death, and self: Fundamental questions of primitive cognition viewed through the lens of body plasticity and synthetic organisms. *Biochem. Biophys. Res. Commun.* **2021**, *564*, 114–133.

entropy

Article

Mechanistic Modelling of Biomass Growth, Glucose Consumption and Ethanol Production by *Kluyveromyces marxianus* in Batch Fermentation

Yolocuauhtli Salazar [1], Paul A. Valle [2,*], Emmanuel Rodríguez [2], Nicolás O. Soto-Cruz [3], Jesús B. Páez-Lerma [3] and Francisco J. Reyes-Sánchez [3]

1 Postgraduate Program in Engineering, Tecnológico Nacional de México/IT Durango, Blvd. Felipe Pescador 1830 Ote., Durango 34080, Mexico
2 Postgraduate Program in Engineering Sciences, BioMath Research Group, Tecnológico Nacional de México/IT Tijuana, Blvd. Alberto Limón Padilla s/n, Tijuana 22454, Mexico
3 Departamento de Ingenierías Química y Bioquímica, Tecnológico Nacional de México/IT Durango, Blvd. Felipe Pescador 1830 Ote., Durango 34080, Mexico
* Correspondence: paul.valle@tectijuana.edu.mx

Abstract: This paper presents results concerning mechanistic modeling to describe the dynamics and interactions between biomass growth, glucose consumption and ethanol production in batch culture fermentation by *Kluyveromyces marxianus* (*K. marxianus*). The mathematical model was formulated based on the biological assumptions underlying each variable and is given by a set of three coupled nonlinear first-order Ordinary Differential Equations. The model has ten parameters, and their values were fitted from the experimental data of 17 *K. marxianus* strains by means of a computational algorithm design in Matlab. The latter allowed us to determine that seven of these parameters share the same value among all the strains, while three parameters concerning biomass maximum growth rate, and ethanol production due to biomass and glucose had specific values for each strain. These values are presented with their corresponding standard error and 95% confidence interval. The goodness of fit of our system was evaluated both qualitatively by in silico experimentation and quantitative by means of the coefficient of determination and the Akaike Information Criterion. Results regarding the fitting capabilities were compared with the classic model given by the logistic, Pirt, and Luedeking–Piret Equations. Further, nonlinear theories were applied to investigate local and global dynamics of the system, the Localization of Compact Invariant Sets Method was applied to determine the so-called localizing domain, i.e., lower and upper bounds for each variable; whilst Lyapunov's stability theories allowed to establish sufficient conditions to ensure asymptotic stability in the nonnegative octant, i.e., $\mathbf{R}^3_{+,0}$. Finally, the predictive ability of our mechanistic model was explored through several numerical simulations with expected results according to microbiology literature on batch fermentation.

Keywords: asymptotic stability; batch fermentation; in silico experimentation; *Kluyveromyces marxianus*; nonlinear data fitting; nonlinear mechanistic model

Citation: Salazar, Y.; Valle, P.A.; Rodríguez, E.; Soto-Cruz, N.O.; Páez-Lerma, J.B.; Reyes-Sánchez, F.J. Mechanistic Modelling of Biomass Growth, Glucose Consumption and Ethanol Production by *Kluyveromyces marxianus* in Batch Fermentation. *Entropy* **2023**, *25*, 497. https://doi.org/10.3390/e25030497

Academic Editor: Pavel Kraikivski

Received: 12 January 2023
Revised: 8 March 2023
Accepted: 10 March 2023
Published: 14 March 2023

1. Introduction

Alcoholic fermentation is an anaerobic process that transforms sugars like glucose or fructose into ethanol and carbon dioxide. Several yeast species are used commonly in this process, e.g., *Kloeckera, Hanseniaspora, Candida, Pichia, Kluyveromyces*, and *Saccharomyces* among others. The growth rate of these microorganisms has an ultimate effect on the sensorial characteristics of the final product, which can be positive or negative depending on the yeast used [1].

Overall, yeasts are indispensable for biotechnological processes such as wine and beer production [2]. In this research, we focus on investigating glucose consumption and ethanol production from several strains of *Kluyveromyces marxianus* (*K. marxianus*). This yeast has a great potential for alcoholic fermentation due to its intraspecific characteristics such as higher specific growth rates, the ability to grow on various substrates, and tolerance to high temperatures [3–5]. Further, *Kluyveromyces* sp. produces aromatic compounds such as fruity esters, carboxylic acids, ketones, furans, and other alcohols in liquid fermentation such as 2-phenyl ethanol whose sensorial characteristics can influence the quality of wine, distilled drinks, and fermented foods [6]; refer to Fonseca et al. for an extensive review on the biotechnological potentials of *K. marxianus* [4,6].

Concerning industrial applications, fermentation is commonly performed in batch culture, which brings certain advantages such as the reduction of contamination risk, in addition to the fact that a large capital investment is not necessary since high-priced production equipment is not required compared to a continuous culture process [7]. Batch process implies that yeasts are incubated in a closed container under controlled conditions with a culture medium composed of the necessary nutrients [8]. Hence, biomass cannot grow indefinitely and four phases have been identified in its dynamics, i.e., lag phase, exponential growth, stationary state, and death phase. While this process is carried out, the substrate is consumed and converted into the product, e.g., ethanol produced by sugars such as glucose [9]. Therefore, properly identifying the time interval of these phases as well as predicting the maximum product concentration that could be produced from the initial concentrations of both substrate and biomass may help to optimize production costs on the resulting product of several applications. The latter may be achieved by mechanistic modeling through predictive microbiology, which can be considered a powerful tool to investigate and summarize the overall effects of varying conditions and environmental factors within food formulation and processing [10]. Further, mathematical models could aid in gaining insights concerning microbial food safety and quality assurance of increasingly complex food products [11,12], as well as estimating shelf life and forecasting food spoilage [13,14].

Mathematical models in predictive microbiology can be classified according to different criteria, uses, and functionalities that are not mutually exclusive. Based on the type and number of variables, models are classified into primary, secondary, and tertiary; they can also be differentiated on the basis of their mathematical background as mechanistic or empirical [15], and they can be categorized into structured and unstructured conforming to the complexity of the chemical compounds of the biomass [16]. Primary models are those that represent biomass growth dynamics as a function of time, the main equations in the literature are the exponential functions of Gompertz [17] and Vazquez Murado [18], the logarithmic function of Baranyi et al. [19] and the cubic model of Garcia et al. [20]. All models are described by parameters such as maximum growth rate $[\mu_{\max}]$, lag time $[L]$, and both initial $[X_0]$ and maximum biomass $[X_{\max}]$ Concentrations, while secondary models relate to the latter with environmental conditions such as temperature and pH, and other variables such as substrate and product concentrations over time, e.g., equations of Monod [21], Teisser [22], Haldane [23], and Moser [24], which aim to describe biomass growth dynamics as a function of substrate concentration and have been widely used to investigate bacterial growth [25]. Tertiary models are the result of combining primary and secondary models through the use of computer tools that allow predictions regarding the growth or death of microorganisms in food when different environmental conditions are combined [26]. Concerning the second classification mentioned, mechanistic models are formulated by means of theoretical bases and provide an interpretation of microbial growth in terms of known processes and empirical models are usually composed of polynomials of the first or second degree and pragmatically describe the data with convenient mathematical relationships, this does not usually give information on precise responses of microorganisms, because they do not take into account known processes [27]. Finally, according to the third category described, unstructured models consider biomass only as a chemical compound in a culture and its dynamics is described by simple models, while structured models also take into account changes in the internal cellular structure of biomass in terms such as the content of RNA, enzymes, reagents, metabolism and products [28]. The Gompertz, Vazquez-Murado, Baranyi and Garcia models, mentioned above, are also classified as unstructured models since biomass is considered a variable described only by its concentration. Mathematical models used by Sansonetti [29], Lei [30], and Steinmeyer [31] are classified as structured because they describe the growth of biomass considering the intracellular reactions produced by its metabolism.

Thus, it is important to highlight that in a batch fermentation process, multiple reactions occur, so the adaptability and evolution of microorganisms in short periods and changes in environmental conditions usually characterize this type of process, consequently, the modeling of these systems is complex due to time-varying characteristics of biological systems, resulting in nonlinear systems dynamics [28]. Hence, a mathematical model formulated from a system of nonlinear differential equations will allow the application of nonlinear systems control methods to optimize the process so that the characteristics of the final product can be predicted when the environmental conditions of the culture are controlled and the initial conditions of biomass, substrate, and product values are known. It is worth mentioning that most of the models found in the literature focus on the yeast *Saccharomyces cerevisiae* since it is one of the most used in the industry; however, biotechnological opportunities have been found in non-Saccharomyces yeasts because they have metabolic characteristics that lead to the production of compounds of interest. Therefore, it is important to model the growth of *K. marxianus* because of the great potential in the production of esters compounds of industrial importance [32]. Thus, in this paper, we applied mechanistic and computational modeling to formulate a system of three coupled nonlinear first-order Ordinary Differential Equations (ODEs) that describe dynamics between biomass, glucose (substrate), and ethanol (product) concentrations over time. Mechanistic modeling allowed us to provide both qualitative and quantitative descriptions concerning the relationships of biomass growth, glucose consumption, and ethanol production from 17 strains of *K. marxianus*, while computational modeling was used to fit experimental data from these three variables and establish numerical values for each parameter of the mathematical model. Further, nonlinear theories such as the Localization of Compact Invariant Sets (LCIS) method and Lyapunov's Stability Theory were applied to provide a complete analysis of the local and global dynamics of our proposed biological system [33].

2. Materials and Methods

This section provides all the information concerning the experimental data of biomass growth, substrate consumption and ethanol production, i.e., karyotypes of the *K. marxianus* strains with identifiable chromosomal differences among them, environmental conditions, chemical characteristics of the medium, lab equipment used for measurements, and periods for each measurement, then the mathematical model is formulated and each equation as well as values and units of parameters are described. This section concludes by describing the procedure to approximate the experimental data and to fit the numerical values of each parameter by designing an algorithm in Matlab.

2.1. Experimental Data: Culture Medium and Analytical Techniques

Experimental data was obtained from alcoholic fermentation in batch culture by *K. marxianus*, 17 strains with different genetic profiles were incubated in 20 g/L of yeast extract peptone dextrose agar at 30 °C in order to study their kinetic growth, glucose consumption and ethanol production. Codification and origin of studied karyotypes of *K. marxianus* are identified by Páez et al. in [34], where 15 strains were obtained from different places of México, and they were isolated from agave fermentation for mezcal production, in addition to 2 reference strains that were isolated from pozol (CBS6556) in México, and from yoghurt (CBS397) in Netherlands.

Characteristics of the chemically defined medium are given as follows: glucose 20 g/L, KH_2PO_4 3 g/L, $(NH_4)_2SO_4$ 3 g/L, Na_2HPO_4 1.49 g/L, glutamic acid 1 g/L, $MgCl_2$ heptahydrate 0.41 g/L, $ZnCl_2$ 0.0192 g/L, $CuCl_2$ 0.0006 g/L, $MnCl_2$ 0.044 g/L, $CoCl_2$ 0.0005 g/L, CaCl 0.0117 g/L, $FeCl_2$ 0.011 g/L, $(NH_4)_6Mo_7O_{24}$ 0.004 g/L, H_3BO_4 0.0030 g/L, aminobenzoic acid 0.0010 g/L, myo-inositol 0.1250 g/L, nicotinic acid 0.0050 g/L, pantothenic acid 0.005 g/L, pyridoxine 0.0050 g/L, thiamin HCl 0.005 g/L, biotin 0.000024 g/L [35]. This medium was used to culture the strains for biomass development with agitation, for the conservation of the strains, plates with the same medium with 20% agar were used and stored at 4 °C.

Biomass concentration was measured with a spectrophotometer UV-VIS DR 6000 (HACH, Loveland, CO, USA) by optical density at 600 nm, values in g/L were obtained relating optical density with a calibration curve of the dry weight of *K. marxianus*. For glucose consumption and ethanol production by High-Performance Liquid Chromatography (HPLC series 1200, Agilent Technology, Palo Alto, CA, USA), a BIORAD HP-87H^+ (8%) ion exchange column was used, in an AGILENT® 1200 series equipment, with H_2SO_4 0.005 N as mobile phase, at a flow of 0.5 mL/min, the column temperature was 60 °C, and the Refractive Index detector temperature was 60 °C. The injection volume of 5 µL, calibration curves were made with glucose and ethanol Sigma Aldrich at 99% purity or higher, and a determination coefficient higher than 0.99 for each compound [36,37].

Fermentation was made in duplicate for every strain and samples were taken each hour for 13 consecutive hours. Two samples were taken every hour for each variable in the time interval of the process where t goes from 0 to 13, then the average value of the two measurements was computed. Therefore, each variable, i.e., biomass $[x(t)]$, glucose $[y(t)]$, and ethanol $[z(t)]$, has 14 observations with a total of 42 experimental data points (n) for each *K. marxianus* strain.

2.2. The KM Mechanistic Model

The KM mechanistic model is proposed to describe the dynamics of alcoholic fermentation. This is a biochemical process carried out by yeasts (also known as biomass), to transform sugars such as glucose into ethyl alcohol, otherwise known as ethanol (main product) and other byproducts. In this case, the alcoholic fermentation is taken in a batch fermentation process with established laboratory conditions of temperature and known initial glucose concentrations (substrate). Our mathematical model describes the relationships between biomass concentration $[x(t)]$, glucose consumption $[y(t)]$, and ethanol production $[z(t)]$ over time. The set of three first-order ODEs is presented below

$$\dot{x} = \frac{\rho_1 xy}{\rho_2 + y} - \rho_3 xz - \rho_4 x, \tag{1}$$

$$\dot{y} = -\rho_5 xy - \rho_6 yz - \rho_7 y, \tag{2}$$

$$\dot{z} = \rho_8 xz + \rho_9 yz - \rho_{10} z, \tag{3}$$

where each state variable $x(t)$, $y(t)$ and $z(t)$ are measured in g/L, and the time unit is given in *hours*. Now, by considering results from Leenheer and Aeyels (see Section II.A in [38]), all solutions with nonnegative initial conditions $[x(0), y(0), z(0) \geq 0,]$ will be located in the nonnegative octant as indicated below

$$\mathbf{R}^3_{+,0} = \{x(t), y(t), z(t) \geq 0\},$$

i.e., each positive half trajectory of the system will be positively forward invariant in $\mathbf{R}^3_{+,0}$. The latter also considers the biological sense of each variable as there is no meaning for negative values of biomass, glucose or ethanol in the scope of the KM system (1)–(3). It is important to mention that variables cannot grow exponentially indefinitely, and they must have biologically feasible limits which will be discussed in the next section. Values and units of each parameter of the KM system (1)–(3) are given in Table 1.

Table 1. Description, values, and units of variables and parameters for the KM mechanistic model.

Variables/ Parameters	Description	Values	Units
$x(t)$	Biomass concentration	–	g/L
$y(t)$	Glucose concentration	–	g/L
$z(t)$	Ethanol concentration	–	g/L
ρ_1	Biomass maximum growth rate	$[289.385, 381.419] \times 10^{-3}$	h^{-1}
ρ_2	Affinity with substrate constant	2.281	g/L
ρ_3	Inhibition rate of biomass growth due to product accumulation	1.066×10^{-3}	$L/(g \times h)$
ρ_4	Biomass death rate	7.275×10^{-3}	h^{-1}
ρ_5	Consumption rate for biomass growth	56.893×10^{-3}	$L/(g \times h)$
ρ_6	Consumption rate for ethanol production	71.842×10^{-3}	$L/(g \times h)$
ρ_7	Glucose spontaneous decomposition rate	824.233×10^{-9}	h^{-1}
ρ_8	Ethanol production associated with the biomass growth rate	$[19.088, 49.816] \times 10^{-3}$	$L/(g \times h)$
ρ_9	Glucose converted in ethanol	$[46.352, 70.349] \times 10^{-3}$	$L/(g \times h)$
ρ_{10}	Ethanol degradation rate	149.899×10^{-3}	h^{-1}

Now, let us describe our mechanistic model based on the experimental data described in the previous section and the following biological assumptions. Biomass growth dynamics is described by Equation (1), where the first term uses the classical Monod form for the growth of microorganisms [21], where ρ_1 is the biomass maximum growth rate (also found in the literature as μ_{max}), and ρ_2 is the affinity or half-velocity constant for glucose consumption. The second term describes biomass death due to ethanol accumulation toxicity by the law of mass action (see Section 2.3 in [39]) with a rate ρ_3. This term is negative because ethanol accumulation increases the membrane fluidity and negatively affects the membrane protein's function, which can lead to cell growth inhibition or even death [40,41]. The third term represents the natural yeast death rate $[\rho_4]$ mainly due to environmental resources depletion [42]. Glucose dynamics is formulated in Equation (2) as a decrescent function where the law of mass action gives the first two terms. The first one represents glucose consumption to support biomass growth. In contrast, the second term accounts for the glucose consumption used for ethanol production, with rates ρ_5 and ρ_6, respectively. The third term represents the spontaneous decomposition rate of glucose $[\rho_7]$ [43]. The latter stems from the fact that the culture medium is placed in a sealed container in batch fermentation, and no other nutrients (primarily glucose) are supplied into the system. Ethanol dynamics is described in Equation (3). The first term represents ethanol production associated with biomass growth. It is due to ethanol being recognized as a primary metabolite, a product obtained from reactions required or cellular growth [44,45]. The second term represents the glucose conversion to ethanol not directly linked to cellular growth, attributed to the need for Gibbs's free energy for cellular maintenance [44,46]. In both cases, terms are formulated by the law of mass action with respective rates ρ_8 and ρ_9. Finally, the third term represents ethanol degradation with a rate ρ_{10}. The flow diagram shown in Figure 1 was constructed to illustrate the dynamics of the system.

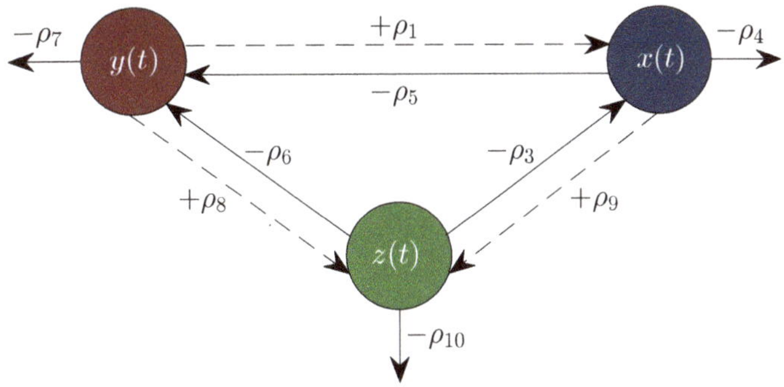

Figure 1. Flow diagram describing interactions between each variable and their corresponding relationship with each parameter.

It should be noted that fixed parameters values were estimated for the 17 *K. marxianus* strains, particularly for ρ_2, ρ_3, ρ_4, ρ_5, ρ_6, ρ_7 and ρ_{10}. Further, concerning the death rate of biomass $[\rho_4]$, spontaneous decomposition rate of glucose $[\rho_7]$, and degradation rate of ethanol $[\rho_{10}]$, one can see that they are in a different order of magnitude and the following constraint is formulated for these three parameters:

$$\rho_{10} > \rho_4 > \rho_7. \tag{4}$$

Now, concerning the equilibrium points of the KM system, Equations (1)–(3) have a unique biologically meaningful equilibrium point in the domain $\mathbf{R}^3_{+,0}$ given by

$$(x_0^*, y_0^*, z_0^*) = (0, 0, 0). \tag{5}$$

Another set of five equilibria with at least one negative value is shown in Appendix A; therefore, these equilibrium points are discarded from the biologically meaningful dynamics of the system. Further, from the biological characteristics of each variable the following can be stated with respect to each solution as time increases

$$\lim_{t\to\infty} x(t) = \lim_{t\to\infty} y(t) = \lim_{t\to\infty} z(t) = 0,$$

due to the eventual death of microorganisms, glucose consumption and ethanol degradation [47], asymptotic stability of trajectories is discussed in the next section.

2.3. Parameter Value Estimation

First, let us compute the glucose decomposition rate $[\rho_7]$ by assuming a first-order kinetics [48] for glucose dynamics, and considering a half-life $[t_{1/2}]$ of 96 years [43]. Then, ρ_7 can be computed from the next equation

$$y(t) = y_0 e^{-\rho_7 t},$$

as follows

$$\frac{y_0}{2} = y_0 e^{-\rho_7 t_{1/2}},$$

where y_0 is the glucose initial concentration, i.e., $y(0)$; hence

$$\rho_7 = \frac{\ln 2}{t_{1/2}} = 824.233 \times 10^{-9}\, h^{-1}.$$

Now, in order to determine the numerical values of parameters ρ_i, $i = 1, 2, 3, 4, 5, 6, 8, 9, 10$; the computational model of Equations (1)–(3) was formulated as follows

$$
\begin{aligned}
x_{i+1} &= x_i + \left(\frac{\rho_1 x_i y_i}{\rho_2 + y_i} - \rho_3 x_i z_i - \rho_4 x_i\right)\Delta_t, &(6)\\
y_{i+1} &= y_i + (-\rho_5 x_i y_i - \rho_6 y_i z_i - \rho_7 y_i)\Delta_t, &(7)\\
z_{i+1} &= z_i + (\rho_8 x_i z_i + \rho_9 y_i z_i - \rho_{10} z_i)\Delta_t, &(8)
\end{aligned}
$$

by applying Euler's method (see Section 1.7 in [49]) where Δ_t was set to 1×10^{-5}. Then, an algorithm was formulated in Matlab 2022b with the *lsqcurvefit* function from the optimization toolbox as its core [50] (initial points were set as 1×10^{-1} for each parameter $[\rho_1, \rho_8, \rho_9]$, i.e., $x_0 = [1 \times 10^{-1}; 1 \times 10^{-1}; 1 \times 10^{-1}]$, and optimotions of the function were set as follows: Max Function Evaluations $= 1 \times 10^3$, Max Iterations $= 1 \times 10^3$, and Function Tolerance $= 1 \times 10^{-9}$). This allowed us to establish a fixed value for parameters $\rho_j, j = 2,3,4,5,6,10$; by averaging the corresponding values for each of the 17 strains, these results are shown in Table 1. However, this procedure was not applicable for parameters ρ_1, ρ_8 and ρ_9, as it was expected that each strain of *K. marxianus* will have its own biomass growth rate $[\rho_1]$, and its corresponding ethanol production rates (ρ_8 and ρ_9), this is directly linked to the chromosomal differences among the strains affecting their growth kinetics. Hence, the main algorithm was redesigned to fit these three parameters and consider the others as fixed constants. Overall results are shown in Table 2 with their corresponding standard error (SE), and 95% confidence interval (CI). These two statistics allow us to establish that estimates for parameters ρ_1, ρ_8, and ρ_9 in the 17 strains are statistically significant. The latter follows from the fact that each $SE(\rho_k) < \rho_k/2$, $k = 1,8,9$; i.e., the value of the SE is less than half of the value fitted for each parameter, thus, the null hypothesis $[\rho_k = 0]$ can be rejected (see Section 5.2.8 from Koutsoyiannis [51]). Furthermore, both the lower and upper limit of the 95% CI of all fitted values are positive, hence, as there is no change in the sign of the bounds, this implies that the value of the null hypothesis is excluded, and one can conclude that all P-values are less than 0.05 (see Chapter 17 from Motulsky [52]).

Table 2. Fitted values, their standard error $[SE(\rho_k)]$, and 95% confidence intervals $[CI(\rho_k)]$ for the biomass growth rate $[\rho_1]$, and ethanol production rates $[\rho_8, \rho_9]$, where all values are written with a magnitude of 10^{-3}. Thus, it is possible to identify both lower and upper bounds for the values of the three fitted parameters as follows $\rho_1 \in [289.385, 381.419] \times 10^{-3}$, $\rho_8 \in [19.088, 49.816] \times 10^{-3}$, and $\rho_9 \in [46.352, 70.349] \times 10^{-3}$.

Strain	ρ_1	$SE(\rho_1)$	95% $CI(\rho_1)$	ρ_8	$SE(\rho_8)$	95% $CI(\rho_8)$	ρ_9	$SE(\rho_9)$	95% $CI(\rho_9)$
1	312.378	14.457	(283.137, 341.620)	31.563	3.669	(24.140, 38.985)	60.706	1.151	(58.378, 63.033)
2	320.848	8.407	(303.842, 337.853)	40.073	2.395	(35.228, 44.919)	51.253	0.650	(49.937, 52.569)
3	315.502	9.267	(296.756, 334.247)	42.560	2.657	(37.186, 47.935)	52.146	0.797	(50.533, 53.759)
4	319.364	8.448	(302.276, 336.453)	38.795	2.368	(34.005, 43.585)	54.576	0.763	(53.032, 56.119)
5	312.618	13.016	(286.291, 338.945)	30.025	3.939	(22.056, 37.993)	57.305	1.125	(55.030, 59.580)
6	318.556	12.204	(293.870, 343.242)	25.252	3.360	(18.456, 32.048)	60.595	1.126	(58.316, 62.874)
7	336.119	7.376	(321.200, 351.038)	29.110	1.793	(25.483, 32.737)	56.806	0.627	(55.538, 58.075)
8	326.840	11.939	(302.691, 350.989)	25.102	2.794	(19.450, 30.754)	64.465	1.004	(62.434, 66.497)
9	322.375	17.757	(286.457, 358.293)	29.314	4.383	(20.448, 38.180)	61.948	1.570	(58.771, 65.125)
10	381.419	15.931	(349.195, 413.642)	19.088	3.051	(12.915, 25.260)	69.068	1.414	(66.208, 71.928)
11	307.816	9.485	(288.631, 327.002)	48.803	3.070	(42.593, 55.014)	46.352	0.810	(44.714, 47.990)
12	289.385	10.222	(268.709, 310.060)	49.816	3.380	(42.978, 56.654)	48.047	0.887	(46.252, 49.841)

Table 2. *Cont.*

Strain	ρ_1	$SE(\rho_1)$	95% $CI(\rho_1)$	ρ_8	$SE(\rho_8)$	95% $CI(\rho_8)$	ρ_9	$SE(\rho_9)$	95% $CI(\rho_9)$
13	309.540	17.851	(273.434, 345.647)	28.475	4.498	(19.376, 37.574)	62.432	1.481	(59.436, 65.428)
14	312.244	12.998	(285.953, 338.536)	25.011	3.442	(18.050, 31.973)	61.883	1.121	(59.615, 64.151)
15	298.551	11.822	(274.638, 322.464)	37.245	3.440	(30.287, 44.202)	57.670	1.036	(55.574, 59.766)
16	335.425	16.486	(302.078, 368.773)	20.480	3.266	(13.873, 27.088)	70.349	1.295	(67.728, 72.969)
17	310.122	9.556	(290.793, 329.451)	44.563	2.980	(38.534, 50.592)	50.780	0.755	(49.252, 52.307)

Finally, it should be noted that the in silico experimentation performed in this research was done on a high-end desktop computer with a Ryzen 9 5950X CPU, 128 GB of RAM DDR4 CL18, a 12 GB GPU NVIDIA GeForce RTX 3080, and 1 TB Samsung 980 Pro Gen 4 NVMe M.2. The complete algorithm that was designed to fit the numerical values of parameters $[\rho_1, \rho_8, \rho_9]$, and to determine results concerning the statistics and goodness of fit can be found in the Supplementary Materials.

3. Results

In this section, the in silico experimentation is performed by means of several numerical simulations, and results relating to the nonlinear analysis of the system are derived, i.e., bounds for the localizing domain, asymptotic stability, and existence and uniqueness for all solutions of our model in the nonnegative octant $\mathbf{R}^3_{+,0}$.

3.1. In Silico Experimentation and Goodness of Fit

First, qualitative results are illustrated by means of numerical simulations. For the sake of simplicity, the strains were clustered in groups of four from strain 1 to the 16 (see Figures 2–5, respectively), and results concerning only for the strain 17 are shown in Figure 6. In all panels, the $\times$ green marker represents the average value for the two experimental data measurements for each variable, i.e., biomass $[x(t)]$, glucose $[y(t)]$, and ethanol $[z(t)]$, whilst the blue continuous line represents the approximated value given by the KM system (1)–(3) when is solved by means of Equations (6)–(8) with $\Delta_t = 1 \times 10^{-5}$. The time units are given in *hours* and the concentration for each variable is measured in g/L as indicated in each axis. Values for all ten parameters corresponding to each strain are shown in Tables 1 and 2.

Now, let us provide a quantitative measure of the fitting capabilities of the KM mechanistic model (1)–(3), thus, the coefficient of determination $[R^2]$ is calculated for each variable with results shown in Table 3.

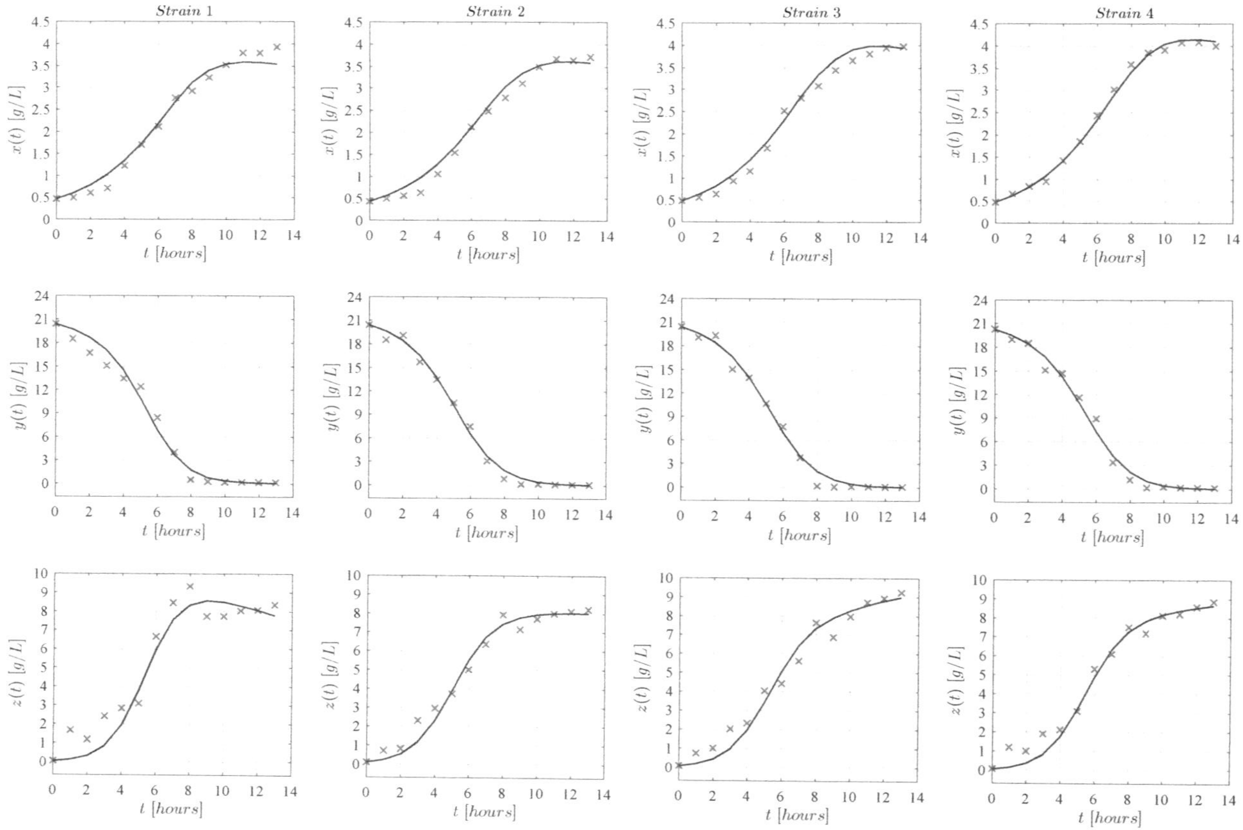

Figure 2. Each column from left to right (in landscape orientation) illustrates both the experimental data (× green marker), and the approximated values obtained with the KM system (continuous blue line) for each corresponding strain 1–4; the top row shows results for biomass $[x(t)]$, the middle row for glucose $[y(t)]$, and the lower row for ethanol $[z(t)]$. The × green marker represents the average value calculated from the two measurements that were made for each variable in every strain.

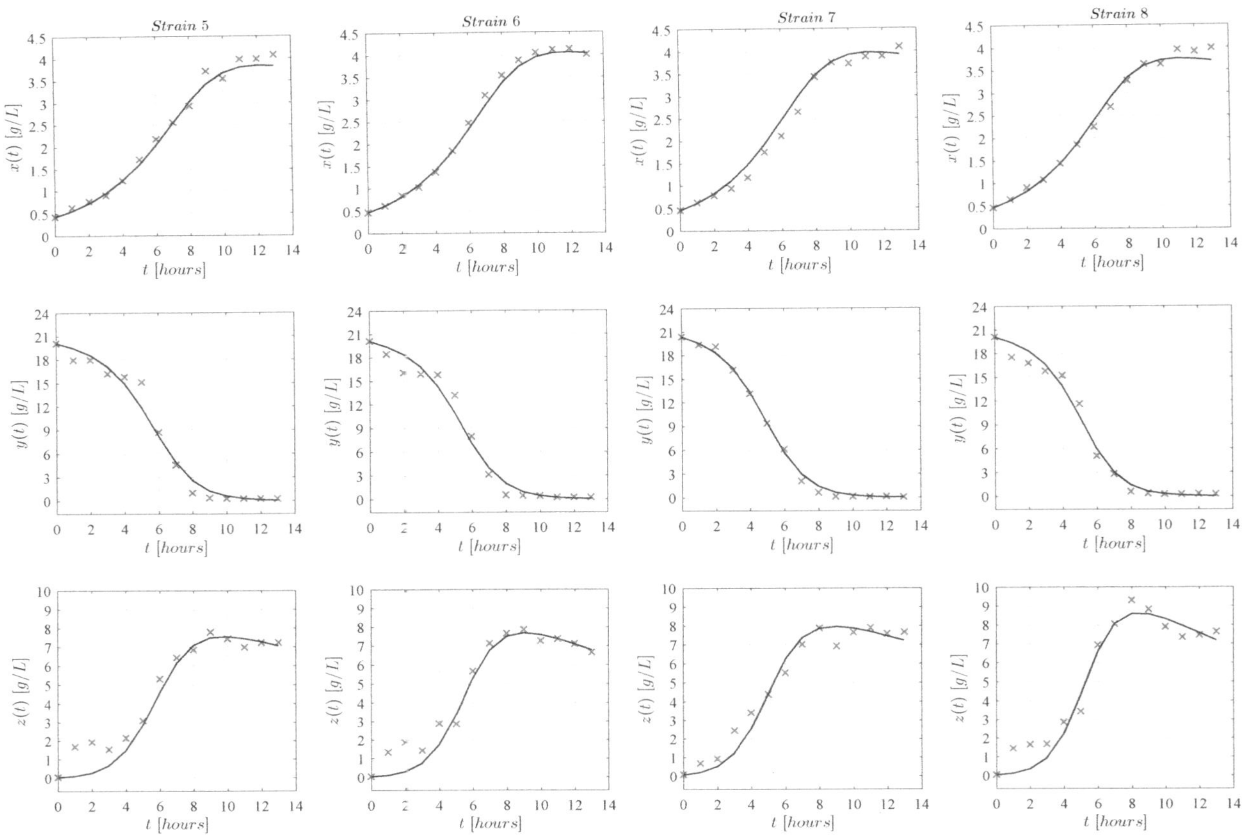

Figure 3. Each column from left to right (in landscape orientation) illustrates both the experimental data (× green marker), and the approximated values obtained with the KM system (continuous blue line) for each corresponding strain 5–8; the top row shows results for biomass $[x(t)]$, the middle row for glucose $[y(t)]$, and the lower row for ethanol $[z(t)]$. The × green marker represents the average value calculated from the two measurements that were made for each variable in every strain.

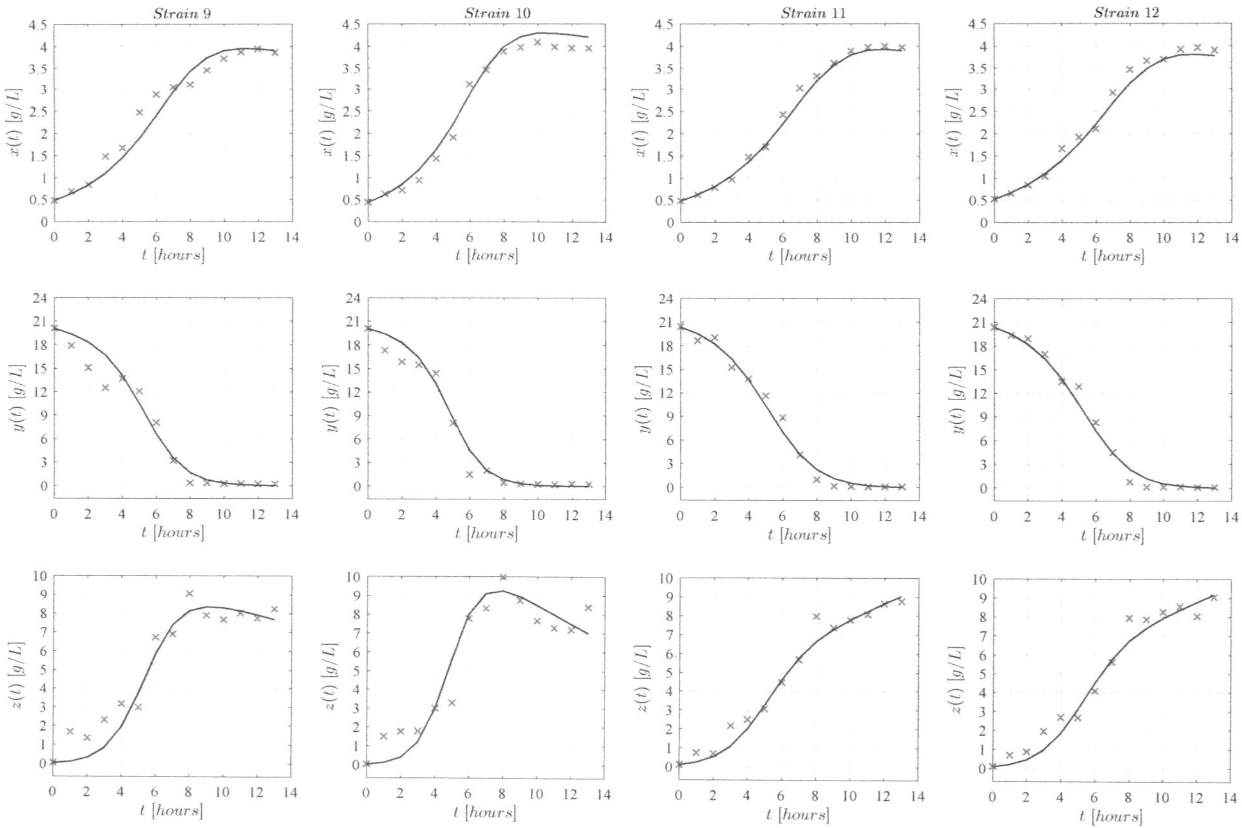

Figure 4. Each column from left to right (in landscape orientation) illustrates both the experimental data ($\times$ green marker), and the approximated values obtained with the KM system (continuous blue line) for each corresponding strain 9–12; the top row shows results for biomass $[x(t)]$, the middle row for glucose $[y(t)]$, and the lower row for ethanol $[z(t)]$. The $\times$ green marker represents the average value calculated from the two measurements that were made for each variable in every strain.

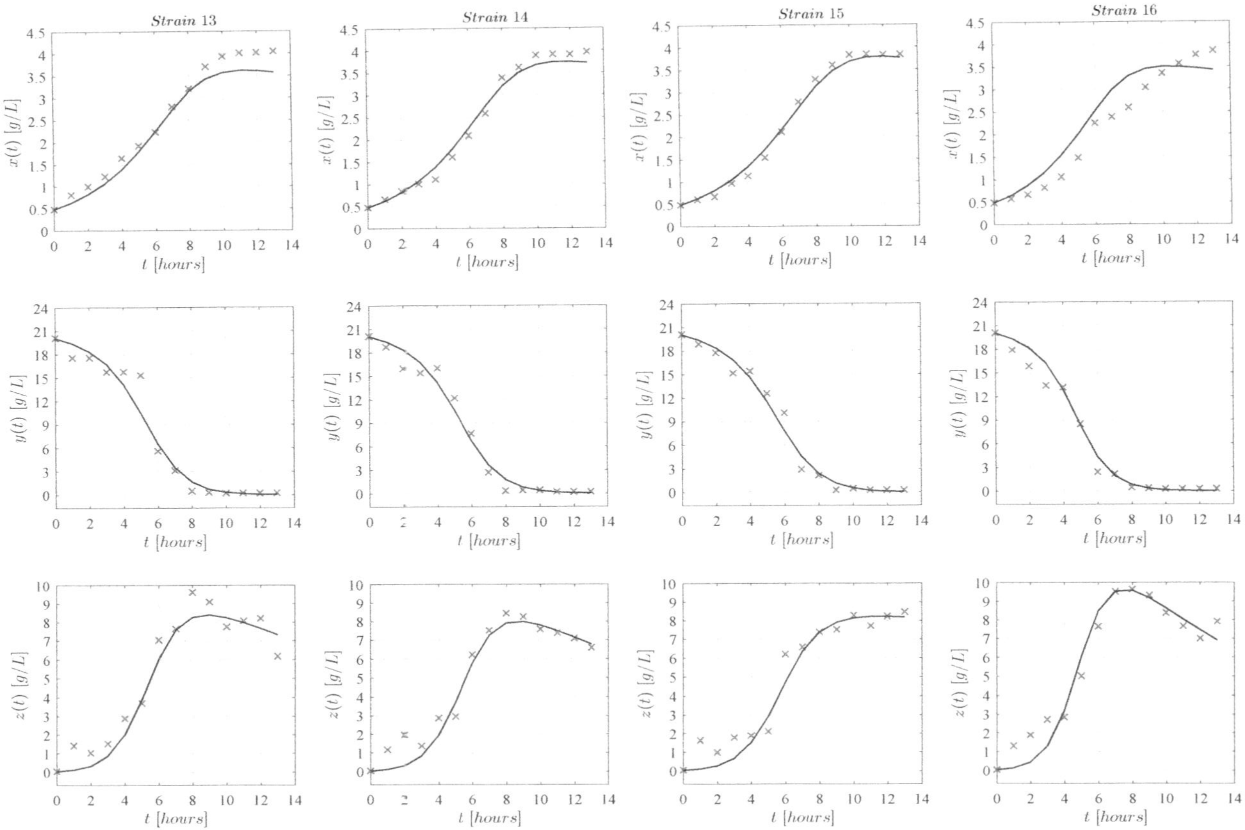

Figure 5. Each column from left to right (in landscape orientation) illustrates both the experimental data (× green marker), and the approximated values obtained with the KM system (continuous blue line) for each corresponding strain 13–16; the top row shows results for biomass $[x(t)]$, the middle row for glucose $[y(t)]$, and the lower row for ethanol $[z(t)]$. The × green marker represents the average value calculated from the two measurements that were made for each variable in every strain.

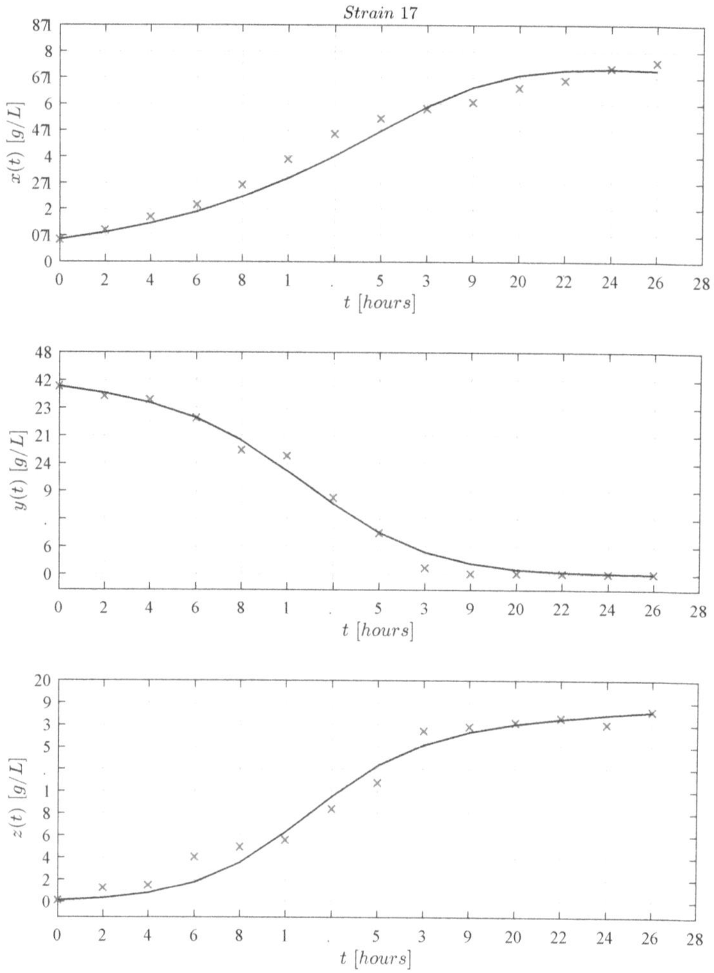

Figure 6. Experimental data (×green marker), and approximated values obtained with the KM system (continuous blue line) for strain 17; the top panel shows results for biomass $[x(t)]$, the panel row for glucose $[y(t)]$, and the lower panel for ethanol $[z(t)]$. The × green marker represents the average value calculated from the two measurements that were made for each variable in every strain.

Table 3. The R^2 provides a measure of how well the experimental data are replicated by the KM mathematical model (1)–(3) for each strain. This coefficient was computed independently for each variable, i.e., biomass $[x(t)]$, glucose $[y(t)]$, and ethanol $[z(t)]$. One can see that the values for R^2 ranges between 0.902 to 0.997 which allows us to conclude an overall well goodness of fit for the model.

Strain	Biomass	Glucose	Ethanol
1	0.980	0.979	0.923
2	0.983	0.994	0.974
3	0.983	0.990	0.967
4	0.996	0.990	0.974

Table 3. *Cont.*

Strain	Biomass	Glucose	Ethanol
5	0.989	0.979	0.926
6	0.995	0.979	0.940
7	0.982	0.997	0.960
8	0.990	0.985	0.951
9	0.953	0.950	0.918
10	0.979	0.972	0.905
11	0.992	0.988	0.974
12	0.985	0.988	0.967
13	0.960	0.962	0.942
14	0.986	0.979	0.947
15	0.992	0.983	0.945
16	0.902	0.974	0.940
17	0.965	0.991	0.973

Further, the Akaike Information Criterion (AIC) [53–55] was computed by considering a small sample relative to the number of parameters ($n/K < 40$) with a bias correction as indicated below

$$AIC = n \ln\left(\frac{\sum_{i=1}^{n}[f_e(i) - f_a(i)]^2}{n}\right) + 2K + \frac{2K(K+1)}{n-K-1},$$

where n is the total number of experimental data points; f_e the experimental data and f_a the approximated value for the residual sum of squares (RSS); and K the number of parameters of the system; therefore, $n/K = (3 \times 14)/10 = 4.2$. Results, including RSS, AIC and R^2, for the complete trajectory of the system, i.e., $\phi(x, y, z)$ for the total of 42 experimental points (14 for each variable) are summarized in Table 4.

Table 4. In order to provide overall measures for the fitting capabilities of our mathematical model, i.e., the KM system (1)–(3), values were calculated for the RSS, the AIC, and the R^2 to estimate and describe the dynamics between the three variables $\phi(x, y, z)$, where $x(t)$, $y(t)$ and $z(t)$ represent, respectively, the evolution of biomass, glucose, and ethanol.

Strain	RSS	AIC	R^2
1	28.375	+10.626	0.976
2	9.233	−36.530	0.993
3	13.679	−20.020	0.989
4	12.267	−24.596	0.990
5	26.471	+7.709	0.979
6	24.403	+4.292	0.980
7	7.825	−43.478	0.994
8	19.638	−4.832	0.983
9	48.551	+33.184	0.954
10	37.663	+22.520	0.966

Table 4. *Cont.*

Strain	RSS	AIC	R^2
11	13.978	−19.111	0.989
12	16.201	−12.912	0.988
13	43.020	+28.105	0.966
14	24.183	+3.912	0.980
15	21.902	−0.248	0.983
16	30.653	+13.869	0.972
17	13.047	−22.005	0.990

The AIC yields a value that relates the amount of information that our model loses when approximating the experimental data. Hence, one can compare the capabilities of the model to estimate the concentrations over time of biomass $[x(t)]$, glucose $[y(t)]$, and ethanol $[z(t)]$ among the 17 *K. marxianus* strains while providing a statistical measured for the quality of the KM system (1)–(3).

3.2. Nonlinear Analysis: Localizing Domain, Asymptotic Stability, Existence and Uniqueness

The localizing domain can be determined by computing the upper bounds for all variables of the KM mechanistic model (1)–(3), the lower bounds are given by the boundary of the domain $\mathbf{R}^3_{+,0}$, i.e., $\{x_{\text{inf}} = 0, y_{\text{inf}} = 0, z_{\text{inf}} = 0\}$. The latter is achieved by means of integration and the LCIS method [56]. Within the localizing domain, one may find all biologically meaningful dynamics of the system, i.e., compact invariant sets such as equilibrium points, periodic orbits, limit cycles and chaotic attractors (see Section 3 in [57]), among others.

First, in order to find the upper bound for the glucose concentration $[y(t)]$, Equation (2) is integrated as follows

$$\int \frac{dy}{y} = -\int_0^t f(x,z)dt,$$

where

$$f(x,z) = \rho_5 x + \rho_6 z + \rho_7 > 0,$$

by considering $\rho_7 > 0$ and $x(t), z(t) \geq 0$ from the domain $\mathbf{R}^3_{+,0}$. Then,

$$y(t) = y(0)\exp\left[-\int_0^t f(x,z)dt\right],$$

with $y(0) \in \mathbf{R}_{+,0}$. Therefore, all solutions with nonnegative initial conditions will be bounded as indicated below

$$K_y = \{0 \leq y(t) \leq y_{\text{sup}} = y(0)\},$$

hence, any upper bound for $x(t)$ and $z(t)$ depending on K_y will be directly related to the glucose initial concentration $[y(0)]$, which is expected as biomass and ethanol production over time is directly related to glucose dynamics.

Now, let us provide the mathematical background that allows us to compute a localizing domain where all compact invariant sets of a nonlinear dynamical system are located. The General Theorem concerning the LCIS method was formalized by Krishchenko and Starkov (see Section 2 in [58]) and it states the following: *Each compact invariant set* Γ *of* $\dot{x} = f(x)$ *is contained in the localizing domain:*

$$K(h) = \{h_{\text{inf}} \leq h(x) \leq h_{\text{sup}}\}.$$

From the latter $f(x)$ is a C^∞−differentiable vector function where $x \in \mathbf{R}^n$ is the state vector. $h(x) : \mathbf{R}^n \to \mathbf{R}$ is a C^∞−differentiable function called localizing function, $h|_S$ denotes the restriction of $h(x)$ on a set $S \subset \mathbf{R}^n$ with $S(h) = \left\{x \in \mathbf{R}^n \mid L_f h(x) = 0\right\}$, and $L_f h(x) = (\partial h/\partial x) f(x)$ is the Lie derivative of $f(x)$. Hence, one can define $h_{\inf} = \inf\{h(x) \mid x \in S(h)\}$ and $h_{\sup} = \sup\{h(x) \mid x \in S(h)\}$. Furthermore, if all compact invariant sets are contained in the set $K(h_i)$ and in the set $K(h_j)$ then they are contained in $K(h_i) \cap K(h_j)$ as well. The nonexistence of compact invariant sets can be considered for a given set $\Lambda \subset \mathbf{R}^n$ if $\Lambda \cap K(h) = \emptyset$, then the system $\dot{x} = f(x)$ has no compact invariant sets located in Λ.

Following the LCIS method, one can explore the next localizing function

$$h_1 = x + \alpha y; \ \alpha > 0,$$

then, the Lie derivative may be written as follows

$$L_f h_1 = -\rho_4 x - \alpha \rho_7 y - \frac{\alpha \rho_5 y + \alpha \rho_2 \rho_5 - \rho_1}{\rho_2 + y} xy - \rho_3 xz - \alpha \rho_6 yz,$$

and the set $S(h_1) = \left\{L_f h_1 = 0\right\}$ is given by

$$S(h_1) = \left\{\rho_4 x = -\alpha \rho_7 y - \frac{\alpha \rho_5 y + \alpha \rho_2 \rho_5 - \rho_1}{\rho_2 + y} xy - \rho_3 xz - \alpha \rho_6 yz\right\},$$

where $x = h_1 - \alpha y$, therefore set $S(h_1)$ is rewritten as indicated below

$$S(h_1) = \left\{h_1 = \frac{\alpha(\rho_4 - \rho_7)}{\rho_4} y - \frac{\alpha \rho_5 y + \alpha \rho_2 \rho_5 - \rho_1}{\rho_4(\rho_2 + y)} xy - \frac{\rho_3}{\rho_4} xz - \frac{\alpha \rho_6}{\rho_4} yz\right\},$$

and the next two conditions are formulated

$$\rho_4 - \rho_7 > 0, \quad (9)$$

$$\alpha > \frac{\rho_1}{\rho_2 \rho_5}, \quad (10)$$

where (9) is directly fulfilled by (4). Now, let us apply the Iterative Theorem in order to find an upper bound for the localizing function

$$S(h_1) \cap K_y \subset \left\{h_1 \le \frac{\alpha(\rho_4 - \rho_7)}{\rho_4} y_{\sup}\right\},$$

then

$$K(h_1) = \left\{x(t) + \alpha y(t) \le \frac{\alpha(\rho_4 - \rho_7)}{\rho_4} y_{\sup}\right\},$$

from the latter, the upper bound for the biomass concentration $[x(t)]$ may be written in terms of the parameters and the initial glucose concentration $[y(0)]$ as follows

$$K_x = \left\{0 \le x(t) \le x_{\sup} = \frac{\alpha(\rho_4 - \rho_7)}{\rho_4} y_{\sup}\right\}.$$

Now, an upper bound for the ethanol concentration $[z(t)]$ can be determined by the following localizing function

$$h_2 = \beta_1 x + \beta_2 y + z; \ \beta_1, \beta_2 > 0,$$

whose Lie derivative is computed as indicated below

$$L_f h_2 = -\beta_1\rho_4 x - \beta_2\rho_7 y - \rho_{10} z - \frac{\beta_2\rho_5 y + \beta_2\rho_2\rho_5 - \beta_1\rho_1}{\rho_2 + y} xy - (\beta_1\rho_3 - \rho_8)xz - (\beta_2\rho_6 - \rho_9)yz,$$

and at this step, the following conditions are formulated

$$\beta_1 > \frac{\rho_8}{\rho_3}, \tag{11}$$

$$\beta_2 > \max\left\{\frac{\rho_9}{\rho_6}, \frac{\beta_1\rho_1}{\rho_2\rho_5}\right\}, \tag{12}$$

then, set $S(h_2) = \left\{L_f h_2 = 0\right\}$, can be written as follows

$$S(h_2) = \left\{\rho_{10} z = -\beta_1\rho_4 x - \beta_2\rho_7 y - \frac{\beta_2\rho_5 y + \beta_2\rho_2\rho_5 - \beta_1\rho_1}{\rho_2 + y} xy - (\beta_1\rho_3 - \rho_8)xz - (\beta_2\rho_6 - \rho_9)yz\right\},$$

hence, as $z = h_2 - \beta_1 x - \beta_2 y$, then set $S(h_2)$ is rewritten as indicated below

$$S(h_2) = \left\{h_2 = \frac{\beta_1(\rho_{10} - \rho_4)}{\rho_{10}} x + \frac{\beta_2(\rho_{10} - \rho_7)}{\rho_{10}} y - \frac{\beta_2\rho_5 y + \beta_2\rho_2\rho_5 - \beta_1\rho_1}{\rho_2 + y} xy - (\beta_1\rho_3 - \rho_8)xz - (\beta_2\rho_6 - \rho_9)yz\right\},$$

where the next condition is formulated

$$\rho_{10} > \max\{\rho_4, \rho_7\}, \tag{13}$$

and it holds by (4). Then, the Iterative Theorem is applied to get the following result

$$S(h_2) \cap K_x \cap K_y \subset \left\{h_2 \leq \frac{\beta_1(\rho_{10} - \rho_4)}{\rho_{10}} x_{\sup} + \frac{\beta_2(\rho_{10} - \rho_7)}{\rho_{10}} y_{\sup}\right\},$$

then, the upper bound for the localizing function h_2 is derived as follows

$$K(h_2) = \left\{\beta_1 x(t) + \beta_2 y(t) + z(t) \leq \frac{\beta_1(\rho_{10} - \rho_4)}{\rho_{10}} x_{\sup} + \frac{\beta_2(\rho_{10} - \rho_7)}{\rho_{10}} y_{\sup}\right\},$$

now, from the latter one can get the upper bound for ethanol concentration $[z(t)]$ over time in terms of the parameters, the initial glucose concentration $[y(0)]$ and the upper bound of biomass $[x_{\sup}]$ as given below

$$K_z = \left\{0 \leq z(t) \leq z_{\sup} = \frac{\beta_1(\rho_{10} - \rho_4)}{\rho_{10}} x_{\sup} + \frac{\beta_2(\rho_{10} - \rho_7)}{\rho_{10}} y_{\sup}\right\}.$$

Results shown above allow us to conclude the following regarding the boundedness of the KM system (1)–(3) solutions:

Theorem 1. ***Localizing domain.*** *If conditions (9)–(13) are fulfilled, then all compact invariant sets of the KM mechanistic model (1)–(3) are located either at the boundaries or within the following domain*

$$K_\Gamma = K_x \cap K_y \cap K_z,$$

where $K_\Gamma \subset \mathbf{R}^3_{+,0}$, and the ultimate bounds for biomass $[x(t)]$, glucose $[y(t)]$, and ethanol $[z(t)]$ concentrations over time are given below

$$\begin{aligned} K_x &= \left\{0 \leq x(t) \leq x_{\sup} = \frac{\alpha(\rho_4 - \rho_7)}{\rho_4} y_{\sup}\right\}, \\ K_y &= \{0 \leq y(t) \leq y_{\sup} = y(0)\}, \\ K_z &= \left\{0 \leq z(t) \leq z_{\sup} = \frac{\beta_1(\rho_{10} - \rho_4)}{\rho_{10}} x_{\sup} + \frac{\beta_2(\rho_{10} - \rho_7)}{\rho_{10}} y_{\sup}\right\}. \end{aligned}$$

Now, let us briefly provide the mathematical background concerning the stability theory in the sense of Lyapunov, particularly the direct method where it is necessary to formulate a Lyapunov candidate function, which is usually denoted as $V(x) : \mathbf{R}^n \rightarrow \mathbf{R}$, a continuously differentiable function whose temporal derivative is given by $\dot{V}(x) = (\partial V/\partial x)f(x)$. This function must be positive definite, i.e., $V(0) = 0$ and $V(x) > 0$ for $x \neq 0$, whilst a negative definite function is also $V(0) = 0$ but $V(x) < 0$ for $x \neq 0$. Further, function $V(x)$ is said to be radially unbounded if $V(x) \rightarrow \infty$ as $\|x\| \rightarrow \infty$. The latter allows the formulation of the Global Asymptotic Stability Theorem (see Chapter 4 in [59] and Chapter 2 in [60]) which states the following: *The equilibrium point x^* is globally asymptotically stable if there exists a function $V(x)$ positive definite, radially unbounded and decrescent such that its temporal derivative $\dot{V}(x)$ is negative definite.* A function $V(x)$ satisfying properties of this theorem is called Lyapunov function.

Following the latter, let us explore the next Lyapunov candidate function

$$V(x,y,z) = \gamma_1 x + \gamma_2 y + z,$$

with

$$\gamma_1, \gamma_2 > 0,$$

then, the time derivative is computed as shown below

$$\dot{V}(x,y,z) = \gamma_1\left(\frac{\rho_1 xy}{\rho_2 + y} - \rho_3 xz - \rho_4 x\right) - \gamma_2(\rho_5 xy + \rho_6 yz + \rho_7 y) + \rho_8 xz + \rho_9 yz - \rho_{10} z,$$

which can be rewritten as follows

$$\dot{V}(x,y,z) = -\gamma_1\rho_4 x - \gamma_2\rho_7 y - \rho_{10} z - (\gamma_1\rho_3 - \rho_8)xz - (\gamma_2\rho_6 - \rho_9)yz - \frac{y\gamma_2\rho_5 + \gamma_2\rho_2\rho_5 - \gamma_1\rho_1}{\rho_2 + y}xy,$$

where it is evident that $\dot{V}(0,0,0) = 0$, therefore the following constraints on coefficients γ_1 and γ_2 are formulated to ensure $\dot{V}(x,y,z) < 0$

$$\gamma_1 > \frac{\rho_8}{\rho_3}, \tag{14}$$

$$\gamma_2 > \max\left\{\frac{\rho_9}{\rho_6}, \frac{\gamma_1\rho_1}{\rho_2\rho_5}\right\}, \tag{15}$$

thus, as parameters $\rho_i, i = 1,2,3,5,6,8,9$; in both conditions are different for each term, then it is possible to assume that there exists a set of solutions that satisfies (14) and (15). Hence, the following result can be concluded:

Theorem 2. ***Asymptotic stability.*** *If conditions (14) and (15) are fulfilled, then the KM mechanistic model (1)–(3) is asymptotically stable and all trajectories will go to the equilibrium point $(x_0^*, y_0^*, z_0^*) = (0,0,0)$.*

The latter implies that any given trajectory $[\phi(x(t), y(t), z(t))]$ with nonnegative initial conditions $[x(0), y(0), z(0) \geq 0]$ passing through any point $(x(t), y(t), z(t))^T$ in $\mathbf{R}^3_{+,0}$ its $\omega-$limit set is not empty and it is a compact invariant set, i.e.,

$$\lim_{t\to\infty} \phi(x(t), y(t), z(t)) = (0,0,0)^T,$$

see Lemma 4.1 by Khalil in [59] at Section 4.2 and Theorem 1 by Perko in [61] at Section 3.2.

Concerning the existence and uniqueness of solutions for the KM system (1)–(3), let us introduce the following notations for the sake of simplicity

$$
\begin{aligned}
f_1(t,x,y,z) &= \frac{\rho_1 xy}{\rho_2+y} - \rho_3 xz - \rho_4 x, \\
f_2(t,x,y,z) &= -\rho_5 xy - \rho_6 yz - \rho_7 y, \\
f_3(t,x,y,z) &= \rho_8 xz + \rho_9 yz - \rho_{10} z,
\end{aligned}
$$

and compute the Jacobian matrix $[\partial f/\partial u](t,u)$ (see [49] at Section 7.4) with results shown below for $f_i(t,u), i = 1,2,3;$ and $u = [x,y,z]^T$

$$
J = \begin{bmatrix} \frac{\rho_1 y}{\rho_2+y} - \rho_3 z - \rho_4 & \frac{\rho_1\rho_2 x}{(\rho_2+y)^2} & -\rho_3 x \\ -\rho_5 y & -\rho_5 x - \rho_6 z - \rho_7 & -\rho_6 y \\ \rho_8 z & \rho_9 z & \rho_8 x + \rho_9 y - \rho_{10} \end{bmatrix}, \tag{16}
$$

and it is evident that $f_i(t,u)$ and $[\partial f/\partial u](t,u)$ are continuous and exist on the domain $\Omega = \left[t_0, t_f\right] \times K_\Gamma$ with $\left[t_0, t_f\right] \in [t_0, \infty]$ and $K_\Gamma \subset \mathbf{R}^3_{+,0}$ [33]. Hence, the latter implies that $f_i(t,u)$ is locally Lipschitz in u on Ω (see Lemma 3.2 by Khalil in [59] at Section 3.1). Further, each element of (16) is bounded by Theorem 1. Thus, the following can be concluded:

Theorem 3. ***Existence and uniqueness.*** *There is a Lipschitz constant $\ell \geq 0$ such that $\|[\partial f/\partial u](t,u)\| \leq \ell$ on Ω. Then, $f_i(t,u)$ satisfies the Lipschitz condition*

$$
\|f(t,u_1) - f(t,u_2)\| \leq \ell\|u_1 - u_2\|,
$$

and there exists some $\delta > 0$ such that the KM mechanistic model (1)–(3), given as $\dot{u} = f_i(t,u)$ with $u(t_0) = u_0$, has a unique solution over $[t_0, t_0 + \delta]$.

Although conditions for asymptotic stability of the equilibrium point $(x_0^*, y_0^*, z_0^*) = (0,0,0)$ in $\mathbf{R}^3_{+,0}$ were established in Theorem 2, it is straightforward to demonstrate its local asymptotic stability by evaluating (5) in (16) as follows

$$
J_{(0,0,0)} = \begin{bmatrix} -\rho_4 & 0 & 0 \\ 0 & -\rho_7 & 0 \\ 0 & 0 & -\rho_{10} \end{bmatrix},
$$

where the eigenvalues $[\lambda_i, i = 1,2,3]$ are given by each element of the diagonal. Thus, $\lambda_1 = -\rho_4$, $\lambda_2 = -\rho_7$, and $\lambda_3 = -\rho_{10}$. Therefore, Theorem 4.7 by Khalil in [59] allows us to conclude the next additional result to Theorem 2:

Corollary 1. ***Local stability.*** *The equilibrium point $(x_0^*, y_0^*, z_0^*) = (0,0,0)$ of the KM mechanistic model (1)–(3) is locally asymptotically stable in $\mathbf{R}^3_{+,0}$.*

4. Discussion

The KM mechanistic model (1)–(3) was formulated by considering the biological relationships between each variable in a controlled batch fermentation where concentrations in g/L were measured for biomass $[x(t)]$, glucose $[y(t)]$, and ethanol $[z(t)]$ over 13 consecutive *hours*. Then, by means of the *lsqcurvefit* function, an algorithm was developed in Matlab to approximate the experimental data from the 17 *K. marxianus* strains discussed at Section 2; both qualitative (see Figures 2–6) and quantitative (see Tables 3 and 4) results were shown in Section 3. The in silico experimentation illustrates the capabilities of the system to approximate the experimental data of each strain, whilst both the R^2 and the AIC provide a value for the goodness of fit of the model to each set of data. In Table 4, one can see that R^2 values range from 0.955 to 0.994, and AIC from -43.478 to 33.184, these values are for strains 7 and 9, respectively.

Now, it should be noted that the dynamics between biomass growth, substrate consumption and product generation have been modeled before by means of the logistic growth law [62], the Pirt Equation [63], and Luedeking–Piret Equation [64] as indicated below in Equations (17)–(19), respectively:

$$\dot{X} = \mu_{max} X \left(1 - \frac{X}{X_{max}}\right), \tag{17}$$

$$\dot{S} = -\frac{1}{Y_{X/S}} \dot{X} - mX, \tag{18}$$

$$\dot{P} = \alpha \dot{X} + \beta X, \tag{19}$$

where μ_{max} is the biomass maximum growth rate, this parameter is equivalent to ρ_1 in our mathematical model; X_{max} the maximum concentration value of biomass in the experimental data set for the time-interval of the process being observed; $Y_{X/S}$ the biomass/substrate yield; m is the maintenance coefficient; α is the growth-associated coefficient for the product; and β is the non-growth-associated coefficient for the product. Our algorithm was applied to approximate the experimental data of the 17 *K. marxianus* strains with overall results shown in Table 5.

Table 5. The logistic, Pir, and Luedeking–Piret Equations (17)–(19) provides valuable information concerning biomass growth $[\mu_{max}]$, biomass/substrate yield $[Y_{X/S}]$, and product generation $[\alpha]$; estimated numerical values are given in their respective columns. Concerning the goodness of fit, results regarding the RSS, AIC, and R^2 are provided in the following columns.

Strain	$\mu_{max}\,[\times 10^{-3}]$	$Y_{X/S}\,[\times 10^{-3}]$	α	RSS	AIC	R^2
1	442.669	159.066	2.648	35.905	+7.815	0.970
2	451.284	150.295	2.551	19.641	−17.523	0.984
3	437.516	158.265	1.903	23.501	−9.987	0.982
4	428.145	164.594	2.185	25.859	−5.971	0.979
5	417.879	166.067	2.296	50.863	+22.441	0.960
6	434.735	169.347	2.253	48.715	+20.629	0.959
7	482.158	168.159	2.248	21.992	−12.774	0.982
8	463.808	165.934	2.505	47.813	+19.844	0.959
9	455.745	163.157	2.530	40.974	+13.362	0.961
10	544.441	175.477	2.344	55.484	+26.094	0.950
11	422.190	158.260	2.120	23.199	−10.530	0.981
12	394.033	151.888	2.241	34.084	+5.629	0.974
13	433.945	165.334	2.562	83.294	+43.158	0.935
14	437.632	161.878	2.422	52.732	+23.957	0.956
15	394.825	151.920	2.692	42.685	+15.079	0.966
16	523.096	162.308	2.643	41.717	+14.116	0.962
17	421.708	150.435	2.352	28.825	−1.411	0.978

The main comparison between the KM system (1)–(3) and Equations (17)–(19) is performed with respect to the biomass maximum growth rate, given by ρ_1 and μ_{max}, respectively. Tables 2 and 5 show that estimated values of ρ_1 are on average $\sim$0.717 smaller than those estimated for μ_{max}. The latter is a direct consequence of the biological assumptions on which each mechanistic model was formulated. The KM system (1)–(3) was constructed by considering interactions between the three variables as illustrated in the flow diagram of Figure 1, whilst the logistic, Pirt, and Luedeking–Piret Equations (17)–(19) are constructed by only assuming a logistic growth for biomass without taking into account the overall effect of ethanol production over the entire system as well as the death rate of biomass $[x(t)]$, decomposition rate of glucose $[y(t)]$, and degradation of ethanol $[z(t)]$. Further, the in silico experimentation concerning Equations (17)–(19) illustrated in Figures A1–A5 at Appendix B shows that approximated values for substrate $[S(t)]$, i.e., glucose, becomes negative as time increases, which is not biologically possible for this variable. Further, one can see from the experimental data that ethanol production does not follow a smooth sigmoidal growth, the data even illustrates degradation among some strains, which is better approximated by our model as it is shown in the lower panels of Figures 2–6.

When comparing the goodness of fit by computing the AIC and R^2, it is evident that the KM system (1)–(3) had overall better results than the logistic, Pirt, and Luedeking–Piret Equations (17)–(19). Although the latter has fewer parameters than ours (six and ten, respectively) and the AIC penalizes a model with more parameters to be fitted, results for the RSS were lower for the KM system which ultimately worked in our favor. Further, the capabilities of the KM mechanistic model may extend beyond its ability to approximate experimental data and estimate the biomass maximum growth rate, in Appendix C the in silico experimentation illustrates the dynamics for $t \in [0, 39]$, i.e., three times the period for the experimental data. Figures A6–A10 show that as time increases and the substrate is no longer added into the system, then the death of biomass and degradation of ethanol begins to take over the system. The latter was expected from the asymptotic stability results of Section 3, particularly Theorem 2 and Corollary 1, as these state that the concentration of all variables will eventually be zero, i.e., both biomass $[x(t)]$ and ethanol $[z(t)]$ concentrations are going to be depleted. Additionally, it is important to note that all solutions of the KM system are bounded from above, which is consistent with the localizing domain results of Theorem 1.

Regarding the values of parameters m and β, our algorithm yielded results in the magnitude of 10^{-14} for m in all strains; in fact, setting m to zero does not affect the ultimate results for the other parameters $[\mu_{\max}, Y_{X/S}, \alpha$, and $\beta]$ which may allow us to completely disregard this term $[-mX]$ from Equations (17)–(19). Concerning β, values for 12 strains were in the same order of magnitude $[10^{-14}]$, however, the following results were determined for strains 3, 4, 11, 12, and 17: 87.633×10^{-3}, 38.660×10^{-3}, 51.050×10^{-3}, 53.813×10^{-3}, 37.702×10^{-3}, respectively. Hence, the non-growth-associated coefficient for the product may influence the dynamics in some karyotypes of *K. marxianus*.

5. Conclusions

Mechanistic modeling has proven to be a powerful tool capable of describing the relationships between different variables in the dynamics of biological systems when considering assumptions based on scientific principles underlying the phenomenon being modeled. In this work, a set of three coupled first-order ODEs was formulated which can approximate experimental changes over time of alcoholic fermentation in batch culture by 17 different strains of *K. marxianus*.

The KM mechanistic model (1)–(3) describes biomass growth $[x(t)]$, glucose consumption $[y(t)]$, and ethanol production $[z(t)]$ in concentrations of g/L per hour. The parameter values of the system were estimated through a nonlinear curve-fitting algorithm in Matlab with the experimental data of each batch culture fermentation described in Section 2. The latter allowed us to conclude that seven parameters have the same numerical value for the dynamics observed in the 17 strains, particularly the affinity with substrate constant $[\rho_2]$, inhibition rate of biomass growth due to product accumulation $[\rho_3]$, biomass death rate $[\rho_4]$, consumption rates for biomass growth and ethanol production $[\rho_5$ and $\rho_6]$, glucose spontaneous decomposition rate $[\rho_7]$, and ethanol degradation rate $[\rho_{10}]$; these values are shown in Table 1. However, the biomass maximum growth rate $[\rho_1]$, ethanol production associated with biomass growth $[\rho_8]$, and glucose converted in ethanol $[\rho_9]$ parameters have specific values for each strain, results are shown in Table 2 with a 95% confidence interval that gives us the margin of error for each parameter value estimation.

As predictive microbiology establishes, mathematical models must be simplified until measurable parameters can be obtained, the KM mechanistic model successfully achieves this with ρ_1, ρ_8, and ρ_9 as the main parameters that describe the overall dynamics of the batch fermentation process under study in this research. The biomass growth rate is a very specific value for each strain that must be as high as possible. Ethanol production with respect to biomass growth represents the fermentative capacity of each strain, and the concentration of glucose converted to ethanol is directly related to these rates. It should be noted that in batch culture the latter requires high sugar concentrations to achieve alcoholic fermentation.

Further, the in silico experimentation illustrates that our model may be able to accurately predict the concentration of each variable as it is shown in Appendix C; nonetheless, further experimental data are needed to properly validate this assessment. One can see in Figures A6–A10 that when no more substrate is added to the culture, then biomass growth goes into the death phase, and ethanol degradation begins to happen in the system. This behavior is to be expected as the nonlinear analysis of the system allowed us to conclude that all concentrations will eventually go to zero in the absence of glucose, i.e., the asymptotic stability of the equilibrium (5) $[(x_0^*, y_0^*, z_0^*) = (0,0,0)]$ by Theorem 2 and Corollary 1; further, concentrations over time of all variables are bounded by the Localizing Domain Theorem 1. The latter is illustrated in all panels for the predictions of biomass growth $[x(t)]$, glucose consumption $[y(t)]$, and ethanol production $[z(t)]$.

Finally, the KM mechanistic model may be useful in the field of predictive microbiology, particularly in alcoholic fermentation through yeast and sugar, such as *K. marxianus* and glucose as only three parameters of our system needs to be fitted for different strains. Furthermore, when comparing the results of the biomass maximum growth rate of our model with the classic logistic, Pirt, and Luedeking–Piret Equations (17)–(19), our values are on average 71.7% smaller as the KM system (1)–(3) takes into account the effect of both substrate and product concentrations in the batch culture over the biomass growth phases.

Supplementary Materials: The following supporting information can be downloaded at: https://www.mdpi.com/article/10.3390/e25030497/s1. this algorithm can be applied to fit the numerical values of parameters $[\rho_1, \rho_8, \rho_9]$, and to determine results concerning the statistics and goodness of fit.

Author Contributions: Conceptualization, Y.S. and P.A.V.; methodology, J.B.P.-L., N.O.S.-C. and P.A.V.; software, P.A.V.; validation, J.B.P.-L. and N.O.S.-C.; formal analysis, P.A.V. and E.R.; investigation, Y.S. and E.R.; resources, J.B.P.-L., N.O.S.-C. and P.A.V.; data curation, J.B.P.-L., N.O.S.-C. and F.J.R.-S.; writing—original draft preparation, P.A.V., Y.S. and E.R.; writing—review and editing, P.A.V., Y.S., J.B.P.-L. and N.O.S.-C.; visualization, P.A.V. and Y.S.; supervision, Y.S. and P.A.V.; project administration, Y.S. and P.A.V.; funding acquisition, Y.S., P.A.V., J.B.P.-L. and N.O.S.-C. All authors have read and agreed to the published version of the manuscript.

Funding: Authors of this work were supported by different projects indicated as follows: Yolocuauhtli Salazar by the TecNM project 14166.22-P: Modelizado de la producción y vida de anaquel de microorganismos en productos fermentados; Paul A. Valle by the TecNM project: Gemelos digitales para el análisis y control de sistemas biológicos; Nicolás O. Soto-Cruz and Jesús B. Páez-Lerma by the CONACyT project Ciencia Básica 252465, and the TecNM project 10197-21-P.

Institutional Review Board Statement: Not applicable.

Informed Consent Statement: Not applicable.

Data Availability Statement: The data supporting the findings of the present work are available through authors Nicolás O. Soto-Cruz and Jesús B. Páez-Lerma upon reasonable request.

Acknowledgments: Emmanuel Rodríguez and Francisco J. Reyes-Sánchez express their gratitude to CONACyT for the financial support during their postgraduate studies under their scholarship agreements 813941 and 465646, respectively.

Conflicts of Interest: The authors declare no conflict of interest. Funders had no role in the design of the study; in the collection, analysis, or interpretation of data; in the writing of the manuscript; or in the decision to publish the results.

Abbreviations

The following abbreviations are used in this manuscript:

AIC	Akaike Infoirmation Criterion
CI	Confidence Interval
K. marxianus	*Kluyveromyces marxianus*
LCIS	Localization of Compact Invariant Sets
RSS	Residual Sum of Squares
R^2	Coefficient of Determination
SE	Standard Error

Appendix A. Equilibria

In order to further discuss the nonlinear mathematical analysis of Section 3, all equilibrium points of the system are calculated and it becomes evident that the origin is the only biologically meaningful result. The set of equilibria of the KM system (1)–(3) is determined by solving the next system of equations

$$\begin{aligned} f_1(t,x,y,z) &= x\left(\frac{\rho_1 y}{\rho_2+y}-\rho_3 z-\rho_4\right)=0, \\ f_2(t,x,y,z) &= -y(\rho_5 x+\rho_6 z+\rho_7)=0, \\ f_3(t,x,y,z) &= z(\rho_8 x+\rho_9 y-\rho_{10})=0, \end{aligned}$$

from which one can compute the following

$$\begin{aligned} (x_0^*,y_0^*,z_0^*) &= (0,0,0), \\ (x_1^*,y_1^*,z_1^*) &= \left(0,\frac{\rho_{10}}{\rho_9},-\frac{\rho_7}{\rho_6}\right), \\ (x_2^*,y_2^*,z_2^*) &= \left(\frac{\rho_{10}}{\rho_8},0,-\frac{\rho_4}{\rho_3}\right), \\ (x_3^*,y_3^*,z_3^*) &= \left(-\frac{\rho_7}{\rho_5},\frac{\rho_2\rho_4}{\rho_1-\rho_4},0\right), \end{aligned}$$

and

$$\begin{aligned} &(x_4^*,y_4^*,z_4^*), \\ &(x_5^*,y_5^*,z_5^*), \end{aligned}$$

where

$$\begin{aligned} x_4^* &= \frac{-\rho_1\rho_6\rho_8+\rho_3\rho_5\rho_{10}-\rho_3\rho_7\rho_8+\rho_4\rho_6\rho_8+\rho_2\rho_3\rho_5\rho_9-\sqrt{\rho}}{2\rho_3\rho_5\rho_8}, \\ y_4^* &= \frac{\rho_1\rho_6\rho_8+\rho_3\rho_5\rho_{10}+\rho_3\rho_7\rho_8-\rho_4\rho_6\rho_8-\rho_2\rho_3\rho_5\rho_9+\sqrt{\rho}}{2\rho_3\rho_5\rho_9}, \\ z_4^* &= \frac{\rho_1\rho_6\rho_8-\rho_3\rho_5\rho_{10}-\rho_3\rho_7\rho_8-\rho_4\rho_6\rho_8-\rho_2\rho_3\rho_5\rho_9+\sqrt{\rho}}{2\rho_3\rho_6\rho_8}, \end{aligned}$$

$$
\begin{aligned}
x_5^* &= \frac{-\rho_1\rho_6\rho_8 + \rho_3\rho_5\rho_{10} - \rho_3\rho_7\rho_8 + \rho_4\rho_6\rho_8 + \rho_2\rho_3\rho_5\rho_9 + \sqrt{\rho}}{2\rho_3\rho_5\rho_8}, \\
y_5^* &= \frac{\rho_1\rho_6\rho_8 + \rho_3\rho_5\rho_{10} + \rho_3\rho_7\rho_8 - \rho_4\rho_6\rho_8 - \rho_2\rho_3\rho_5\rho_9 - \sqrt{\rho}}{2\rho_3\rho_5\rho_9}, \\
z_5^* &= \frac{\rho_1\rho_6\rho_8 - \rho_3\rho_5\rho_{10} - \rho_3\rho_7\rho_8 - \rho_4\rho_6\rho_8 - \rho_2\rho_3\rho_5\rho_9 - \sqrt{\rho}}{2\rho_3\rho_6\rho_8},
\end{aligned}
$$

with

$$
\rho = (\rho_3\rho_7\rho_8 + \rho_1\rho_6\rho_8 - \rho_4\rho_6\rho_8 - \rho_3\rho_5\rho_{10} - \rho_2\rho_3\rho_5\rho_9)^2 + 4\rho_3\rho_5\rho_8(\rho_1\rho_6\rho_{10} + (\rho_{10} + \rho_2\rho_9)(\rho_3\rho_7 - \rho_4\rho_6)).
$$

Now, it is evident that the equilibrium points $(x_i^*, y_i^*, z_i^*), i = 1,2,3;$ have at least one negative term. However, although is not straightforward, the same can be concluded regarding equilibriums (x_4^*, y_4^*, z_4^*) and (x_5^*, y_5^*, z_5^*), as these are computed by disregarding the common term in each equation as follows

$$
\begin{aligned}
\frac{\rho_1 y}{\rho_2 + y} - \rho_3 z - \rho_4 &= 0, \\
\rho_5 x + \rho_6 z + \rho_7 &= 0, \\
\rho_8 x + \rho_9 y - \rho_{10} &= 0,
\end{aligned}
$$

and equality $\rho_5 x + \rho_6 z + \rho_7 = 0$ can only be fulfilled when either x_j^* or $z_j^*, j = 4,5;$ are negative. Therefore, the KM mechanistic model (1)–(3) has unique biologically meaningful equilibrium given by

$$
(x_0^*, y_0^*, z_0^*) = (0,0,0).
$$

Appendix B. Logistic, Pirt, and Luedeking–Piret Equations

This appendix presents results concerning the in silico experimentation when fitting the experimental data to the logistic, Pirt, and Luedeking–Piret Equations (17)–(19). Figures A1–A5 are aiming to qualitative compare the proposed mathematical model with the classic model of biomass-substrate-product. Further, a quantitative comparison is carried out through the coefficient of determination and the Akaike Information Criterion at Section 4, see Tables 2, 4, and 5.

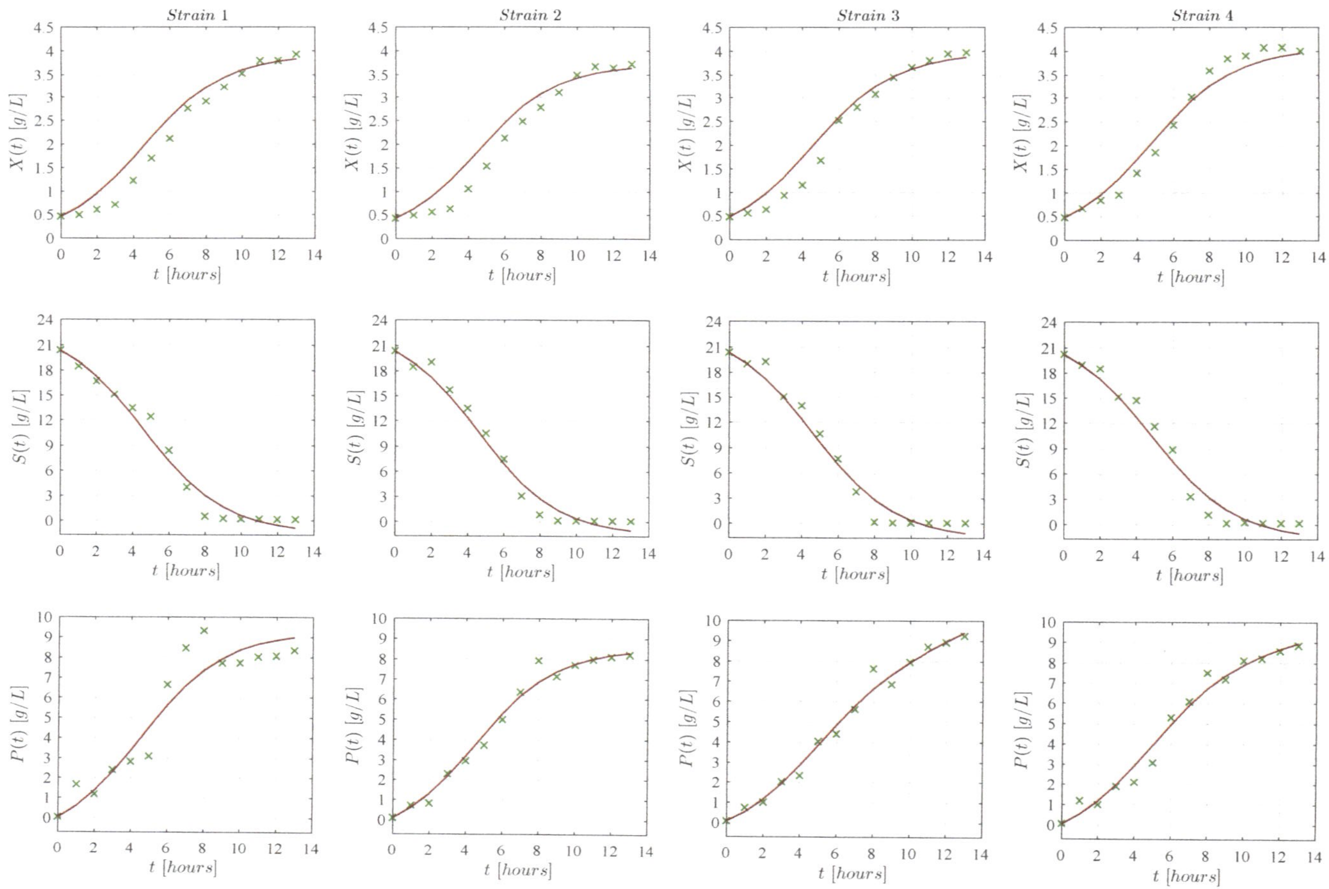

Figure A1. Observed data (×green marker), and approximated values (continuous red line) for strains 1–4 with the Logistic, Pirt, and Luedeking–Piret Equations (17)–(19).

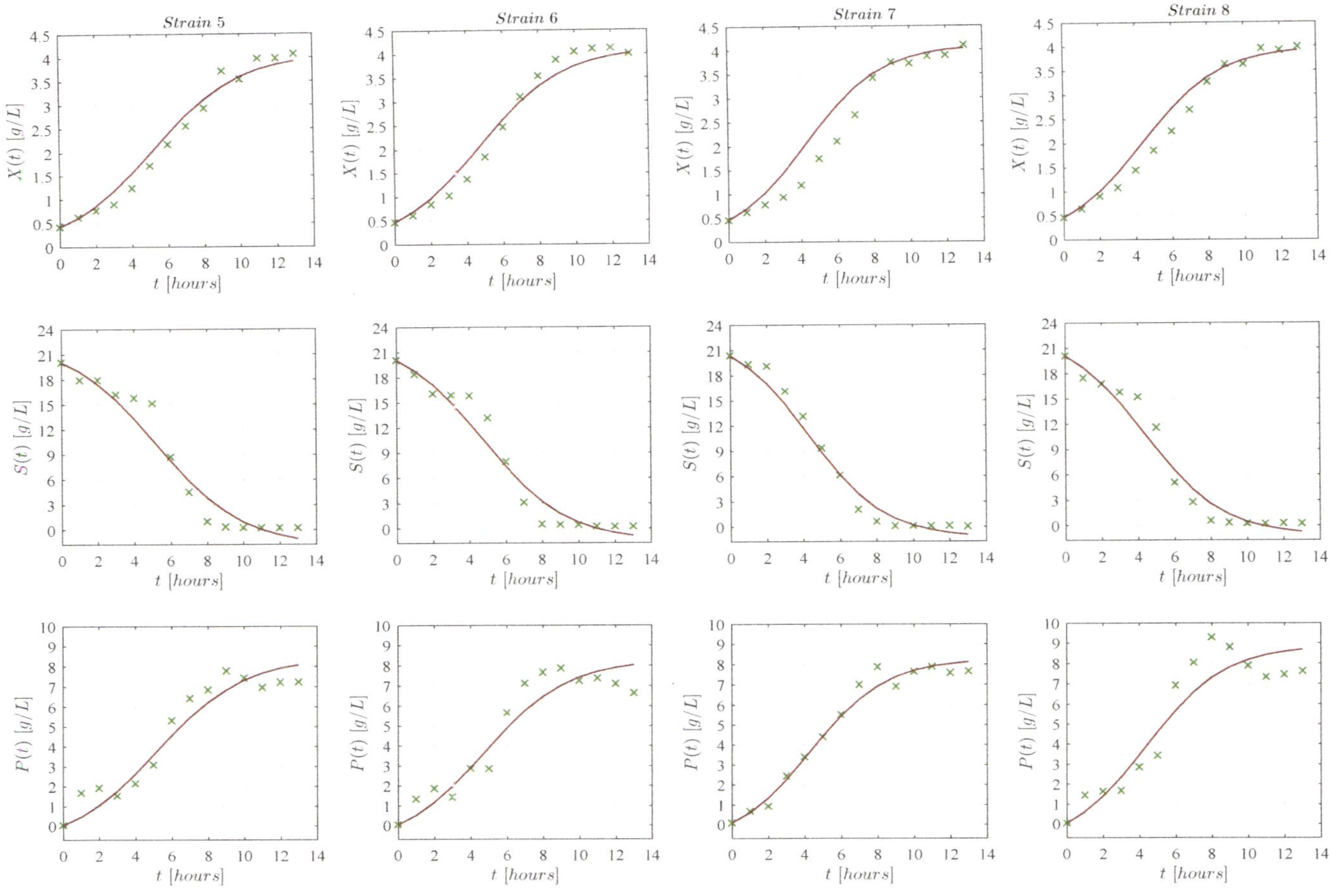

Figure A2. Observed data (×green marker), and approximated values (continuous red line) for strains 5–8 with the Logistic, Pirt, and Luedeking–Piret Equations (17)–(19).

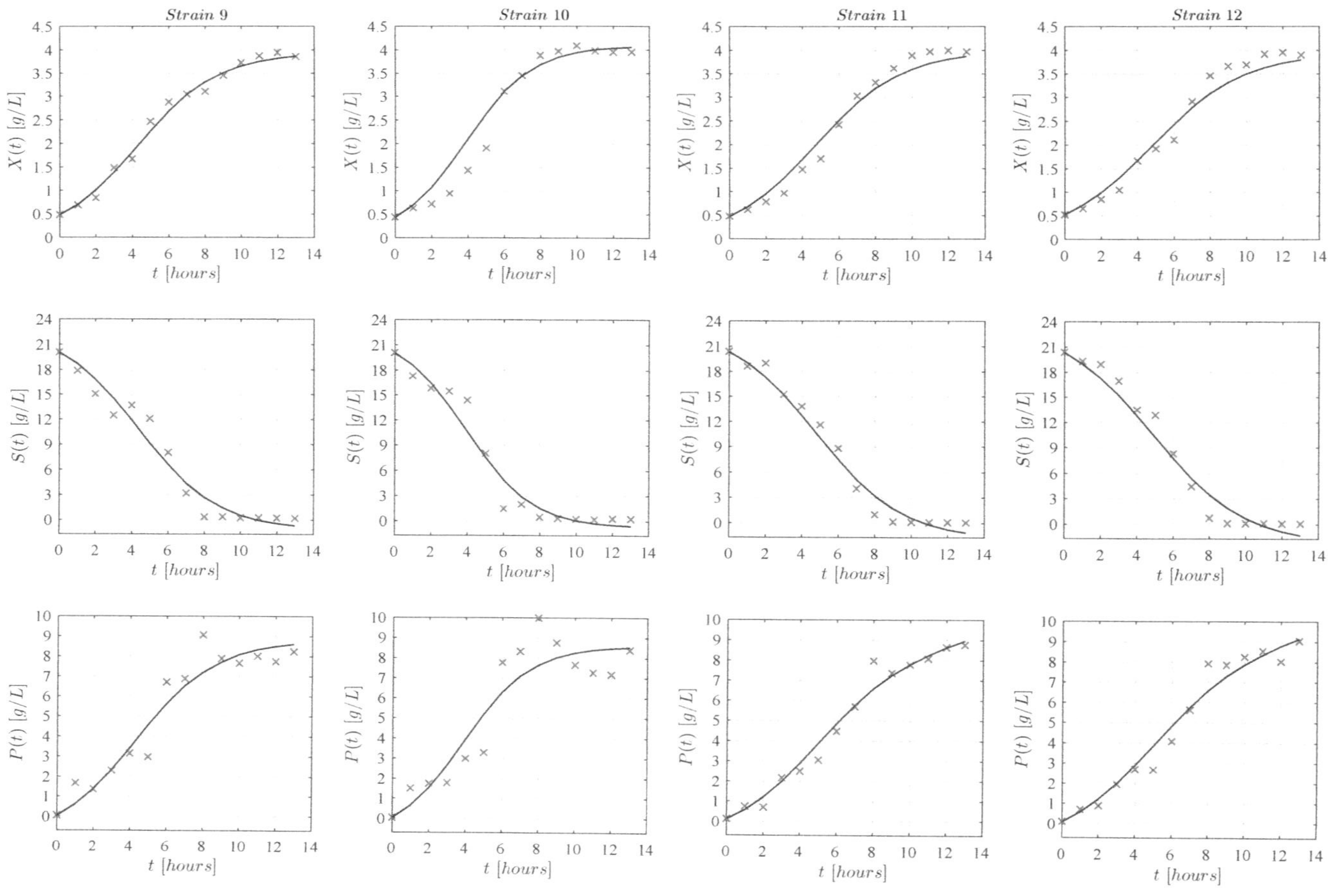

Figure A3. Observed data (×green marker), and approximated values (continuous red line) for strains 9–12 with the Logistic, Pirt, and Luedeking–Piret Equations (17)–(19).

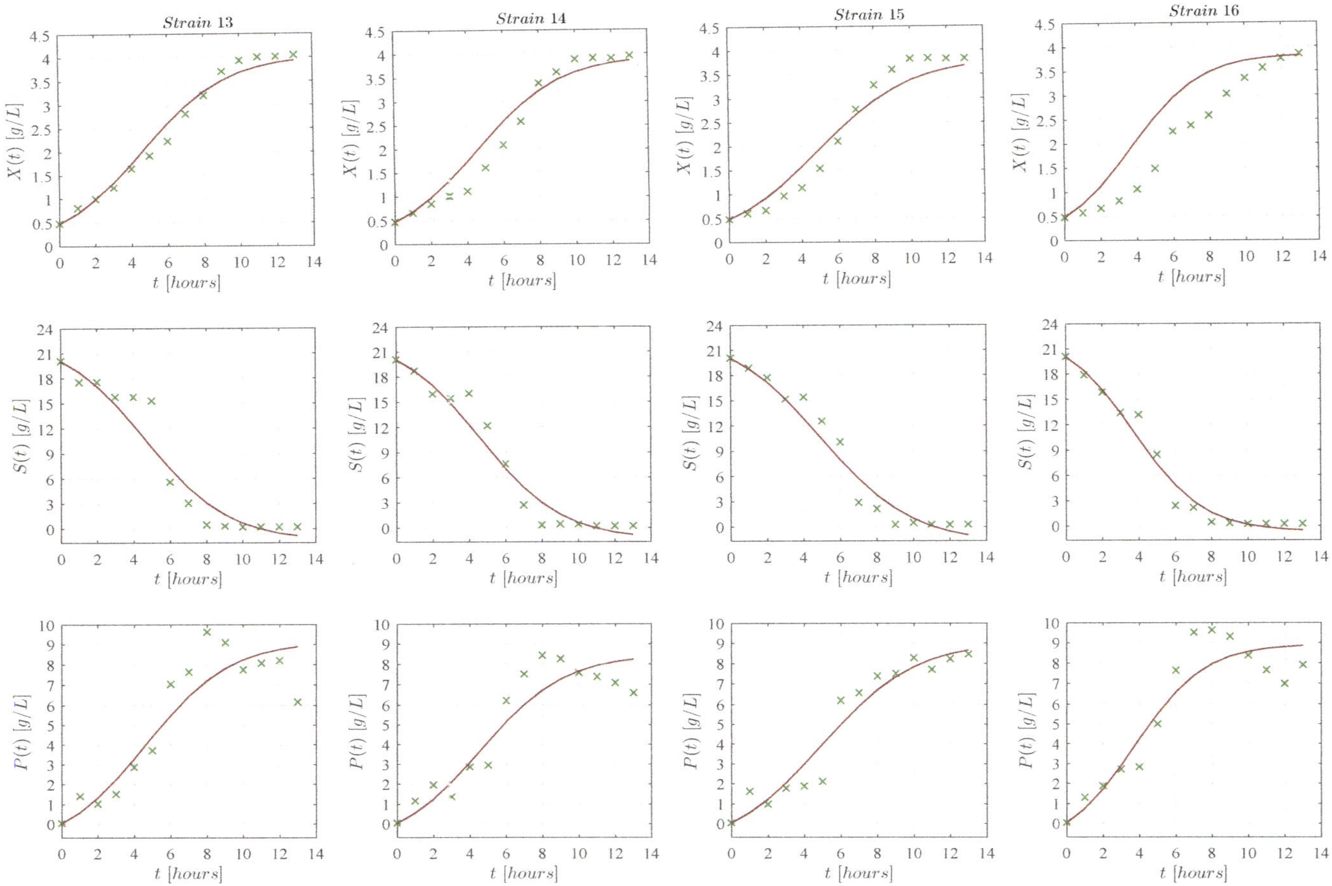

Figure A4. Observed data (×green marker), and approximated values (continuous red line) for strains 13–16 with the Logistic, Pirt, and Luedeking–Piret Equations (17)–(19).

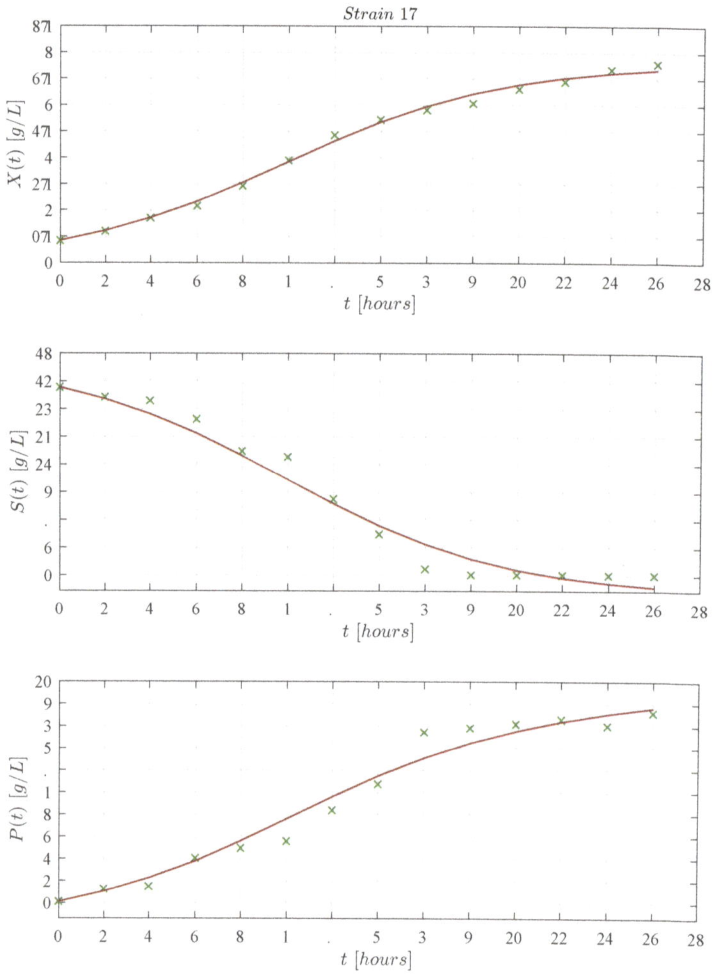

Figure A5. Observed data (×green marker), and approximated values (continuous red line) for strain 17 with the Logistic, Pirt, and Luedeking–Piret Equations (17)–(19).

Appendix C. Predictive Ability of the KM Mechanistic Model

This appendix presents results concerning the in silico experimentation when solving the KM mechanistic model (1)–(3) for a time interval of $t \in [0, 39]$ in order to illustrate its ability to predict the dynamics of the three variables, i.e., the concentration in g/L over time between biomass $[x(t)]$, glucose $[y(t)]$, and ethanol $[z(t)]$, after the last experimental data point without further substrate addition into the batch. It should be noted that at this stage of the research there is not available data to validate if the model is be able to accurately predict the evolution of both biomass and ethanol in the system, further experimental data points could be helpful to better fit the values of biomass death rate $[\rho_4]$, and ethanol degradation rate $[\rho_{10}]$. However, Figures A6–A10 allow us to illustrate results concerning Theorem 2 and Corollary 1 from the asymptotic stability analysis.

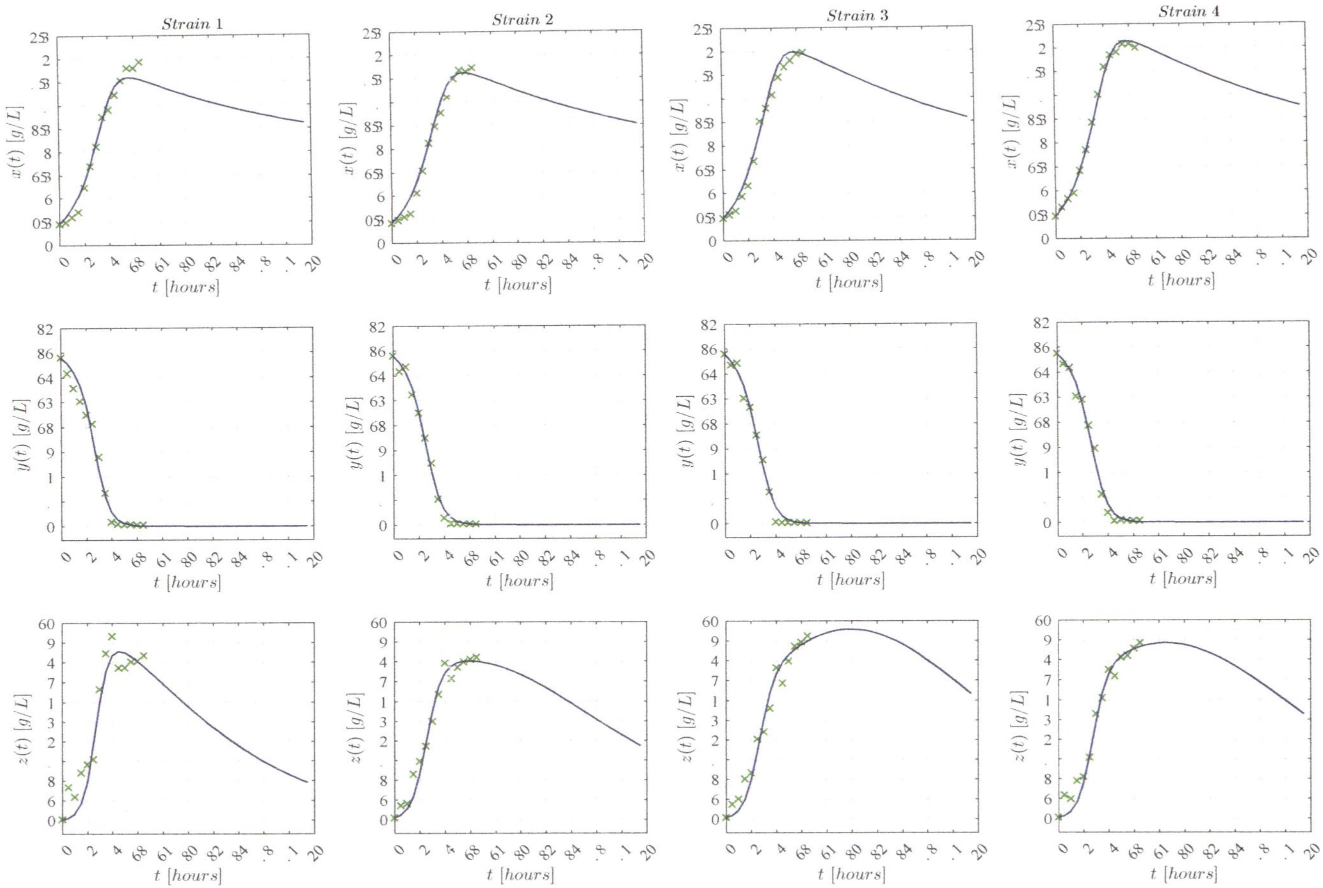

Figure A6. Dynamics prediction for strains 1–4 with the KM system (1)–(3). Observed data are given by the × green marker, and approximated values by the continuous blue line.

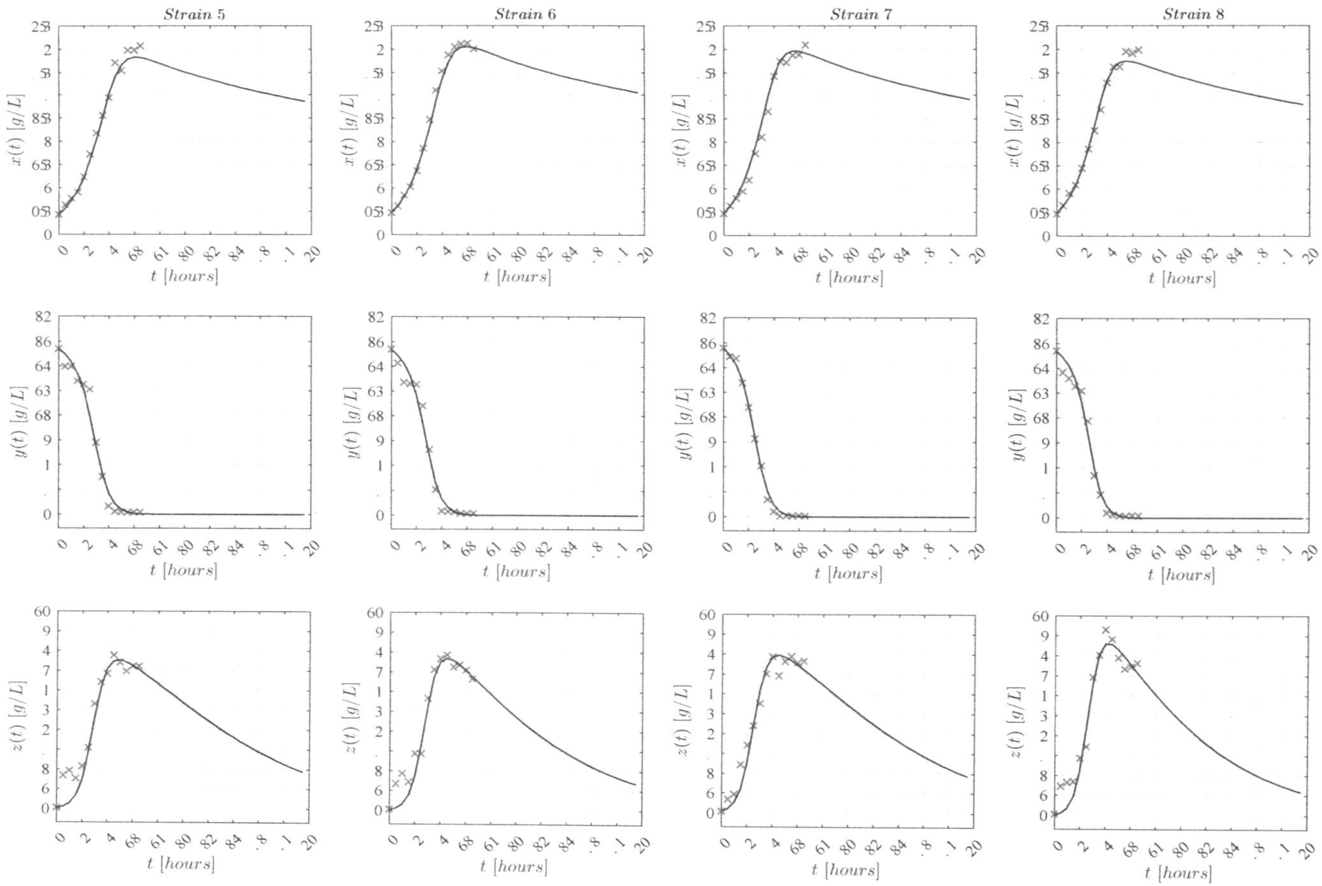

Figure A7. Dynamics prediction for strains 5–8 with the KM system (1)–(3). Observed data are given by the × green marker, and approximated values by the continuous blue line.

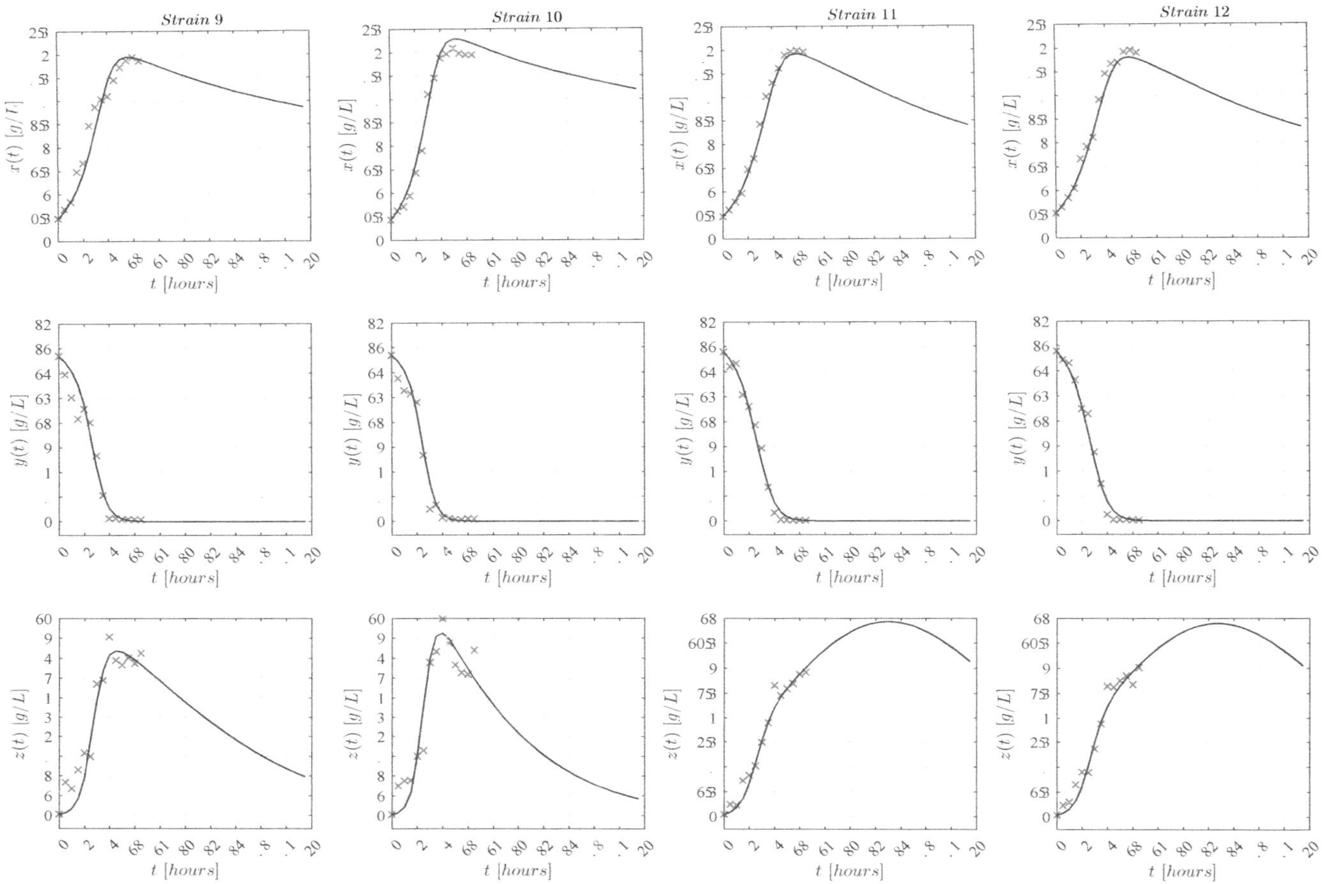

Figure A8. Dynamics prediction for strains 9–12 with the KM system (1)–(3). Observed data are given by the × green marker, and approximated values by the continuous blue line.

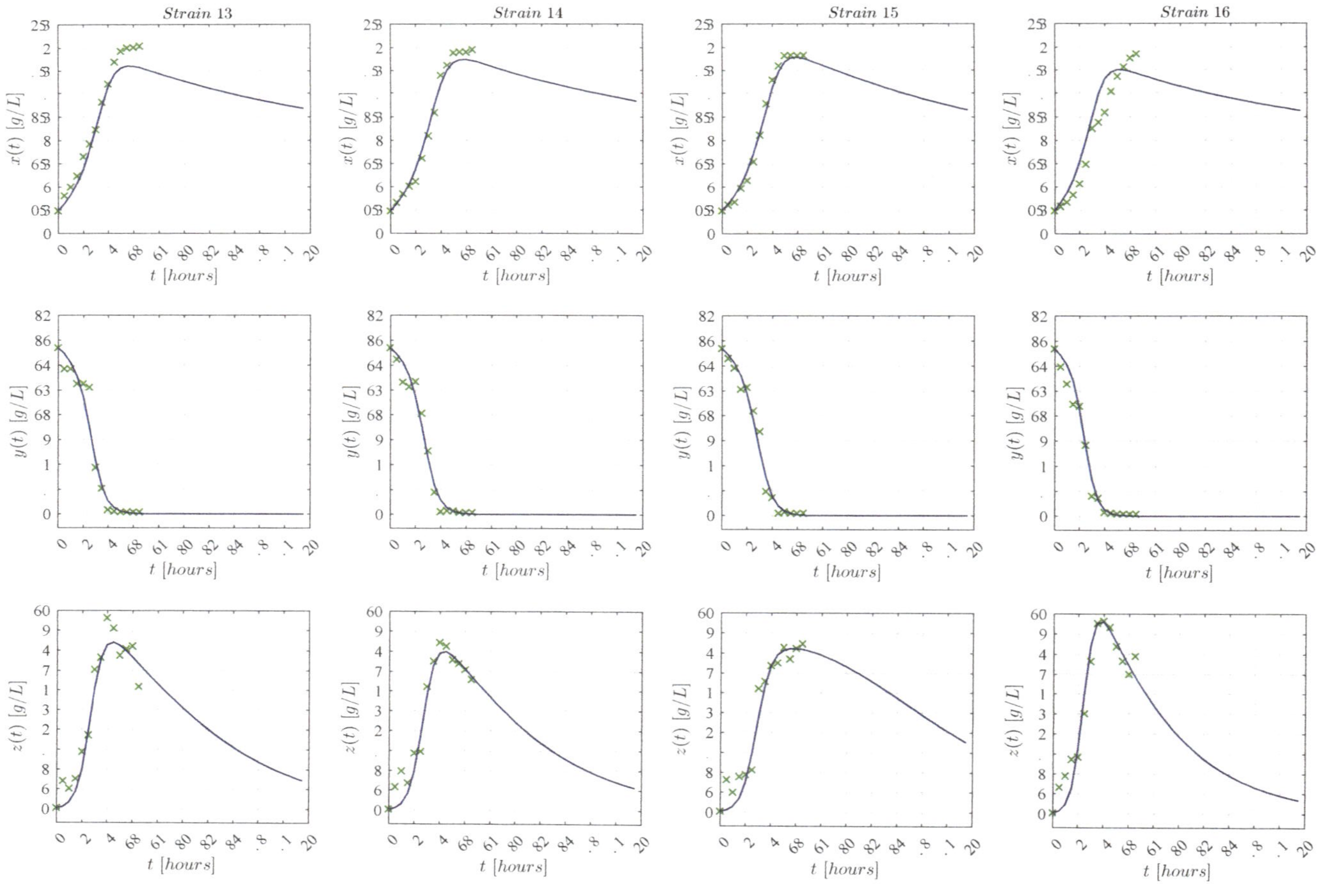

Figure A9. Dynamics prediction for strains 13–16 with the KM system (1)–(3). Observed data are given by the × green marker, and approximated values by the continuous blue line.

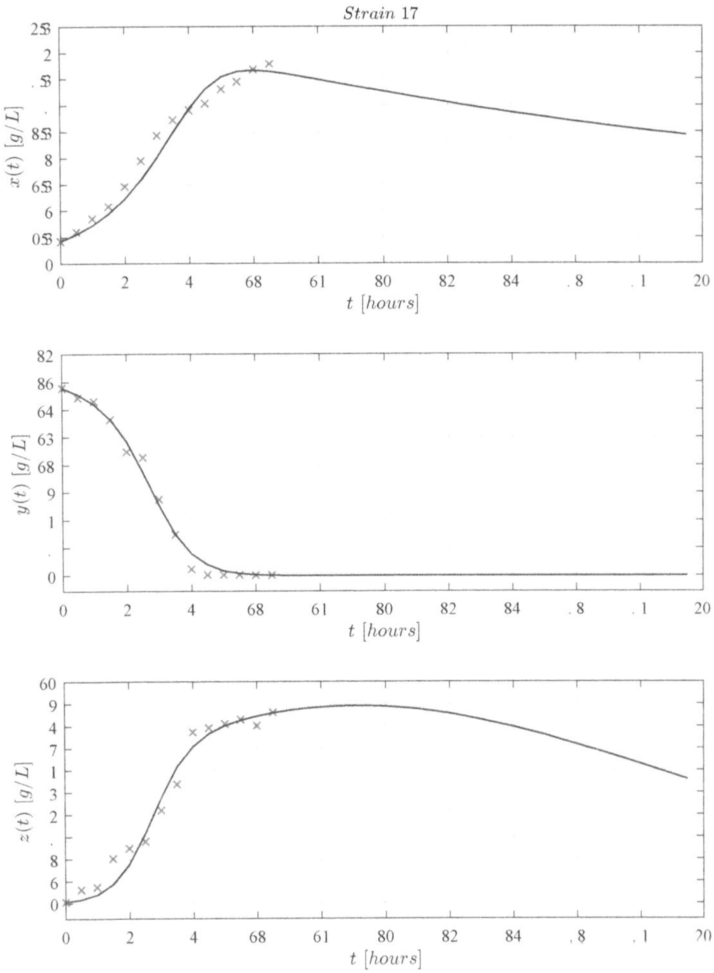

Figure A10. Dynamics prediction for strain 17 with the KM system (1)–(3). Observed data are given by the × green marker, and approximated values by the continuous blue line.

References

1. Zamora, F. Biochemistry of alcoholic fermentation. In *Wine Chemistry and Biochemistry*; Springer: Berlin/Heidelberg, Germany, 2009; pp. 3–26. [CrossRef]
2. Dashko, S.; Zhou, N.; Compagno, C.; Piškur, J. Why, when, and how did yeast evolve alcoholic fermentation? *FEMS Yeast Res.* **2014**, *14*, 826–832. [CrossRef] [PubMed]
3. Kourkoutas, Y.; Dimitropoulou, S.; Kanellaki, M.; Marchant, R.; Nigam, P.; Banat, I.; Koutinas, A. High-temperature alcoholic fermentation of whey using Kluyveromyces marxianus IMB3 yeast immobilized on delignified cellulosic material. *Bioresour. Technol.* **2002**, *82*, 177–181. [CrossRef] [PubMed]
4. Fonseca, G.G.; Heinzle, E.; Wittmann, C.; Gombert, A.K. The yeast Kluyveromyces marxianus and its biotechnological potential. *Appl. Microbiol. Biotechnol.* **2008**, *79*, 339–354. [CrossRef]
5. Morrissey, J.P.; Etschmann, M.M.; Schrader, J.; de Billerbeck, G.M. Cell factory applications of the yeast Kluyveromyces marxianus for the biotechnological production of natural flavour and fragrance molecules. *Yeast* **2015**, *32*, 3–16. [CrossRef]
6. Wittmann, C.; Hans, M.; Bluemke, W. Metabolic physiology of aroma-producing Kluyveromyces marxianus. *Yeast* **2002**, *19*, 1351–1363. [CrossRef]
7. Zentou, H.; Zainal Abidin, Z.; Yunus, R.; Awang Biak, D.R.; Abdullah Issa, M.; Yahaya Pudza, M. A new model of alcoholic fermentation under a byproduct inhibitory effect. *ACS Omega* **2021**, *6*, 4137–4146. [CrossRef]
8. Madigan, M.; Bender, K.; Buckley, D.; Sattley, W.; Stahl, D. *Brock Biology of Microorganisms*; Pearson: London, UK, 2018.

9. Van Maris, A.J.; Abbott, D.A.; Bellissimi, E.; van den Brink, J.; Kuyper, M.; Luttik, M.A.; Wisselink, H.W.; Scheffers, W.A.; van Dijken, J.P.; Pronk, J.T. Alcoholic fermentation of carbon sources in biomass hydrolysates by Saccharomyces cerevisiae: Current status. *Antonie Van Leeuwenhoek* **2006**, *90*, 391–418. [CrossRef] [PubMed]
10. Valdramidis, V. 1-Predictive Microbiology. In *Modeling in Food Microbiology*; Elsevier: Oxford, UK, 2016; pp. 1–15. [CrossRef]
11. Perez-Rodriguez, F.; Valero, A. Predictive microbiology in foods. In *Predictive Microbiology in Foods*; Springer: Berlin/Heidelberg, Germany, 2013; pp. 1–10. [CrossRef]
12. Ross, T.; McMeekin, T.; Baranyi, J. Predictive Microbiology and Food Safety. In *Encyclopedia of Food Microbiology*, 2nd ed.; Academic Press: Oxford, UK, 2014; pp. 59–68. [CrossRef]
13. Zhi, N.N.; Zong, K.; Thakur, K.; Qu, J.; Shi, J.J.; Yang, J.L.; Yao, J.; Wei, Z.J. Development of a dynamic prediction model for shelf-life evaluation of yogurt by using physicochemical, microbiological and sensory parameters. *CyTA-J. Food* **2018**, *16*, 42–49. [CrossRef]
14. Ontiveros, M.F.A. Modelizado de la dinámica de la vida de anaquel de microorganismos en leche fermentada. *Rev. Aristas* **2022**, *9*, 219–225.
15. Buchanan, R.L. Predictive food microbiology. *Trends Food Sci. Technol.* **1993**, *4*, 6–11. [CrossRef]
16. Sharma, V.; Mishra, H.N. Unstructured kinetic modeling of growth and lactic acid production by Lactobacillus plantarum NCDC 414 during fermentation of vegetable juices. *LWT-Food Sci. Technol.* **2014**, *59*, 1123–1128. [CrossRef]
17. Gompertz, B. XXIV. On the nature of the function expressive of the law of human mortality, and on a new mode of determining the value of life contingencies. In a letter to Francis Baily, Esq. FRS &c. *Philos. Trans. R. Soc. Lond.* **1825**, 513–583. [CrossRef]
18. Vázquez, J.A.; Murado, M.A. Unstructured mathematical model for biomass, lactic acid and bacteriocin production by lactic acid bacteria in batch fermentation. *J. Chem. Technol. Biotechnol. Int. Res. Process. Environ. Clean Technol.* **2008**, *83*, 91–96. [CrossRef]
19. Baranyi, J.; Roberts, T.; McClure, P. A non-autonomous differential equation to model bacterial growth. *Food Microbiol.* **1993**, *10*, 43–59. [CrossRef]
20. Garcia, B.E.; Rodriguez, E.; Salazar, Y.; Valle, P.A.; Flores-Gallegos, A.C.; Rutiaga-Quiñones, O.M.; Rodriguez-Herrera, R. Primary Model for Biomass Growth Prediction in Batch Fermentation. *Symmetry* **2021**, *13*, 1468. [CrossRef]
21. Monod, J. The growth of bacterial cultures. *Annu. Rev. Microbiol.* **1949**, *3*, 371–394. [CrossRef]
22. Teissier, G. Growth of bacterial populations and the available substrate concentration. *Rev. Sci. Instrum.* **1942**, *3208*, 209–214.
23. Haldane, J. *Enzymes Longmans*; Green and Co. Oxford, UK, 1930.
24. Moser, H. *The Dynamics of Bacterial Populations Maintained in the Chemostat*; Carnegie Institution of Washington: Washington, DC, USA, 1958.
25. Muloiwa, M.; Nyende-Byakika, S.; Dinka, M. Comparison of unstructured kinetic bacterial growth models. *S. Afr. J. Chem. Eng.* **2020**, *33*, 141–150. [CrossRef]
26. Li, H.; Xie, G.; Edmondson, A.S. Review of secondary mathematical models of predictive microbiology. *J. Food Prod. Mark.* **2008**, *14*, 57–74. [CrossRef]
27. Fakruddin, M.; Mazumder, R.M.; Mannan, K.S.B. Predictive microbiology: Modeling microbial responses in food. *Ceylon J. Sci.* **2011**, *40*, 121–131. [CrossRef]
28. Zentou, H.; Zainal Abidin, Z.; Yunus, R.; Awang Biak, D.R.; Zouanti, M.; Hassani, A. Modelling of molasses fermentation for bioethanol production: A comparative investigation of Monod and Andrews models accuracy assessment. *Biomolecules* **2019**, *9*, 308. [CrossRef]
29. Sansonetti, S.; Hobley, T.J.; Calabrò, V.; Villadsen, J.; Sin, G. A biochemically structured model for ethanol fermentation by Kluyveromyces marxianus: A batch fermentation and kinetic study. *Bioresour. Technol.* **2011**, *102*, 7513–7520. [CrossRef]
30. Lei, F.; Rotbøll, M.; Jørgensen, S.B. A biochemically structured model for Saccharomyces cerevisiae. *J. Biotechnol.* **2001**, *88*, 205–221. [CrossRef] [PubMed]
31. Steinmeyer, D.; Shuler, M. Structured model for Saccharomyces cerevisiae. *Chem. Eng. Sci.* **1989**, *44*, 2017–2030. [CrossRef]
32. Reyes-Sánchez, F.J.; Páez-Lerma, J.B.; Rojas-Contreras, J.A.; López-Miranda, J.; Soto-Cruz, N.Ó.; Reinhart-Kirchmayr, M. Study of the Enzymatic Capacity of Kluyveromyces marxianus for the Synthesis of Esters. *Microb. Physiol.* **2019**, *29*, 1–9. [CrossRef] [PubMed]
33. Valle, P.A.; Coria, L.N.; Plata, C.; Salazar, Y. CAR-T Cell Therapy for the Treatment of ALL: Eradication Conditions and In Silico Experimentation. *Hemato* **2021**, *2*, 28. [CrossRef]
34. Páez-Lerma, J.B.; Arias-García, A.; Rutiaga-Quiñones, O.M.; Barrio, E.; Soto-Cruz, N.O. Yeasts isolated from the alcoholic fermentation of Agave duranguensis during mezcal production. *Food Biotechnol.* **2013**, *27*, 342–356. [CrossRef]
35. Paredes-Ortíz, A.; Olvera-Martínez, T.; Páez-Lerma, J.; Rojas-Contreras, J.; Moreno-Jiménez, M.; Aguilar, C.; Soto-Cruz, N. Isoamyl acetate production during continuous culture of Pichia fermentans. *Rev. Mex. Ing. Química* **2022**, *21*, Bio2654. [CrossRef]
36. da Silva, B.; Gonzaga, L.V.; Fett, R.; Costa, A.C.O. Simplex-centroid design and Derringer's desirability function approach for simultaneous separation of phenolic compounds from Mimosa scabrella Bentham honeydew honeys by HPLC/DAD. *J. Chromatogr. A* **2019**, *1585*, 182–191. [CrossRef] [PubMed]
37. Almeida, I.C.; Pacheco, T.F.; Machado, F.; Gonçalves, S.B. Evaluation of different strains of Saccharomyces cerevisiae for ethanol production from high-amylopectin BRS AG rice (*Oryza sativa* L.). *Sci. Rep.* **2022**, *12*, 1–15. [CrossRef]
38. De Leenheer, P.; Aeyels, D. Stability properties of equilibria of classes of cooperative systems. *IEEE Trans. Autom. Control.* **2001**, *46*, 1996–2001. [CrossRef]

39. Britton, N.F.; Britton, N. *Essential Mathematical Biology*; Springer: Berlin/Heidelberg, Germany, 2003; Volume 453. [CrossRef]
40. Ding, J.; Huang, X.; Zhang, L.; Zhao, N.; Yang, D.; Zhang, K. Tolerance and stress response to ethanol in the yeast Saccharomyces cerevisiae. *Appl. Microbiol. Biotechnol.* **2009**, *85*, 253–263. [CrossRef] [PubMed]
41. Kubota, S.; Takeo, I.; Kume, K.; Kanai, M.; Shitamukai, A.; Mizunuma, M.; Miyakawa, T.; Shimoi, H.; Iefuji, H.; Hirata, D. Effect of ethanol on cell growth of budding yeast: Genes that are important for cell growth in the presence of ethanol. *Biosci. Biotechnol. Biochem.* **2004**, *68*, 968–972. [CrossRef]
42. Fonseca, G.G.; Gombert, A.K.; Heinzle, E.; Wittmann, C. Physiology of the yeast Kluyveromyces marxianus during batch and chemostat cultures with glucose as the sole carbon source. *FEMS Yeast Res.* **2007**, *7*, 422–435. [CrossRef] [PubMed]
43. Wolfenden, R.; Yuan, Y. Rates of spontaneous cleavage of glucose, fructose, sucrose, and trehalose in water, and the catalytic proficiencies of invertase and trehalas. *J. Am. Chem. Soc.* **2008**, *130*, 7548–7549. [CrossRef] [PubMed]
44. Rodrussamee, N.; Lertwattanasakul, N.; Hirata, K.; Limtong, S.; Kosaka, T.; Yamada, M. Growth and ethanol fermentation ability on hexose and pentose sugars and glucose effect under various conditions in thermotolerant yeast Kluyveromyces marxianus. *Appl. Microbiol. Biotechnol.* **2011**, *90*, 1573–1586. [CrossRef]
45. Wu, Z.; Song, L.; Liu, S.Q.; Huang, D. Independent and additive effects of glutamic acid and methionine on yeast longevity. *PLoS ONE* **2013**, *8*, e79319. [CrossRef] [PubMed]
46. Amin, G.; Standaert, P.; Verachtert, H. Effects of metabolic inhibitors on the alcoholic fermentation by several yeasts in batch or in immobilized cell systems. *Appl. Microbiol. Biotechnol.* **1984**, *19*, 91–99. [CrossRef]
47. Willey, J. *Prescott's Microbiology*; McGraw-Hill Education: New York, NY, USA, 2019.
48. Byers, J.P.; Sarver, J.G. Pharmacokinetic modeling. In *Pharmacology*; Elsevier: Amsterdam, The Netherlands, 2009; pp. 201–277. [CrossRef]
49. Garfinkel, A.; Shevtsov, J.; Guo, Y. *Modeling Life: The Mathematics of Biological Systems*; Springer: Cham, Switzerland, 2017.
50. MathWorks. lsqcurvefit. 2022. Available online: https://www.mathworks.com/help/optim/ug/lsqcurvefit.html (accessed on 7 December 2022).
51. Koutsoyiannis, A. *Theory of Econometrics. An Introductory Exposition of Econometric Methods*; The Macmillan Press LTD: New York, NY, USA, 1977.
52. Motulsky, H. *Intuitive Biostatistics: A Nonmathematical Guide to Statistical Thinking*; Oxford University Press: New York, NY, USA, 2018.
53. Akaike, H. Canonical correlation analysis of time series and the use of an information criterion. *Math. Sci. Eng.* **1976**, *126*, 27–96. [CrossRef]
54. Hu, S. Akaike information criterion. *Cent. Res. Sci. Comput.* **2007**, *93*. Available online: https://www.researchgate.net/publication/267201163 (accessed on 11 January 2023).
55. Slavkova, K.P.; Patel, S.H.; Cacini, Z.; Kazerouni, A.S.; Gardner, A.; Yankeelov, T.E.; Hormuth, D.A. Mathematical modelling of the dynamics of image-informed tumor habitats in a murine model of glioma. *Sci. Rep.* **2023**, *13*, 2916. [CrossRef] [PubMed]
56. Krishchenko, A.P. Localization of invariant compact sets of dynamical systems. *Differ. Equations* **2005**, *41*, 1669–1676. [CrossRef]
57. Valle, P.A.; Coria, L.N.; Gamboa, D.; Plata, C. Bounding the Dynamics of a Chaotic-Cancer Mathematical Model. *Math. Probl. Eng.* **2018**, *2018*, 14. [CrossRef]
58. Krishchenko, A.P.; Starkov, K.E. Localization of compact invariant sets of the Lorenz system. *Phys. Lett. A* **2006**, *353*, 383–388. [CrossRef]
59. Khalil, H.K. *Nonlinear Systems*, 3rd ed.; Prentice-Hall: Hoboken, NJ, USA, 2002.
60. Hahn, W.; Hosenthien, H.H.; Lehnigk, S.H. *Theory and Application of Liapunov's Direct Method*; Dover Publications, Inc.: Mineola, NY, USA, 2019.
61. Perko, L. *Differential Equations and Dynamical Systems*; Springer Science & Business Media: Berlin/Heidelberg, Germany, 2013; Volume 7.
62. Jiménez-Islas, D.; Páez-Lerma, J.; Soto-Cruz, N.O.; Gracida, J. Modelling of ethanol production from red beet juice by Saccharomyces cerevisiae under thermal and acid stress conditions. *Food Technol. Biotechnol.* **2014**, *52*, 93–100. Available online: https://hrcak.srce.hr/118561 accessed on 11 March 2023).
63. Pirt, S.J. *Principles of Microbe and Cell Cultivation*; Blackwell Scientific Publications: Hoboken, NJ, USA, 1975.
64. Luedeking, R.; Piret, E.L. A kinetic study of the lactic acid fermentation. Batch process at controlled pH. *J. Biochem. Microbiol. Technol. Eng.* **1959**, *1*, 393–412. [CrossRef]

entropy

MDPI

Review

Mathematical Models of Death Signaling Networks

Madhumita Srinivasan [1], Robert Clarke [2] and Pavel Kraikivski [3,*]

1 College of Architecture, Arts, and Design, Virginia Polytechnic Institute and State University, Blacksburg, VA 24061, USA
2 The Hormel Institute, University of Minnesota, Austin, MN 55912, USA
3 Academy of Integrated Science, Division of Systems Biology, Virginia Polytechnic Institute and State University, Blacksburg, VA 24061, USA
* Correspondence: pavelkr@vt.edu

Abstract: This review provides an overview of the progress made by computational and systems biologists in characterizing different cell death regulatory mechanisms that constitute the cell death network. We define the cell death network as a comprehensive decision-making mechanism that controls multiple death execution molecular circuits. This network involves multiple feedback and feed-forward loops and crosstalk among different cell death-regulating pathways. While substantial progress has been made in characterizing individual cell death execution pathways, the cell death decision network is poorly defined and understood. Certainly, understanding the dynamic behavior of such complex regulatory mechanisms can be only achieved by applying mathematical modeling and system-oriented approaches. Here, we provide an overview of mathematical models that have been developed to characterize different cell death mechanisms and intend to identify future research directions in this field.

Keywords: cell death; apoptosis; necroptosis; ferroptosis; pyroptosis; immunogenic cell death; regulatory networks; death execution pathways; mathematical models; computational analysis

Citation: Srinivasan, M.; Clarke, R.; Kraikivski, P. Mathematical Models of Death Signaling Networks. *Entropy* **2022**, *24*, 1402. https://doi.org/10.3390/e24101402

Academic Editor: Yoh Iwasa

Received: 7 September 2022
Accepted: 28 September 2022
Published: 1 October 2022

1. Introduction

Mathematical modeling is a powerful tool that allows one to connect molecular biology to cell physiology by associating the qualitative and quantitative features of dynamical molecular networks with signal–response curves measured by cell biologists [1]. Mathematical and systems-oriented approaches have been successfully applied to describe the dynamics of complex molecular networks that control cell cycle [2,3], nutrient signaling [4], checkpoints [5], signaling dysregulation in cancer [6], and cell death [7–13]. Systems-oriented mathematical approaches are especially useful for analyzing complex systems that cannot be understood by intuitive reasoning. Undoubtedly, cell death regulation is one such molecular mechanism that cannot be fully understood without mathematical modeling. Here, we provide an overview of mathematical models that have been successfully applied to quantitatively characterize death signaling networks.

Cell death mechanisms are directly involved in regulations of tissue homeostasis, inflammation, immunity, development and other physiological processes [14]. Characterization of new genes and molecular components, involved in signaling pathways by regulating cell death, continues to progress. A detailed characterization of cell death regulation can help identify novel targets and develop effective therapeutic protocols to strike acquired drug resistance in cancer cells. Accurate predictive mechanistic models of complex molecular networks regulating cell death can be used to test the effects of new drugs on the system, and to search for synergistic drug combinations and effective treatment protocols. Different modeling approaches have been already successfully applied to model extensive cell death molecular networks. Ordinary differential equations (ODEs), Boolean logic, pharmacokinetic-pharmacodynamic (PK-PD), Petri nets, agent-based modeling (ABM), cellular automata and hybrid approaches are the common choices available to model molecular mechanisms involved in cell death control, decisions and execution [6,12,13,15–19].

Cell death execution is an all-or-none, irreversible process [20]. Mathematically the activation of irreversible cell death can be described by an irreversible bistable switch with a stable survival steady state, a stable death steady state, and a third unstable steady state separating the survival and death states [21–24]. A pro-death signal can induce cell death by driving the bistable system from the survival to the death state. The transition occurs when the pro-death signal reaches a threshold that corresponds to the limit point bifurcation. Transition in the reverse direction, from death to survival, is impossible because the second limit point bifurcation, where the death steady state vanishes, occurs in the biologically irrelevant negative signal values (i.e., the concentration of a death-inducing ligand or stressor cannot be negative). Therefore, the activation of the cell death execution in such a bistable system cannot be reversed, even if the initial cell death trigger is removed. This mathematical description of the cell death activation is consistent with a threshold mechanism for cell death induction [25] and an all-or-none death decision [22,26,27]. Importantly, understanding how cells control the cell death/survival switch can help to identify targets that can force cancer cells to flip the switch to activate the irreversible cell death execution.

Complexity of cell death regulatory networks, a requirement to account for all important regulating molecular details and pathways, availability of merely small sets of sparse data for model calibration, as well as under- and over-fitting of the model are issues that must be routinely solved in order to develop a predictive model of cell death [6,15]. This review describes mathematical models that have been successfully applied to quantitatively characterize such cell death control mechanisms as apoptosis, necroptosis, ferroptosis, pyroptosis and immunogenic cell death.

2. Apoptosis

Apoptosis is one of the most well-studied and characterized programmed cell death mechanisms. The detailed characterization of molecular interactions involved in apoptosis, and the growing amount of related quantitative data, has encouraged computational and systems biologists to develop mathematical models of apoptosis [12,13,17]. Over the last twenty years, several dozen mathematical models of apoptosis regulation have been described. These apoptosis models aim to explain different data or effects of different treatments on cell death. While the core molecular components regulating apoptosis are shared by all models, variations in molecular circuit designs, components, data, mathematical approaches, and study goals make each model a unique tool to study apoptosis. Most often, molecular mechanisms of apoptosis are mathematically represented using ODEs [7,21,22,25–33], Boolean logics [34–36], and Petri nets [16]; other computational approaches have also been applied [18,37,38].

The execution core of apoptosis regulation involves a family of proteases termed caspases. Caspases can be separated into the following two groups: effector or executioner caspases (caspase-3, -6, -7), and active initiator caspases (e.g., caspases-8, -9). Activation of the caspases initiates the cleavage of several important cellular proteins, such as actin and nuclear lamins, which results in cell body and nuclear shrinkage and cell death [39]. Apoptosis can be processed through mitochondria-dependent (intrinsic apoptosis) and mitochondria-independent (extrinsic apoptosis) caspase-3 activation pathways [14]. The core components involved in these two pathways are commonly included in all mathematical models of apoptosis and can be found in the earliest mathematical models of apoptosis [7].

Extrinsic apoptosis is characterized by high amounts of active caspase-8 that activates the downstream effectors caspase-3, caspase-6, and caspase-7. The activation of caspase-8 is receptor-mediated, which occurs upon receipt of a death signal that is processed by a surface death receptor such as FAS (a member of the tumor necrosis factor gene superfamily) [14]. Therefore, extrinsic apoptosis is a receptor-mediated cell death mechanism, as shown in Figure 1 (left). By contrast, intrinsic apoptosis can be executed even in cells with lower levels of active caspase-8 but requires an additional amplification that involves activation of the pro-apoptotic functions of the mitochondria. For example, stress-related factors (e.g., DNA damage) can induce activation of the executioner caspases via a mitochondria-dependent pathway in the absence of an external death signal [40] (Figure 1, right panel). The mitochondria-dependent pathway begins with the cleavage of anti-apoptotic Bcl-2 family members, which causes the aggregation of pro-apoptotic proteins such as Bax and Bak. Aggregation of pro-apoptotic proteins is followed by the release of cytochrome *c* from the mitochondria, which induces the formation of a large protein complex known as the apoptosome. The apoptosome recruits and activates caspase-9, allowing it to cleave the downstream effectors pro-caspase-3, pro-caspase-6, and pro-caspase-7. Notably, the expression of anti-apoptotic Bcl-2 family members can block the intrinsic apoptosis signaling in cells. By contrast, extrinsic apoptosis cannot be blocked by the expression of high levels of Bcl-2 proteins because large amounts of caspase-8 are already generated.

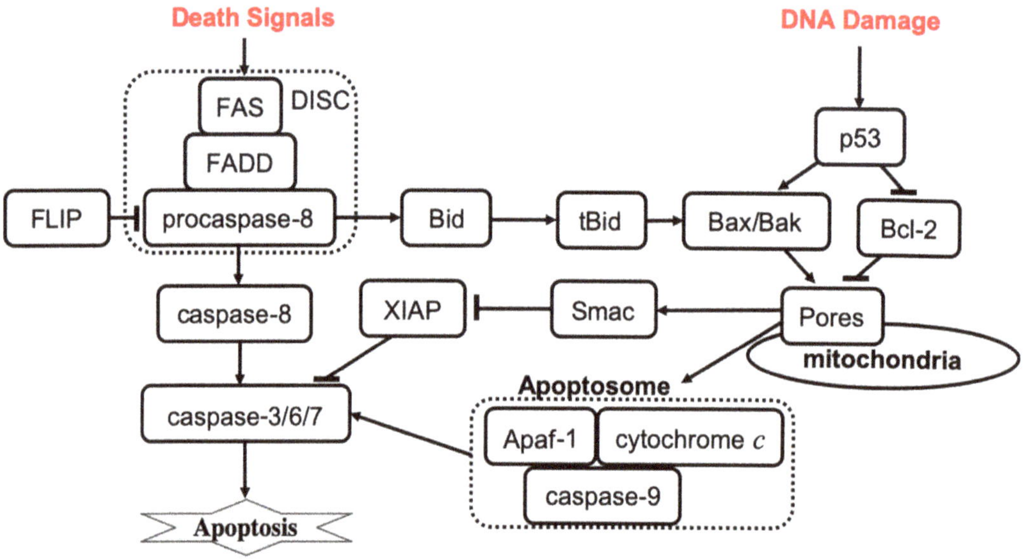

Figure 1. The mechanisms of receptor-induced apoptosis (left) and stress-mediated, mitochondria-dependent apoptosis (right). Solid lines represent activation (arrowhead) and inhibition (bar head) influences.

The earliest mathematical models of apoptosis described both mitochondria-dependent and independent death activation pathways. In early 2000, Fussenegger et al. published a mechanistic ODE-based mathematical model of apoptosis that describes both receptor-mediated and stress-induced caspase activation mechanisms [7]. The receptor-mediated feature of the model describes the FAS surface receptor that activates procaspase-8. Activation of apoptosis initiator caspases involves the following reactions: the binding of an extracellular death ligand to the FAS receptor, the binding of FAS-associated death domain (FADD) protein to the FAS death domain, and the binding of caspase-8 to a domain on FADD that enables caspase-8 activation by proteolytic cleavage. Each binding process is described by a specific rate parameter in the model. Simulation results show that about 50% of procaspase-8 is activated within two hours after the death signal is received. After procaspase-8 activation, the executioner caspase is activated within minutes, and then the initiation of procaspase-9 occurs with the lag time ~20–30 min. The activation curves have a sigmoidal shape indicating, that the transition between the inactive to the active state is characterized by a threshold. If the binding between FADD and clustered FAS death domains is disrupted, then only <0.1% of active caspase-8 is observed upon receipt of the death signal, which is consistent with experimental observations [41].

Fussenegger's model of stress-mediated apoptosis regulation describes the activation of procaspase-9 by cytosolic cytochrome *c*, and the apoptotic protease-activating factor 1 (Apaf-1) complex. Activated caspase-9 then activates apoptosis executioner caspases at some specific rate. Formation of the Apaf-1–cytochrome *c* complex is inhibited by antiapoptotic Bcl-2 family members such as Bcl-x_L. Proapoptotic Bcl-2 family members (e.g., Bax, Bak) can bind to antiapoptotic family members and remove their inhibitory effect. The ratio of anti- versus pro-apoptotic Bcl-2 family members is controlled by the p53 transcription factor that is activated in cells under stress conditions. Simulation results of stress-induced caspase activation dynamics were consistent with experimental observations [42]. Specifically, the model shows that cytochrome *c* is released within 10 min after a stress death signal is received, which results in procaspase-9 activation, 35–40% of the executioner caspase being active within 1 h, and 70% of the executioner caspase being active at 2 h. In addition, simulations revealed that the active fraction of both initiator and executioner caspases is reduced in p53 mutant cells as compared to wild-type cells. Overexpression of antiapoptotic Bcl-2 family members is predicted to block the activation of procaspase-9. The model also confirms that the ratio of anti- versus pro-apoptotic Bcl-2 family members determines whether or not executioner caspases will be activated. The model was then used to predict the effects of combined therapies based on simultaneous receptor- and stress-induced caspase activation.

The model developed by Fussenegger et al. was successful in explaining qualitative experimental observations. However, more quantitative data would be required to complete the model calibration. Quantitative information on reaction rates and molecular concentrations is required to perform reliable mathematical simulations of signal transduction in the apoptosis regulatory network. In 2004, Eissing et al. developed a reduced receptor-induced apoptosis, using parameter values from the literature to evaluate the system behavior within a wide range of parameters [21]. The model revealed that caspase activity remains low for a time that is inversely proportional to the stimulus strength, followed by a steep rise in activity when the input exceeds the threshold; caspase activity then ceases at some maximum level. Bifurcation analysis of the model confirmed that the apoptosis regulation system exhibits a bistable behavior. The same year, Bentele et al. developed a data-based model of receptor-induced apoptosis with parameters estimated on the basis of quantitative experimental data [25]. The time series data for concentrations of 15 different molecules after activation of FAS receptors were used to calibrate the core model of the FAS-induced apoptosis. In addition, data from distinct apoptosis activation scenarios in response to different initial values of ligand concentration were used to improve the estimation of model parameters. The model predicted that apoptosis is not executed when a ligand–receptor concentration ratio is below a critical value, which was also confirmed by experimental observations. In conclusion, Bentele et al. proposed a threshold mechanism for induction of receptor-induced apoptosis. A year later, Hua et al. published a FAS-induced apoptosis model to investigate the effects of altering the level of Bcl-2 on the kinetics of caspase-3 activation [43]. The model predicts that Bcl-2 blocks the mitochondrial pathway by binding to proapoptotic Bax, Bak, and tBid proteins. Further, the model predicts that apoptosis signaling flow can be switched between mitochondria-dependent and mitochondria-independent pathways by varying molecular component levels without changing network structure.

In 2006, Legewie et al. developed a quantitative kinetic model of intrinsic (stress-induced) apoptosis, which displays an all-or-none behavior of caspase activation in response to an apoptotic stimulus [22]. The model helped to identify the positive feedback mechanism that allows cells to achieve ultrasensitivity and bistability in cell death decision making. The pathway molecular regulators that control the apoptotic threshold stimulus and integrate multiple inputs into an all-or-none caspase output were also determined. Time-course simulation results agreed with experimental observations that the induction of maximal caspase-3 cleavage after exogenous addition of cytochrome *c* occurs within ~15–60 min. Furthermore, cytochrome *c*-induced activation of caspase-3 was observed to be bistable and irreversible. The bistable and irreversible caspase-3 activation arises in the system due to XIAP-mediated feedback that cooperates with caspase-9 cleavage by caspase-3. X-linked inhibitor of apoptosis (XIAP) inhibits the catalytic activities of caspase-9 and caspase-3 through reversible binding. The feedback cleavage of caspase-9 by caspase-3 leads to autoamplification of the apoptotic signal. Simulation results show that XIAP-mediated feedback is observed only if caspase-9 and caspase-3 compete for binding to XIAP. Depletion and re-addition experiments using different Apaf-1, caspase-3, caspase-9, and/or XIAP concentrations were proposed to test the all-or-none caspase activation.

Also in 2006, Rehm et al. published a computational model of apoptosome-dependent caspase activation based on biochemical data from HeLa cells [26]. The model predicts that the all-or-none apoptotic response depends on caspase-3-dependent feedback signaling and XIAP, which was then verified quantitatively using single-cell experiments with a caspase fluorescence resonance energy transfer substrate. A concentration threshold of XIAP between 0.15 and 0.30 μM, controlling the substrate cleavage by effector caspases, was identified. The model suggested that high levels of XIAP may promote apoptosis resistance and sublethal caspase activation. This result agrees with a computational analysis that was performed earlier, which also suggested that the inhibitor of apoptosis plays an important role in both the induction and prevention of apoptosis [44]. Conversely, Bagci et al. proposed a mathematical model of mitochondria-dependent apoptosis to study both the role of Bax and Bcl-2 synthesis, degradation rates and the number of mitochondrial permeability transition pores involved in the cell response to a death signal [23]. The main finding was that the transition from bistable to monostable (survival) cell behavior is controlled by the synthesis and degradation rates of Bax and Bcl-2 and by the number of mitochondrial permeability transition pores. Also, the model results suggested that cooperative apoptosome formation is a much more robust mechanism to induce bistability than feedback mechanisms involving, for example, the inhibition of caspase-3 by the inhibitor of apoptosis. Later, Chen and Cui et al. analyzed the robustness of Bax and Bcl-2 apoptotic switches using both deterministic and stochastic models [38,45,46]. These mechanisms were confirmed to be bistable and robust to noise and wide ranges of parameter variation.

Albeck et al. developed a mathematical model of extrinsic, receptor-induced apoptosis to explain the molecular mechanism of the variable-delay, snap-action switch function that determines the cell choice between life and death [27]. The model was calibrated by experimental data collected from live-cell imaging, flow cytometry, and immunoblotting of cells perturbed by protein depletion and overexpression. The model was then used to reveal the mechanism by which a steady and gradual increase in caspase-8 activity is converted into a snap-action downstream signal. Permeabilization of the mitochondrial membrane and relocalization of proteins are the key factors in the extrinsic apoptosis network by which a graded signal that activates caspase-8 and promotes the formation of pores in the mitochondrial membrane is transformed into an all-or-none death decision. Importantly, such snap-action behavior at the level of the mitochondrial outer membrane permeabilization occurs independently of caspase-dependent feedback mechanisms. The formation of pores in the mitochondrial membrane involves the pore-forming proteins Bax and Bak that can self-assemble into transmembrane pores, which are antagonized by anti-apoptotic Bcl-2 proteins [47]. Cytochrome *c* is released into the cytosol when the level of active pore-forming proteins exceeds the threshold set by anti-apoptotic Bcl-2 proteins. Using experimental and modeling techniques, Spencer et al. demonstrated that cell-to-cell variability in time-to-death significantly depends on the activation rate of the tBid protein that activates the pore-forming proteins, Bax and Bak [33]. Therefore, in the case of receptor-mediated apoptosis, the timing and probability of death relies on the differences in the protein levels that can be caused, for example, by noise in gene expression. Furthermore, the stochastic protein turnover in a receptor-mediated apoptosis model can result in fractional killing [48].

Later models were developed to investigate crosstalk between apoptosis regulation and NF-κB pathways [32], the estrogen signaling network [31], endoplasmic pathways [28], and autophagy regulation [29]. Neumann et al. described a model of the crosstalk between receptor-mediated apoptosis regulation and NF-κB signaling that are activated by the same receptor in parallel to the apoptotic signaling and on a similar time scale [32]. Model and experimental analysis suggested that the balance between apoptotic and NF-κB signaling is shaped by the proteins that regulate the assembly dynamics of the death-inducing signaling complex (DISC). Therefore, the assembly of DISC acts as a signal processor, determining life/death decisions in a nonlinear manner. Tyson et al. provided a roadmap for a detailed mathematical model that would allow researchers to characterize the crosstalk among the estrogen signaling network, apoptosis, autophagy, and cell cycle regulations in breast epithelial cells [31]. Later, the same research lab published a detailed mathematical model to examine the decision process that moves a cell from autophagy to apoptosis [29]. The model was successful in explaining quantitative time-course data of autophagy and apoptosis under cisplatin treatment. Further, the model allows for characterization of the prosurvival and prodeath cell responses to cytotoxic stress. Also, in 2012, Hong et al. published a model of cisplatin-induced apoptosis that integrates the death receptor pathway, and mitochondrial and endoplasmic reticulum stress response mechanisms [28]. The model predicts the relative contribution of each signaling pathway to apoptosis. Simulation results revealed that the mitochondrial and death receptor pathways as well as crosstalk among pathways make the greatest contribution to the level of apoptosis, whereas the contribution of the endoplasmic reticulum stress pathway is negligible.

The Role of p53 in Apoptosis

The tumor suppressor gene p53 (TP53) has been reported as an upregulated modulator of apoptosis and as a driver of cell fate transition from cell cycle arrest to apoptosis [49]. Mathematical models that characterize the p53 contribution to apoptosis have been developed by several groups [7,23,28,30,50]. p53 targets many genes regulating cell apoptosis, including BCL2 and BAX genes [51]. Computational study of apoptosis regulation shows that the balance between anti- and proapoptotic Bcl-2 family members is altered in p53 mutant cells [7]. Also, the active fraction of both initiator and executioner caspases is reduced in p53 mutant cells as compared with wild-type cells. The mathematical model also predicts that overexpression of the death ligand and the FAS receptor can be used to initiate executioner caspase activation in p53 mutant cells [7]. Bagci et al. have shown that apoptosis is not sensitive to caspase-3 activation when p53 expression is low, and that bistability to apoptotic stimuli is observed when p53 level is high [23]. Predictions from this apoptosis model agree with experimental data [52]. Another study reported that inhibition of p53 protects against cisplatin-induced apoptosis [28]. Cisplatin induces DNA damage that results in the phosphorylation and activation of p53. There, the activation of Bax by p53 induces mitochondrial membrane permeabilization and apoptosis [53]. Also, p53 mediates caspase-2 activation and the mitochondrial release of apoptosis-inducing factor. The model predicts time courses for p53, caspase-2, Bax activation, apoptosis-inducing factor release and apoptosis activation. Simulation results agree with experimental data that p53 inhibition prevents the mitochondrial release of apoptosis-inducing factor and cisplatin-induced apoptosis [54]. Overexpression of p53 results in caspase-2 activation and also the mitochondrial release of apoptosis-inducing factors [54].

Ballweg et al. developed a mathematical model that integrates p53 signaling, cisplatin-induced events, and apoptosis regulation that was used to study the dynamics of fractional killing induced by cytotoxic drugs [30]. Many drugs activate not only apoptosis execution signaling but also expression of anti-apoptotic genes, which results only in fractional killing amongst a population of treated cells [55]. Thus, fractional killing may occur due to crosstalk between the apoptosis and survival pathways [56]. The model predicts that the probability of apoptosis depends on the dynamics of p53 and the rate of p53 activation determines the cell fate [30]. Slow activation of p53 results in cell survival, whereas fast p53 activation induces cell death. This result also agrees with the experimental observation showing that apoptotic cells accumulate p53 much earlier than cells that survive the treatment [55]. In the model, activation of Bax and subsequent execution of apoptosis occur when the level of p53 exceeds a threshold value. However, the apoptosis initiation threshold depends on the inhibitor of apoptosis, cIAP. Cells with an elevated level of cIAP require a higher level of p53 to induce apoptosis. Because the level of apoptosis regulator cIAP increases with time, the rate of p53 activation plays an important role in the determination of cell fate. Cell-to-cell variability due to stochastic gene expression and environmental noise can also set different apoptosis initiation thresholds in different cells, resulting in fractional killing.

Up to this point, we have reviewed mathematical models of apoptosis that use ODEs to describe the mechanism of cell death (apoptosis) regulation. However, other mathematical approaches have been also used to study apoptosis regulation [16,18,34–37]. Several apoptosis models have been developed using a Boolean (logical) approach that can analyze extensive regulatory networks with many molecular components and their interactions [34–36]. Schlatter et al. developed an apoptosis regulation model that comprises 86 nodes and 125 interactions [34]. Mai et al. developed a model that describes 37 internal states of signaling molecules involved in apoptosis regulation, 2 extracellular signal inputs, and the DNA damage event as an output [35]. Calzone et al. developed a model to study crosstalk between receptor-mediated apoptosis regulation, NFκB pro-survival pathways, and RIP1-dependent necroptosis regulation [36]. These models were used to characterize feedback loops in the apoptosis regulation network structure.

While Boolean models are excellent tools to reproduce the *qualitative* behavior of a regulatory network, they are weak at addressing detailed *quantitative* questions about molecular mechanisms [19]. Petri nets have been applied to analyze and validate a qualitative model of extensive apoptosis regulation [16]. Agent-based modeling turned out to be a more appropriate approach for modeling the death-inducing signaling complex assembly than an ODE-based model that must describe a large number of intermediate products involved in DISC assembly [37]. A cellular automata approach has been applied to study apoptosis blocking in the immunological response of T cells by varying the inhibitor actions such as FLIP and IAP [18]. The model predicts that only joint suppression of both FLIP and IAP apoptosis inhibitors can effectively act to kill cancer cells through apoptosis.

In conclusion, comprehensive data and extensive experimental characterization of apoptosis allowed computational and systems biologists to develop several mathematical models of apoptosis regulation. These models not only increase our understanding of mechanisms of apoptosis execution induced by stress or signals, but also predict perturbations that can prevent or enhance apoptosis. An accurate mathematical model of apoptosis can help find novel combinations of existing therapies that can induce the death of cancer cells using low doses. Further studies that integrate apoptosis with other cell death regulations will help to understand the cell death decision mechanism that determines the execution of a specific cell death fate.

3. Necroptosis

Necroptosis is a regulated cell death that can be initiated by changes in extracellular or intracellular homeostasis, detected by specific death receptors [14]. Triggering necroptosis primarily involves the receptor-interacting protein kinase 1 (RIP1), RIP3, and mixed lineage kinase domain-like protein (MLKL). Necroptosis can be induced by the binding of tumor necrosis factor (TNF) or other ligands to cell surface receptors that trigger the sequential phosphorylation of receptor-interacting protein kinases. At a cell physiology level, necroptosis results in cell volume expansion, cell membrane rupture, and intracellular material overflow that leads to a local inflammatory reaction and immune response activation. Necroptosis-inhibiting drugs can be used to treat inflammatory diseases [57]. Necroptosis-promoting drugs are potential anticancer therapies [58]. Studies of necroptosis regulation can help to identify molecular targets that can be used to reprogram the necroptosis execution in a desired direction. While many molecular components involved in necroptosis regulation are known, the precise interaction network, signaling spread, dynamic behavior of necroptotic regulation, and the decision-making processes within the molecular network, remain poorly understood. Several mathematical models have been developed recently to characterize the dynamics of necroptosis regulation [8,59].

Xu et al. have developed a computational model of the cellular necroptosis signaling network [8], to study the necroptosis signaling dynamics that lead to cell death in the form of oscillation-induced trigger waves. The study focused on the core cellular necroptosis signaling module that includes four components: TRADD, RIP1, caspase-8, and RIP3. The activities of key components are regulated either by phosphorylation, dephosphorylation, or cleavage. The corresponding mathematical model described 4 variables and involved 10 interaction terms. Xu et al. used a Latin hypercube sampling method to randomly scan the model network parameters and evaluate the stable oscillation behavior of the cellular necroptosis signaling circuit. Bifurcation analysis and potential landscape theory were applied to explore oscillation modes in different cellular necroptosis signaling circuits. The results indicate that the cellular necroptosis signaling circuit more likely produces oscillations when the total amount of RIP1 or caspase-8 is high, while fluctuations in the value of RIP3 have no significant effects on the oscillation probability. Also, oscillations are often obtained when the activation of caspase-8 by RIP1 is fast, while RIP3 phosphorylation by RIP1 is relatively slow. Further, oscillations are more robust when the reaction rate constants that describe RIP1 activation by RIP3 are stronger than rate constants describing other interactions. Overall, oscillation robustness analysis revealed three regulatory feedback loops formed by RIP1, caspase-8, and RIP3 interactions. These loops comprise a negative feedback loop: RIP3 activates RIP1, which activates caspase-8, that inhibits RIP3; a positive feedback loop: RIP1 activates RIP3, which inhibits caspase-8, that inhibits RIP1; and an incoherent feedforward loop: RIP1 activates both caspase-8 and RIP3, and caspase-8 inhibits RIP3. Importantly, for oscillations to be robust, the reactions in the positive feedback loop must be slower than reaction rates in the negative feedback loop. Also, a stochastic parameter analysis indicated that the incoherent feedforward loop is the essential molecular mechanism that allows the necroptosis signaling system to generate oscillations.

Xu et al. classified oscillations that occur in cellular necroptosis signaling circuits into four groups according to amplitude and oscillation period. About 50% of observed oscillations had a high-amplitude (above the median value of all the counted amplitudes) and fast period (>100 min based on the oscillation period of NF-κB [60,61]), about 37% of oscillations had a low-amplitude and fast period, ~12% of oscillations had high-amplitude and slow period, and ~1% of oscillations had a slow and low-amplitude period. Further analysis revealed that the inhibition rates of RIP1 and RIP3 by caspase-8 play an important role in determining the amplitude behavior of fast oscillations. In addition, bifurcation analysis revealed that the dynamic behavior of the system can be switched from slow high-amplitude oscillations to slow low-amplitude oscillations by tuning the parameters that describe the activation of caspase-8 by RIP1. However, the transition from fast to slow oscillation behavior cannot be achieved by changing any single reaction rate constant. Also, the system changes dynamics from slow high-amplitude oscillations to fast low- or high-amplitude oscillations when two parameters that describe RIP1 inhibition with caspase-8 and RIP1 phosphorylation with RIP3 are simultaneously tuned. Robustness analysis revealed that the period of fast oscillations was more robust to parameter perturbations than the period of slow oscillations. The amplitude of slow low-amplitude oscillations was robust to parameter perturbations, while the robustness of amplitude of fast high-amplitude oscillations was the weakest. Overall, the study provides a quantitative characterization for the mechanism of oscillation mode-switching behavior in the necroptosis signaling network. Xu et al. proposed that MLKL can decode the information according to the amplitude and period of RIP3, which can be an important mechanism that allows cells to generate different responses in various stressful conditions.

A more recent detailed computational model of tumor necrosis factor (TNF)-induced necroptosis has been developed by Ildefonso et al. [59]. The model was derived from the literature-curated molecular mechanisms of necroptosis regulation, which involves 14 proteins, 37 biochemical species, and 40 reactions. The simplified molecular mechanism that shows key species involved in necroptosis execution is shown in Figure 2. Dynamics of species were described by a set of ordinary differential equations where all reactions were described by the mass action law. The model was calibrated and validated using experimental protein time-course data from a well-established necroptosis-executing cell line. Simulations then confirmed that the model is successful in explaining the dynamics of necroptosis reporter, phosphorylated mixed lineage kinase domain-like protein (pMLKL). Furthermore, four distinct necroptosis execution modes were identified by using a dynamical systems analysis and a spectral clustering algorithm. While the temporal dynamics of pMLKL were similar in each mode of necroptosis execution, the sequences of molecular events that led to MLKL phosphorylation and subsequent necroptotic cell death were different. The modes primarily differed in the values of rate constants across the necroptosis execution pathway. For example, the rate constant for binding of A20 to ubiquitinated RIP1 was significantly smaller in mode 4 than in the other modes, and also smaller in mode 2 relative to modes 1 and 3. Also, mode 4 has a significantly larger activation rate and smaller deactivation rate constant for caspase-8 in complex II. The activation/deactivation of caspase-8 in complex II is a critical step in the pathway for determining whether the cell will progress to necroptosis. Differences in rate constant values create the difference in the action of A20 and CYLD enzymes across four modes that are then able to effectively operate as an inhibitor or activator of necroptosis.

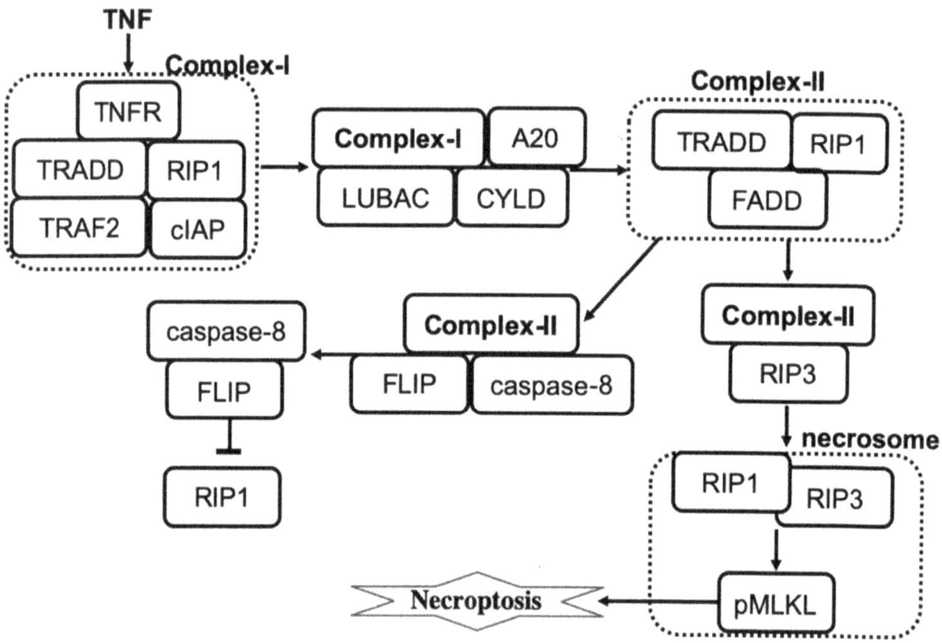

Figure 2. The influence diagram shows key molecular components involved in necroptosis regulation. Lines represent activation (arrowhead) and inhibition (bar head) influences.

Taken together, the computational analysis helped to resolve the controversy in experimental observations by showing that CYLD- and A20-driven deubiquitination of RIP1 may act as pro- and anti-necroptotic in different cell types. According to Ildefonso et al.'s model, knocking out A20 decreases the probability of necroptosis execution (necroptosis sensitivity) in mode 1, and increases the sensitivity to necroptosis in mode 2 [59]. Conversely, knocking out CYLD increases the sensitivity to necroptosis in modes 1 and 4, and decreases the sensitivity to necroptosis in mode 2. Knocking out CYLD or A20 has no effect in mode 3. Also, A20 knockout has no effect in mode 4. These results have been compared to cell phenotype observations in A20 and CYLD knockdown experiments in different cell types. For example, it has been reported that RIP1 is deubiquitinated by both A20 and CYLD in mouse fibrosarcoma cells, but inhibition of CYLD protects cells from necroptosis, while A20 depletion can sensitize cells to death by necroptosis [62]. Thus, A20 and CYLD depletion experiments in mouse fibrosarcoma cells are consistent with the model results obtained for A20 and CYLD knockouts in mode 2.

TNF, TNFR, and MLKL are three common protein modulators of necroptosis across the four modes of necroptosis execution. Furthermore, rate constants that control the association of TNF to TNFR, ubiquitination of RIP1 by cIAP in complex I, and association of LUBAC to complex I can be used to efficiently modulate necroptosis execution across the four modes. Therefore, targeting these modulators can be used to induce or prevent necroptosis, potentially useful for both cancer therapy and treatment of inflammatory diseases.

Apoptosis and necroptosis regulation networks share common nodes and edges and may suppress each other [63]. Either apoptosis or necroptosis can be induced by TNF and the cell death decision depends on the cell state. Complex II can recruit RIP3 to form a necrosome or recruit caspase-8 to stabilize its active conformation, resulting in the release of an activated caspase 8 homodimer that then can induce apoptosis [64]. Li et al. [65] performed a quantitative study of crosstalk between the apoptosis and necroptosis pathways. Specifically, mathematical modeling was used to investigate three possible mechanisms of caspase-8 activation by (i) TRADD, (ii) RIP1, and (iii) TRADD and RIP1 together. The law of mass action was used to convert the proposed molecular mechanisms into a system of ODEs. Simulations of each mechanism were compared with data obtained using the sequential window acquisition of all theoretical fragment ion spectra mass spectrometry methods. All three mechanisms reproduced the amounts of major components in TNFR1, RIP1, and RIP3 complexes. However, only mechanism (ii) could explain a negative regulation of RIP3 phosphorylation by the increase in RIP1 levels. This result was also supported by a sensitivity analysis showing that the most robust negative regulation of RIP3 phosphorylation by RIP1 is achieved when mechanism (ii) is used in the model. To test this prediction, Li et al. experimentally knocked down RIP1 to three different expression levels by using RIP1-specific short hairpin RNA and measured the increase in TNF-induced phosphorylation of RIP3 and MLKL. Deletion of RIP1 completely blocks TNF-induced RIP3 phosphorylation [65]. In addition, simulation results show that pro-caspase-8 activity is necessary for the up-regulation of RIP3 phosphorylation by decreasing RIP1 expression. The mechanism was further refined to make it in agreement with the observation that TNF induces quick caspase-8 activation and apoptosis in RIP1 KO cells [62]. Specifically, TRADD-dependent caspase-8 activation was added to the mechanism (ii). The final model successfully explained both RIP1′s biphasic roles in necroptosis, where RIP1 promotes necroptosis within an extremely low-level range ($<\sim 2\%$ of wildtype) and inhibits necroptosis at higher levels, and the activation level of caspase-8 in RIP1 KO cells. Also, the response of pro-caspase-8 to RIP1 level is linear, whereas RIP3 phosphorylation is determined by the nonlinear (ultrasensitive) threshold pattern.

Overall, a quantitative approach has been applied successfully to describe the roles of RIP1 in cell death determination. In conclusion, Li et al. proposed a "speed competition" decision mechanism in which cells decide to execute apoptosis or necroptosis by the pathway that reaches the final destination first. Interestingly, simultaneous execution of necroptosis and apoptosis has been observed in some individual cells [65].

4. Pyroptosis

The regulated cell death that is associated with the formation of plasma membrane pores by members of the gasdermin protein family is called pyroptosis [14]. The induction of pyroptosis may occur as a consequence of inflammatory caspase activation that can be triggered by pathogen invasion such as Gram-negative bacteria. The critical role of caspase-driven pyroptosis for innate immune responses against invading bacteria has been confirmed in experiments with mice carrying gene mutations that disrupt normal activity of caspase proteins [66]. By killing the host cell, pyroptosis removes the replication compartment of intracellular pathogens and thus prevents their spreading. Hence, pyroptosis has an important role in innate immunity against intracellular pathogens.

Pyroptosis induced by inflammatory caspases is driven by the gasdermin protein GSDMD. Caspases activate GSDMD that then translocates to the plasma membrane where GSDMD induces pore formation and thus rapid plasma membrane permeabilization. The simplified molecular mechanism of the pyroptosis induced by inflammatory caspases is shown in Figure 3. In this scheme, pyroptosis relies on caspase-1 activation.

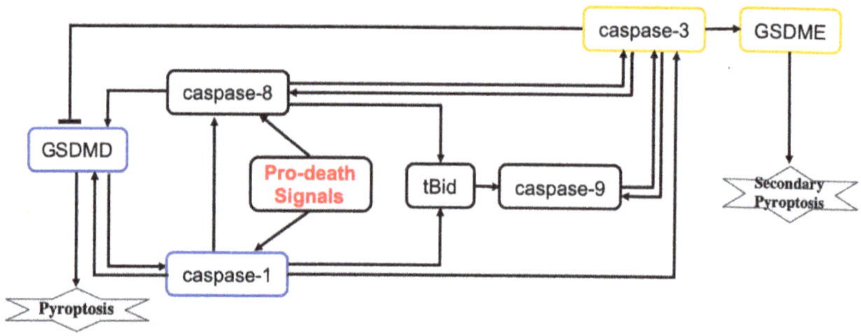

Figure 3. The influence diagram shows key molecular components involved in GSDMD- and GSDME-executed death modes called pyroptosis and secondary pyroptosis respectively. Lines represent activation (arrowhead) and inhibition (bar head) influences.

Beyond inflammatory settings, pyroptotic cell death can be induced by TNF, various DNA-damaging agents, and infection with vesicular stomatitis virus [67,68]. In these cases, pyroptosis is driven by other members of the gasdermin family, specifically GSDME. This form of pyroptosis releases fewer inflammatory cytokines than is observed when pyroptosis is induced by inflammatory caspases. Pyroptotic signaling relies on the activation of caspase-3 that catalyzes proteolytic cleavage of GSDME. The identification of other gasdermin family members that execute pyroptosis in conditions that are beyond inflammatory settings has been significantly expanded [14].

A computational study of the crosstalk between caspase-1- and caspase-3-driven pyroptosis pathways was performed by Zhu et al. [9]. The molecular regulatory network that executes pyroptosis via activation of GSDMD and GSDME is shown in Figure 3. The crosstalk between caspase-1- and caspase-3-driven pyroptosis pathways is realized through tBid, caspase-9, and caspase-8 components. Zhu et al. developed a mathematical model that describes the dynamics of seven molecular components and the dynamics of the cell population governed by cell proliferation and death processes. The model consists of eight coupled ODEs and 83 parameters. Hill functions were used to describe activation and inactivation reactions for molecular components. The values of 44 parameters were estimated from sources available in the literature and 39 parameters were estimated using 138 time-course data points that were measured for eight variables (the death rate and seven molecular components) in wild-type cells and cells with single, double, and triple knockouts of the molecular components.

The pyroptosis decision mechanism was analyzed using bifurcation and sensitivity analysis methods. Bifurcation analysis revealed that the change in expression levels of caspase-1, caspase-3, and GSDMD can switch between GSDMD- and GSDME-executed pyroptosis death modes. Furthermore, the transition between pyroptosis death modes could not be efficiently controlled by varying the expression levels of caspase-8, caspase-9, tBid, or GSDME. According to the model, GSDMD-driven pyroptosis is more likely when the caspase-1 total expression level is below ∼1.5 nM and GSDME-driven pyroptosis occurs when the caspase-1 level is above 14 nM. For caspase-1 levels ranging from 1.5–14 nm, bistability is observed when either GSDMD- or GSDME-driven pyroptosis may occur. Similarly, when GSDMD level is lower than 88 nM, GSDME-driven pyroptosis is induced, whereas cells can selectively execute either pyroptosis mode when the level of GSDMD is between 88 nM and 165 nM. GSDMD-driven pyroptosis occurs when GSDMD level is higher than 165 nM. Also, cells execute GSDMD-driven pyroptosis when caspase-3 level is lower than 250 nM, and selectively induce either GSDMD- or GSDME-executed pyroptosis with higher levels of caspase-3.

Sensitivity analysis confirmed that the expression levels of GSDMD and caspase-1 can efficiently change the pyroptosis death modes. This result agrees with experimental observations [69,70]. In addition, bifurcation analysis predicts that the expression level of caspase-3 can also change the pyroptosis death mode between caspase-1- and caspase-3-driven pyroptosis. Overall, the model predicted 3 molecular components and 12 reactions that can be targeted to control the switch between modes of pyroptosis execution. Drugs that can switch between pyroptosis death modes can help to improve treatment protocols for cancer and inflammasome-mediated diseases. For example, GSDME-induced pyroptosis can act as a tumor suppressor [71,72] and also releases fewer inflammatory cytokines when compared to pyroptosis that is executed by GSDMD.

Li et al. extended the GSDMD-induced pyroptosis model by adding apoptosis regulation [73]. The model allows one to study the crosstalk between pyroptosis and apoptosis and inflammasome-induced cell death under different perturbation conditions. Simulation results reproduce the dynamics of cell death executioners in multiple knockout cells. Pyroptosis and apoptosis events are determined by the level of cleaved GSDMD and cleaved caspase-3, respectively. Sensitivity analysis was performed to determine the molecular components that can significantly affect the occurrence of pyroptosis and apoptosis. The model predicted that caspase-1 and GSDMD are key molecular regulators directing the signal flow that can switch cell death modes between pyroptosis and apoptosis. Decreases in caspase-1 or GSDMD gradually inhibit pyroptosis and enhance apoptosis induction. These model predictions were validated by caspase-1 and GSDMD-knocked down experiments. Furthermore, the model results helped to suggest the death signal propagation pathways, resulting in pyroptosis or apoptosis in cells expressing different levels of caspase-1 or GSDMD. To understand the roles of caspase-1 and GSDMD in triggering the cell death modes, Li et al. employed a potential landscape approach. The cell death landscape was represented by potential wells corresponding to pyroptosis and apoptosis death modes. In the double-well potential landscape, the system evolved into one of the two wells from any initial condition. Caspase-1 or GSDMD could change the potential landscape from monostable to bistable. A monostable landscape corresponding to pyroptosis is obtained in cells with a high expression level of caspase-1 or GSDMD; the potential landscape changes to bistable and then to an apoptotic monostable as the expression level of caspase-1 or GSDMD decreases. Overall, the model helps to understand the inflammasome-induced cell death, crosstalk between pyroptosis and apoptosis, and may be used to determine potential molecular targets for driving cells into a desired death execution mode.

5. Ferroptosis

Ferroptosis is another regulated cell death mechanism that involves iron-catalyzed lipid damage [14,74,75]. Cell death occurring by ferroptosis correlates with the accumulation of markers of lipid peroxidation and can be suppressed by iron chelators, inhibitors of lipid peroxidation, and lipophilic antioxidants [75]. Ferroptotic cell death can be modulated pharmacologically and genetically by perturbing lipid repair systems that involve glutathione and glutathione peroxidase 4 (GPX4) that convert toxic lipid hydroperoxides (L-OOH) into non-toxic lipid alcohols (L-OH) [76]. Depletion or inactivation of GPX4 results in overwhelming lipid peroxidation that causes cell death. Ferroptosis also depends on a set of enzymatic reactions that regulate the biosynthesis of membrane polyunsaturated fatty acids (PUFA)-containing phospholipids, which are the substrates of pro-ferroptotic lipid peroxidation products [75]. Also, the formation of coenzyme-A-derivatives of PUFAs (PUFA-CoA) and their insertion into phospholipids are necessary for the induction of a ferroptotic death signal. Two enzymes, ACSL4 and LPCAT3 are involved in the biosynthesis and remodeling of PUFAs [75,77]. Depletion of PUFAs can suppress the occurrence of ferroptosis, and loss of ACSL4 and LPCAT3 gene products increases resistance to ferroptosis [75].

Iron induces the accumulation of lipid peroxides and thus is important for the execution of ferroptosis. Intracellular iron levels depend on the iron efflux pump ferroportin and the iron importer TFR1 and other proteins that regulate iron import, export, and storage [78–80]. Also, for ferroptosis to start, phospholipid molecules containing polyunsaturated fatty acids (LH-P) are formed from PUFA-CoA, which are then oxidized into lipid hydroperoxides (L-OOH) and eventually into lipid radicals (LO*). LH-P generation is regulated by LPCAT3 and inhibited by monounsaturated fatty acids (MUFAs). Production of MUFAs depends on desaturation of the saturated fatty acids (SFAs) which is catalyzed by the desaturase SCD1 [81]. Formation of lipid radicals LO* is promoted by reactive oxygen species (ROSs) and lipid peroxidation enzymes including ALOX15 [74,82]. The generation of endogenous lipid radicals initiates ferroptosis. In addition, ferroptotic cell-death responses can be modulated by p53 activity [83]. For example, induction of SAT1, a transcription target of p53, is correlated with the expression levels of ALOX15 [83]. The influence diagram that reflects the molecular mechanism of ferroptosis is shown in Figure 4. Overall, ferroptosis is morphologically and mechanistically different from all other types of regulated cell death. Regulation of ferroptosis has great potential for cancer therapy, and molecular targets that promote ferroptosis are being actively explored [84].

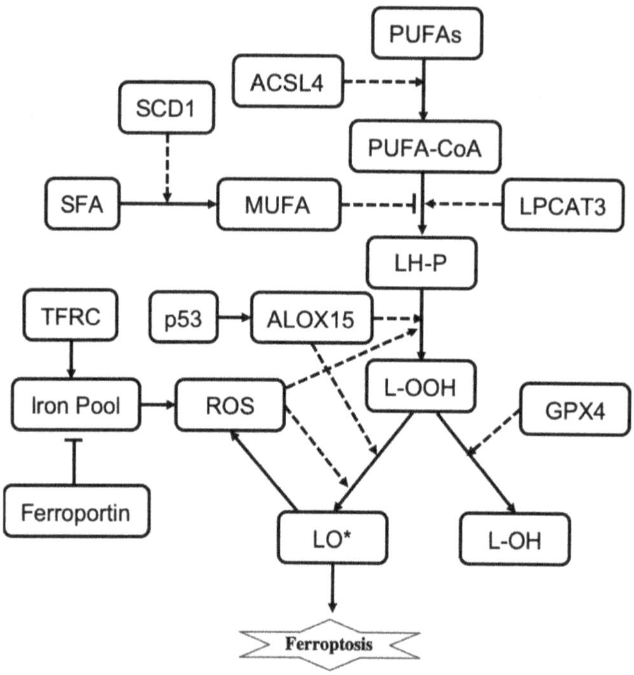

Figure 4. The influence diagram shows key molecular components involved in ferroptosis regulation. Solid arrows represent reactions of transformation, activation (arrowhead), and inhibition (bar head). Dashed lines with arrowheads or bar heads represent activation or inhibition of a reaction, respectively.

Konstorum et al. developed a stochastic, multistate, discrete mathematical model of ferroptosis regulation [10]. The model describes states of eleven variables that represent ALOX15, GPX4, L-HP, LIP, LO*, L-OOH, LPCAT3, MUFA, PUFA-CoA, ROS, and SLC7A11. Each variable can take on three values that respectively represent low, medium, and high molecular species activity or expression level. Variables are updated using updating rules and an asynchronous update scheme at each discrete time step. Five external inputs representing ACSL4, ferroportin, p53, SCD1, and TFRC, which do not change during the course of the simulation, are used to study the sensitivity of ferroptosis induction to different signaling and perturbation conditions.

Konstorum et al. used a system-level analysis to study how different input conditions and parameters alter ferroptosis sensitivity. They found that ferroptosis sensitivity depends on PUFA synthesis and PUFA incorporation into the phospholipid membrane, as well as the balance between levels of pro-oxidant species (ROS, lipoxygenases) and antioxidant factors (GPX4). Ferroptosis sensitivity can be reduced by altering parameters that minimize the production of L-OOH species. High ACSL4 and low SCD1 levels result in high ferroptosis sensitivity. The model also predicted that a high level of SCD1 can inhibit ferroptotic induction even when levels of ACSL4 are high. These model predictions were validated using an in vitro experimental system of an ovarian cancer stem cell culture [10]. Overall, the model allows us to better understand the crosstalk between pathways transmitting signals from different inputs that induce the execution of ferroptosis.

6. Immunogenic Cell Death

Immunogenic cell death (ICD) is a regulated cell death mechanism that induces an immune response in the hosts [14]. Basically, ICD is an immunostimulatory form of apoptosis that is characterized by the ability of dying cells to generate robust adaptive immune responses [85]. The immune response is promoted by damage-associated molecular patterns (DAMPs), which are released by dying cells [86]. DAMPS communicate a state of danger to the organism by activating pattern recognition receptors (PRRs) that are present on the surface of innate immune cells such as monocytes, macrophages, and dendritic cells (DCs) [87]. Activated macrophages and dendritic cells can migrate to the lymph node and pass the antigens to $CD8^+$ and $CD4^+$ T lymphocytes, which results in an adaptive immune response. Tumor cell systems are often used to study ICD regulation and dynamics [88]. The immune responses against cancer- or pathogen-driven antigens that induce ICD are well characterized [85]. Importantly, over the past years, developments of ICD-related cancer immunotherapy approaches are gaining great momentum [89].

To study the ICD dynamics of cancer cells, Checcoli et al. developed a mathematical model that integrates intracellular mechanisms involved in ICD and intercellular interactions among cancer cells, DCs, $CD8^+$, and $CD4^+$ T cells [11]. The modeling approach is based on a continuous time Boolean Kinetic Monte-Carlo formalism that was successfully applied to model different complex molecular mechanisms [90]. The aim of the mathematical characterization of ICD processes was to identify the regulatory molecular targets and combinations of pharmacological compounds that can increase anticancer immunity. The model can predict the time-dependent size of different cell populations involved in ICD that is induced by a treatment exposure.

To determine the role of each of the main cell types involved in ICD, Checcoli et al. first simulated a core ICD mechanism that is merely sufficient to reproduce ICD events observed experimentally [11]. The core regulatory mechanism describes the release of CALR, ATP, and HMGB1 molecules from dying tumor cells, and inner-state activation or evolution of immature DC, activated DC, migrating DC, lymph node DC, T cell, and cytotoxic T lymphocyte cell types. As shown in Figure 5, also included are two processes: tumor cell division, which is inhibited by T cells, and death, which is promoted by cytotoxic T lymphocytes. The states of molecules and cells are described by Boolean variables that can take two values: **1** for active or present and **0** for inactive or absent. The system state is described by a vector of Boolean values that represent each molecule, process, and cell type in the system. In the probabilistic description, the probability is assigned to each system state, such that the sum of probabilities over all possible system states is equal to **1**. Then, to determine the number of cells in a given system state, the system state probability is multiplied by the overall size of the cell population.

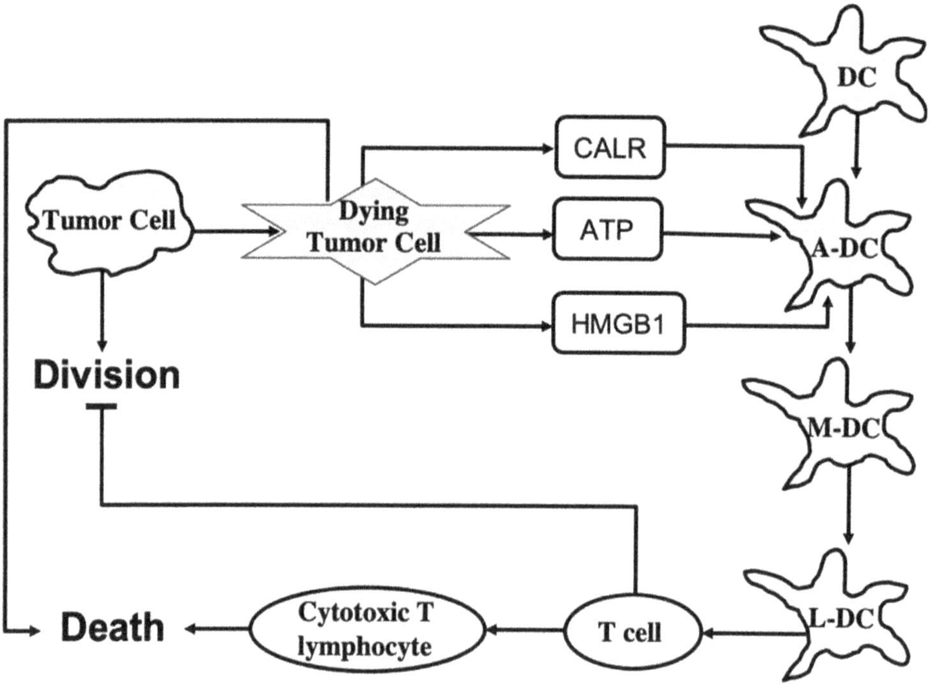

Figure 5. The core mechanism of immunogenic cell death. Lines represent activation (arrowhead) and inhibition (bar head) influences. A-DC, M-DC, and L-DC represent activated DC, migrating DC, and lymph node DC cell types, respectively.

The core model can reproduce the series of events following an ICD-inducing intervention. The release of CALR, ATP, and HMGB1 molecules by dying cancer cells occurs within hours, a slow increase in T cells begins after 100 h, which peaks at 200 h, and the tumor cells are eliminated in about 220 h when a rapid increase in cytotoxic T lymphocyte cell population begins. When the clonal expansion of the cytotoxic T lymphocytes was blocked in the model, tumor cell clearance became less efficient and depended mostly on the direct cytotoxicity of the treatment.

To improve the predictive power of the model, Checcoli et al. extended their core model by including more cell types and molecular components as well as the ligand–receptor dynamics that determines intercellular communication. The extended model describes 57 entities and provides more detailed representations of the series of events that were explored by the core model. Simulation results of the extended model also reproduce the succession of events resulting in ICD. Simulations were performed starting with 80% of tumor cells, 10% of dendritic cells, and 5% of inactive $CD4^+$ and $CD8^+$ cells. The population of tumor cells rapidly decays starting from 250 h when cytotoxic T lymphocytes are engaged to eliminate tumor cells.

To assess the extended model robustness to parameter changes, Checcoli et al. performed a sensitivity analysis measuring the variations in sizes of tumor cell populations within the 220 h and 280 h time frame when the tumor cell population decreases in the standard conditions (WT) of the extended model [11]. The decrease in size of the tumor cell population was seen to be delayed only for a few parameter changes when compared with the WT condition. Changes in parameters that control the number of DCs gave the strongest effect. A lower amount of DCs delayed the time of death, whereas a higher amount enhanced the death process. Changes in parameters controlling the rate of T cell clonal expansion give a similar effect on the cell death process. Sensitivity analysis also suggested the points of intervention that had the strongest effect on ICD. For example, a complete knockout of CD28 or CD80 (costimulatory molecules for T cell activation) resulted in a failure of the ICD-inducing treatment (80% of the tumor cell population persists at t = 280 h). By contrast, an external treatment that increases Interleukin-2 (IL-2) could kill the tumor cells faster, at t = 200 h.

The Boolean approach does not provide quantitative details and different regimens of drug treatments. Nevertheless, the model characterizes ICD events and dynamics in cancer cells and predicts molecular targets that could increase tumor clearance. For future directions, Checcoli et al. suggested to include specific in vitro and in vivo experiments to identify parameter values that will agree with experimentally observed timing of the different events leading to tumor clearance [11]. Further extension of the model including effects of IFNγ or TGFβ on the immune cells, and major signaling pathways inside each cell type, will allow the model to predict more feasible pharmacological interventions that can boost ICD for killing tumor cells.

7. Discussion

The significant progress that has been made in the mathematical characterization of different cell death execution pathways offers quantitative insight into cell death control and mechanistically explains why and how a living cell may die. Table 3 summarizes cell death mathematical model development over a 22-year period. We include the modeled cell death mechanism, methods, a mathematical description of the cell death event used in each model, and the main modeling results obtained in each work. ODE and Boolean logic-based approaches are the most common mathematical techniques used to model cell death mechanisms. However, a physical description based on the potential landscape theory has been recently applied to study stochastic dynamics and global stability of cell death signaling pathways [8,73]. In this approach, the steady state probability distribution of a system P_{ss} and a dimensionless potential function E are related via Boltzmann relation: $E = -\ln(P_{ss})$ [91]. The physical description allows one to employ thermodynamics to analyze cell death regulatory circuits. Conversely, entropy-based approaches have been applied to analyze biological networks [92] and a cell fate selection process [93,94], they have not been yet applied to characterize cell death decision mechanisms. Therefore, one promising future direction is to describe cell death networks using physical approaches that could help to reveal new functional system states and unknown properties of cell death regulatory mechanisms.

Table 1. Summary of cell death mechanism models.

Authors, Year, Cell Death Mechanism Modeled	Methods. Death Rule (DR)	Results
Fussenegger et al., 2000 [7], receptor- and stress-induced apoptosis	ODE approach. DR: the ratio of executioner caspase to free Bcl-x_L is greater than a threshold value	Qualitative explanation of observed caspase activation dynamics
Eissing et al., 2004 [21], receptor-induced apoptosis	ODE approach, stability and bifurcation analysis methods. DR: The bistable system is in apoptotic steady state	Bistable behavior of caspase-3 activation
Bentele et al., 2004 [25], receptor-induced apoptosis	ODE approach, sensitivity analysis. DR: receptor–ligand ratio is greater than a threshold value	A threshold mechanism for induction of receptor-induced apoptosis

Table 2. *Cont.*

Authors, Year, Cell Death Mechanism Modeled	Methods. Death Rule (DR)	Results
Hua et al., 2005 [43], receptor-induced apoptosis	ODE approach, sensitivity analysis. DR: caspase-3 activation	Bcl-2 blocks the mitochondrial apoptosis pathway by binding to proapoptotic proteins
Legewie et al., 2006 [22], intrinsic apoptosis	ODE approach, stability and bifurcation analysis methods. DR: irreversible caspase-3 activation	Bistable and irreversible caspase-3 activation arises in the system due to XIAP-mediated feedback
Rehm et al., 2006 [26], intrinsic apoptosis	ODE approach, sensitivity analysis. DR: complete caspase-dependent substrate cleavage	All-or-none apoptotic response depends on caspase-3-dependent feedback signaling and XIAP
Bagci et al., 2006 [23], mitochondria-dependent apoptosis	ODE approach. DR: caspase-3 activation is above a threshold that depends on Bax degradation and expression rates.	The transition from bistable to monostable (survival) cell behavior is controlled by the number of mitochondrial permeability transition pores
Chen and Cui et al., 2007, 2008 [38,45,46], intrinsic apoptosis	Deterministic and stochastic approaches, robustness analysis. DR: one-way bistable switch of Bax-activation	Apoptotic switches are bistable and robust to noise
Albeck et al., 2008 [27], extrinsic apoptosis	ODE approach, compartmental modeling. DR: mitochondria-to-cytosol cytochrome *c* and Smac translocation in an all-or-none manner	Permeabilization of the mitochondrial membrane and relocalization of proteins are the key factors in all-or-none death decision
Spencer et al., 2009 [33], extrinsic apoptosis	ODE approach. DR: levels of activated tBid, Bax, and Bak exceed a threshold set by inhibitory Bcl-2 proteins	Cell-to-cell variability in time-to-death depends on activation of the pore-forming proteins Bax and Bak
Neumann et al., 2010 [32], crosstalk between receptor-mediated apoptosis and NF-κB signaling	ODE approach, sensitivity analysis. DR: the maximum level of active caspase-8 is used as a readout for apoptosis	Assembly of DISC acts as a signal processor determining life/death decisions in a nonlinear manner
Hong et al., 2012 [28], crosstalk between apoptosis and ER stress response mechanisms	ODE approach, sensitivity analysis. DR: the level of apoptosis is determined by an ODE that depends on caspases-2,3,9,8 and apoptosis-inducing factor	Crosstalks among the mitochondrial, death receptor and ER stress response pathways contribute to the level of apoptosis
Tavassoly et al., 2015 [29], crosstalk between autophagy and apoptosis	ODE approach. DR: apoptosis occurs as soon as proapoptotic BH3 exceeds antiapoptotic Bcl2 protein	Time courses of the relative level of autophagy for different levels of stressor and percentage of apoptotic cells
Ballweg et al., 2017 [30], crosstalk between p53 signaling and apoptosis	ODE approach, dynamical analysis. DR: the level of p53 is elevated higher than a threshold that depends on cIAP level	The probability of apoptosis depends on the dynamics of p53
Schlatter et al., 2009 [34], apoptosis	Boolean logic and multi-value logic approach	High connectivity, crosstalks, and feedback loops in apoptosis regulatory network are significant and essential for apoptosis signaling
Mai et al., 2009 [35], intrinsic and extrinsic apoptosis	Boolean logic approach. DR: the "DNA Damage Event" node has remained in the ON state for 20 successive steps	The feedback loops directly involving the caspase 3 are essential for maintaining irreversibility of apoptosis
Calzone et al., 2010 [36], apoptosis and non-apoptotic cell death (necroptosis)	Boolean logic approach. DR: "Apoptosis" node or "NonACD" node is in ON state	Transient activation of key proteins in necroptosis and mutual inhibitory crosstalks among apoptosis, survival and necroptosis pathways

Table 3. *Cont.*

Authors, Year, Cell Death Mechanism Modeled	Methods. Death Rule (DR)	Results
Xu et al., 2021 [8], cellular necroptosis signaling circuits	ODE approach, sensitivity analysis, bifurcation and potential landscape methods.	The structure and distribution characteristics of all parameters are essential for stable oscillation behavior of necroptosis circuits
Ildefonso et al., 2022 [59], necroptosis regulation	ODE approach, DREAM parameter estimation method, sensitivity analysis. DR: phosphorylated MLKL exceeds a hard threshold of 2772 molecules	Four distinct necroptosis execution modes
Li et al., 2021 [65], crosstalk between apoptosis and necroptosis regulatory networks.	ODE approach. DR: apoptosis occurs when RIP1 level < ∼1000 molecules/cell, co-occurrence of apoptosis and necroptosis when ∼46,000 mpc< RIP1 > ∼1000 mpc, necroptosis alone when RIP1 >∼46,000 mpc	Characterization of RIP1's biphasic roles in necroptosis
Zhu et al. [9], crosstalk between caspase-1 and caspase-3 driven pyroptosis pathways	ODE approach, bifurcation and sensitivity analysis methods. DR: Cell death rate is defined using a ratio of dying cell population to the initial cell population	The change in expression levels of caspase-1, caspase-3, and GSDMD can switch between GSDMD- and GSDME-executed pyroptosis death modes
Li et al., 2022 [73], crosstalk between pyroptosis and apoptosis regulations	ODE and potential energy landscape approaches. DR: by levels of cleaved GSDMD (pyroptosis) and cleaved caspase-3 (apoptosis)	Caspase-1 and GSDMD are key proteins that regulate the switching between pyroptosis and apoptosis
Konstorum et al., 2020 [10], ferroptosis regulation	Stochastic, multistate, discrete mathematical approach. DR: intermediate and high levels of the lipid radical LO*	Ferroptosis sensitivity depends on PUFA synthesis, PUFA incorporation into the phospholipid membrane, and the balance between levels of pro-oxidant species and antioxidant factors
Checcoli et al., 2020 [11], immunogenic cell death (ICD) mechanism	Boolean Kinetic Monte-Carlo approach. DR: Death node is at **1**	The succession of events resulting in ICD. Points of intervention that had the strongest effect on ICD

Importantly, many different cell death pathways share common molecular components, and thus all these pathways can interact together at any time to form a complex mechanism. Therefore, we hypothesize that cell death can be controlled by a singular, highly integrated cell death decision network, see Figure 6. This network enables cells to alter the signal flow through the shared nodes but with different edges and so select alternative cell death execution pathways within a single control network of cell death. A stress death signal can thus initiate multiple death mechanisms but not all reach an execution threshold. Currently, the molecular mechanism that regulates the selection of each specific death execution pathway remains elusive. In addition, mathematical models developed to study crosstalk between necroptosis and apoptosis [65], pyroptosis and apoptosis [73], autophagy and apoptosis [29] support the hypothesis that signals propagating through different cell death pathways are integrated to process the execution of specific cell death. We are developing a mathematical model of the cell death decision network to predict the molecular species and interactions that direct the signal flow towards a specific irreversible cell death fate. Such a model will provide new insights into the integrated control of cell death. Model predictions will help develop new approaches to either block or initiate irreversible cell death and identify which cell death pathways are blocked and which pathways remain accessible to execute cell death. Thus, model predictions will suggest alternative interventions to overcome a block in cell death activation that can occur in cancer cells that acquire drug resistance.

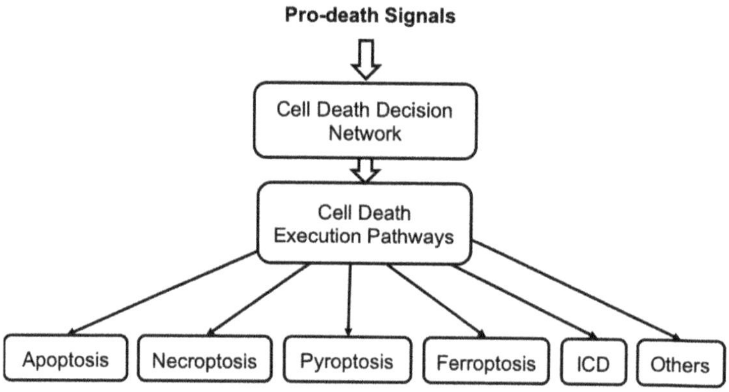

Figure 6. The concept of cell death decision network that controls the irreversible execution of different cell death mechanisms.

Author Contributions: Conceptualization, P.K. and R.C.; writing—original draft preparation, M.S. and P.K.; writing—review and editing, M.S., P.K. and R.C; visualization, M.S. and P.K.; supervision, P.K. All authors have read and agreed to the published version of the manuscript.

Funding: US department of the Army, Breast Cancer Research Program W81XWH-18-1-0722; BC171885, Public Health Service NIH U01CA184902.

Institutional Review Board Statement: Not applicable.

Data Availability Statement: Not applicable.

Conflicts of Interest: The authors declare no conflict of interest.

References

1. Tyson, J.J.; Novak, B. A Dynamical Paradigm for Molecular Cell Biology. *Trends. Cell Biol.* **2020**, *30*, 504–515. [CrossRef] [PubMed]
2. Kraikivski, P.; Chen, K.C.; Laomettachit, T.; Murali, T.; Tyson, J.J. From START to FINISH: Computational analysis of cell cycle control in budding yeast. *NPJ Syst. Biol. Appl.* **2015**, *1*, 15016. [CrossRef] [PubMed]
3. Shafiekhani, S.; Kraikivski, P.; Gheibi, N.; Ahmadian, M.; Jafari, A. Dynamical analysis of the fission yeast cell cycle via Markov chain. *Curr. Genet.* **2021**, *67*, 785–797. [CrossRef] [PubMed]
4. Jalihal, A.P.; Kraikivski, P.; Murali, T.; Tyson, J.J. Modeling and Analysis of the Macronutrient Signaling Network in Budding Yeast. *Mol. Biol. Cell* **2021**, *32*, ar20. [CrossRef] [PubMed]
5. Jung, Y.; Kraikivski, P.; Shafiekhani, S.; Terhune, S.S.; Dash, R.K. Crosstalk between Plk1, p53, cell cycle, and G2/M DNA damage checkpoint regulation in cancer: Computational modeling and analysis. *NPJ Syst. Biol. Appl.* **2021**, *7*, 46. [CrossRef] [PubMed]
6. Clarke, R.; Kraikivski, P.; Jones, B.C.; Sevigny, C.M.; Sengupta, S.; Wang, Y. A systems biology approach to discovering pathway signaling dysregulation in metastasis. *Cancer Metastasis Rev.* **2020**, *39*, 903–918. [CrossRef]
7. Fussenegger, M.; Bailey, J.E.; Varner, J. A mathematical model of caspase function in apoptosis. *Nat. Biotechnol.* **2000**, *18*, 768–774. [CrossRef]
8. Xu, F.; Yin, Z.; Zhu, L.; Jin, J.; He, Q.; Li, X.; Shuai, J. Oscillations Governed by the Incoherent Dynamics in Necroptotic Signaling. *Front. Phys.* **2021**, *9*, 482. [CrossRef]
9. Zhu, L.; Li, X.; Xu, F.; Yin, Z.; Jin, J.; Liu, Z.; Qi, H.; Shuai, J. Network modeling-based identification of the switching targets between pyroptosis and secondary pyroptosis. *Chaos Solitons Fractals* **2022**, *155*, 111724. [CrossRef]
10. Konstorum, A.; Tesfay, L.; Paul, B.T.; Torti, F.M.; Laubenbacher, R.C.; Torti, S.V. Systems biology of ferroptosis: A modeling approach. *J. Theor. Biol.* **2020**, *493*, 110222. [CrossRef]
11. Checcoli, A.; Pol, J.G.; Naldi, A.; Noël, V.; Barillot, E.; Kroemer, G.; Thieffry, D.; Calzone, L.; Stoll, G. Dynamical Boolean Modeling of Immunogenic Cell Death. *Front. Physiol.* **2020**, *11*, 590479. [CrossRef] [PubMed]
12. Lavrik, I.N. Systems biology of apoptosis signaling networks. *Curr. Opin. Biotechnol.* **2010**, *21*, 551–555. [CrossRef] [PubMed]
13. Lavrik, I.N.; Eils, R.; Fricker, N.; Pforr, C.; Krammer, P.H. Understanding apoptosis by systems biology approaches. *Mol. Biosyst.* **2009**, *5*, 1105–1111. [CrossRef]

14. Galluzzi, L.; Vitale, I.; Warren, S.; Adjemian, S.; Agostinis, P.; Martinez, A.B.; Chan, T.A.; Coukos, G.; Demaria, S.; Deutsch, E. Consensus guidelines for the definition, detection and interpretation of immunogenic cell death. *J. Immunother. Cancer* **2020**, *8*, e000337. [CrossRef] [PubMed]
15. Aldridge, B.B.; Burke, J.M.; Lauffenburger, D.A.; Sorger, P.K. Physicochemical modelling of cell signalling pathways. *Nat. Cell Biol.* **2006**, *8*, 1195–1203. [CrossRef] [PubMed]
16. Heiner, M.; Koch, I.; Will, J. Model validation of biological pathways using Petri nets—Demonstrated for apoptosis. *Biosystems* **2004**, *75*, 15–28. [CrossRef] [PubMed]
17. Schleich, K.; Lavrik, I.N. Mathematical modeling of apoptosis. *Cell Commun. Signal.* **2013**, *11*, 44. [CrossRef]
18. Apte, A.; Bonchev, D.; Fong, S. Cellular automata modeling of FASL-initiated apoptosis. *Chem. Biodivers.* **2010**, *7*, 1163–1172. [CrossRef]
19. Tyson, J.J.; Laomettachit, T.; Kraikivski, P. Modeling the dynamic behavior of biochemical regulatory networks. *J. Theor. Biol.* **2019**, *462*, 514–527. [CrossRef]
20. Rehm, M.; Dussmann, H.; Janicke, R.U.; Tavare, J.M.; Kogel, D.; Prehn, J.H. Single-cell fluorescence resonance energy transfer analysis demonstrates that caspase activation during apoptosis is a rapid process: Role of caspase-3. *J. Biol. Chem.* **2002**, *277*, 24506–24514. [CrossRef]
21. Eissing, T.; Conzelmann, H.; Gilles, E.D.; Allgower, F.; Bullinger, E.; Scheurich, P. Bistability analyses of a caspase activation model for receptor-induced apoptosis. *J. Biol. Chem.* **2004**, *279*, 36892–36897. [CrossRef] [PubMed]
22. Legewie, S.; Bluthgen, N.; Herzel, H. Mathematical modeling identifies inhibitors of apoptosis as mediators of positive feedback and bistability. *PLoS Comput. Biol.* **2006**, *2*, e120. [CrossRef] [PubMed]
23. Bagci, E.; Vodovotz, Y.; Billiar, T.; Ermentrout, G.; Bahar, I. Bistability in apoptosis: Roles of bax, bcl-2, and mitochondrial permeability transition pores. *Biophys. J.* **2006**, *90*, 1546–1559. [CrossRef] [PubMed]
24. Ferrell, E.J.; Xiong, W. Bistability in cell signaling: How to make continuous processes discontinuous, and reversible processes irreversible. *Chaos* **2001**, *11*, 227–236. [CrossRef] [PubMed]
25. Bentele, M.; Lavrik, I.; Ulrich, M.; Stosser, S.; Heermann, D.; Kalthoff, H.; Krammer, P.; Eils, R. Mathematical modeling reveals threshold mechanism in CD95-induced apoptosis. *J. Cell Biol.* **2004**, *166*, 839–851. [CrossRef]
26. Rehm, M.; Huber, H.J.; Dussmann, H.; Prehn, J.H. Systems analysis of effector caspase activation and its control by X-linked inhibitor of apoptosis protein. *EMBO J.* **2006**, *25*, 4338–4349. [CrossRef]
27. Albeck, J.G.; Burke, J.M.; Spencer, S.L.; Lauffenburger, D.A.; Sorger, P.K. Modeling a snap-action, variable-delay switch controlling extrinsic cell death. *PLoS Biol.* **2008**, *6*, 2831–2852. [CrossRef]
28. Hong, J.-Y.; Kim, G.-H.; Kim, J.-W.; Kwon, S.-S.; Sato, E.F.; Cho, K.-H.; Shim, E.B. Computational modeling of apoptotic signaling pathways induced by cisplatin. *BMC Syst. Biol.* **2012**, *6*, 122. [CrossRef]
29. Tavassoly, I.; Parmar, J.; Shajahan-Haq, A.; Clarke, R.; Baumann, W.T.; Tyson, J.J. Dynamic Modeling of the Interaction Between Autophagy and Apoptosis in Mammalian Cells. *CPT Pharmacomet. Syst. Pharmacol.* **2015**, *4*, 263–272. [CrossRef]
30. Ballweg, R.; Paek, A.L.; Zhang, T. A dynamical framework for complex fractional killing. *Sci. Rep.* **2017**, *7*, 8002. [CrossRef]
31. Tyson, J.J.; Baumann, W.T.; Chen, C.; Verdugo, A.; Tavassoly, I.; Wang, Y.; Weiner, L.M.; Clarke, R. Dynamic modelling of oestrogen signalling and cell fate in breast cancer cells. *Nat. Rev. Cancer* **2011**, *11*, 523–532. [CrossRef] [PubMed]
32. Neumann, L.; Pforr, C.; Beaudouin, J.; Pappa, A.; Fricker, N.; Krammer, P.H.; Lavrik, I.N.; Eils, R. Dynamics within the CD95 death-inducing signaling complex decide life and death of cells. *Mol. Syst. Biol.* **2010**, *6*, 352. [CrossRef] [PubMed]
33. Spencer, S.L.; Gaudet, S.; Albeck, J.G.; Burke, J.M.; Sorger, P.K. Non-genetic origins of cell-to-cell variability in TRAIL-induced apoptosis. *Nature* **2009**, *459*, 428–432. [CrossRef] [PubMed]
34. Schlatter, R.; Schmich, K.; Avalos Vizcarra, I.; Scheurich, P.; Sauter, T.; Borner, C.; Ederer, M.; Merfort, I.; Sawodny, O. ON/OFF and beyond—A boolean model of apoptosis. *PLoS Comput. Biol.* **2009**, *5*, e1000595. [CrossRef]
35. Mai, Z.; Liu, H. Boolean network-based analysis of the apoptosis network: Irreversible apoptosis and stable surviving. *J. Theor. Biol.* **2009**, *259*, 760–769. [CrossRef]
36. Calzone, L.; Tournier, L.; Fourquet, S.; Thieffry, D.; Zhivotovsky, B.; Barillot, E.; Zinovyev, A. Mathematical modelling of cell-fate decision in response to death receptor engagement. *PLoS. Comput. Biol.* **2010**, *6*, e1000702. [CrossRef]
37. Schleich, K.; Warnken, U.; Fricker, N.; Öztürk, S.; Richter, P.; Kammerer, K.; Schnölzer, M.; Krammer, P.H.; Lavrik, I.N. Stoichiometry of the CD95 death-inducing signaling complex: Experimental and modeling evidence for a death effector domain chain model. *Mol. Cell* **2012**, *47*, 306–319. [CrossRef]
38. Chen, C.; Cui, J.; Lu, H.; Wang, R.; Zhang, S.; Shen, P. Modeling of the role of a Bax-activation switch in the mitochondrial apoptosis decision. *Biophys. J.* **2007**, *92*, 4304–4315. [CrossRef]
39. Krammer, P.H.; Arnold, R.; Lavrik, I.N. Life and death in peripheral T cells. *Nat. Rev. Immunol.* **2007**, *7*, 532–542. [CrossRef]
40. Norbury, J.C.; Zhivotovsky, B. DNA damage-induced apoptosis. *Oncogene* **2004**, *23*, 2797–2808. [CrossRef]
41. Chinnaiyan, A.M.; O'Rourke, K.; Tewari, M.; Dixit, V.M. FADD, a novel death domain-containing protein, interacts with the death domain of Fas and initiates apoptosis. *Cell* **1995**, *81*, 505–512. [CrossRef]
42. Kluck, R.M.; Bossy-Wetzel, E.; Green, D.R.; Newmeyer, D.D. The release of cytochrome c from mitochondria: A primary site for Bcl-2 regulation of apoptosis. *Science* **1997**, *275*, 1132–1136. [CrossRef] [PubMed]
43. Hua, F.; Cornejo, M.G.; Cardone, M.H.; Stokes, C.L.; Lauffenburger, D.A. Effects of Bcl-2 levels on Fas signaling-induced caspase-3 activation: Molecular genetic tests of computational model predictions. *J. Immunol.* **2005**, *175*, 985–995. [CrossRef]

44. Stucki, W.J.; Simon, H.U. Mathematical modeling of the regulation of caspase-3 activation and degradation. *J. Theor. Biol.* **2005**, *234*, 123–131. [CrossRef] [PubMed]
45. Chen, C.; Cui, J.; Zhang, W.; Shen, P. Robustness analysis identifies the plausible model of the Bcl-2 apoptotic switch. *FEBS Lett.* **2007**, *581*, 5143–5150. [CrossRef] [PubMed]
46. Cui, J.; Chen, C.; Lu, H.; Sun, T.; Shen, P. Two independent positive feedbacks and bistability in the Bcl-2 apoptotic switch. *PLoS ONE* **2008**, *3*, e1469. [CrossRef] [PubMed]
47. Youle, R.J.; Strasser, A. The BCL-2 protein family: Opposing activities that mediate cell death. *Nat. Rev. Mol. Cell Biol.* **2008**, *9*, 47–59. [CrossRef] [PubMed]
48. Bertaux, F.; Stoma, S.; Drasdo, D.; Batt, G. Modeling dynamics of cell-to-cell variability in TRAIL-induced apoptosis explains fractional killing and predicts reversible resistance. *PLoS Comput. Biol.* **2014**, *10*, e1003893. [CrossRef]
49. Fischer, M. Census and evaluation of p53 target genes. *Oncogene* **2017**, *36*, 3943–3956. [CrossRef]
50. Kim, E.; Kim, J.Y.; Lee, J.Y. Mathematical Modeling of p53 Pathways. *Int. J. Mol. Sci.* **2019**, *20*, 5179. [CrossRef]
51. Miyashita, T.; Reed, J.C. Tumor suppressor p53 is a direct transcriptional activator of the human bax gene. *Cell* **1995**, *80*, 293–299. [CrossRef]
52. Katiyar, S.K.; Roy, A.M.; Baliga, M.S. Silymarin induces apoptosis primarily through a p53-dependent pathway involving Bcl-2/Bax, cytochrome c release, and caspase activation. *Mol. Cancer Ther.* **2005**, *4*, 207–216. [CrossRef] [PubMed]
53. Chipuk, J.E.; Kuwana, T.; Bouchier-Hayes, L.; Droin, N.M.; Newmeyer, D.D.; Schuler, M.; Green, D.R. Direct activation of Bax by p53 mediates mitochondrial membrane permeabilization and apoptosis. *Science* **2004**, *303*, 1010–1014. [CrossRef] [PubMed]
54. Seth, R.; Yang, C.; Kaushal, V.; Shah, S.V.; Kaushal, G.P. p53-dependent caspase-2 activation in mitochondrial release of apoptosis-inducing factor and its role in renal tubular epithelial cell injury. *J. Biol. Chem.* **2005**, *280*, 31230–31239. [CrossRef] [PubMed]
55. Paek, A.L.; Liu, J.C.; Loewer, A.; Forrester, W.C.; Lahav, G. Cell-to-Cell Variation in p53 Dynamics Leads to Fractional Killing. *Cell* **2016**, *165*, 631–642. [CrossRef] [PubMed]
56. Falschlehner, C.; Emmerich, C.H.; Gerlach, B.; Walczak, H. TRAIL signalling: Decisions between life and death. *Int. J. Biochem. Cell. Biol.* **2007**, *39*, 1462–1475. [CrossRef]
57. Choi, M.E.; Price, D.R.; Ryter, S.W.; Choi, A.M. Necroptosis: A crucial pathogenic mediator of human disease. *JCI Insight* **2019**, *4*, e128834. [CrossRef]
58. Wu, Y.; Dong, G.; Sheng, C. Targeting necroptosis in anticancer therapy: Mechanisms and modulators. *Acta Pharm. Sin. B* **2020**, *10*, 1601–1618. [CrossRef] [PubMed]
59. Ildefonso, G.V.; Oliver-Metzig, M.; Hoffmann, A.; Harris, L.A.; Lopez, C.F. Distinct execution modes of a biochemical necroptosis model explain cell type-specific responses and variability to cell-death cues. *bioRxiv* **2022**. [CrossRef]
60. Nelson, D.; Ihekwaba, A.; Elliott, M.; Johnson, J.; Gibney, C.; Foreman, B.; Nelson, G.; See, V.; Horton, C.; Spiller, D. Oscillations in NF-kappaB signaling control the dynamics of gene expression. *Science* **2004**, *306*, 704–708. [CrossRef]
61. Hoffmann, A.; Levchenko, A.; Scott, M.L.; Baltimore, D. The IkappaB-NF-kappaB signaling module: Temporal control and selective gene activation. *Science* **2002**, *298*, 1241–1245. [CrossRef] [PubMed]
62. Vanlangenakker, N.; Bertrand, M.; Bogaert, P.; Vandenabeele, P.; Vanden Berghe, T. TNF-induced necroptosis in L929 cells is tightly regulated by multiple TNFR1 complex I and II members. *Cell Death Dis.* **2011**, *2*, e230. [CrossRef]
63. Han, J.; Zhong, C.Q.; Zhang, D.W. Programmed necrosis: Backup to and competitor with apoptosis in the immune system. *Nat. Immunol.* **2011**, *12*, 1143–1149. [CrossRef] [PubMed]
64. Brenner, D.; Blaser, H.; Mak, T.W. Regulation of tumour necrosis factor signalling: Live or let die. *Nat. Rev. Immunol.* **2015**, *15*, 362–374. [CrossRef] [PubMed]
65. Li, X.; Zhong, C.-Q.; Wu, R.; Xu, X.; Yang, Z.-H.; Cai, S.; Wu, X.; Chen, X.; Yin, Z.; He, Q. RIP1-dependent linear and nonlinear recruitments of caspase-8 and RIP3 respectively to necrosome specify distinct cell death outcomes. *Protein Cell* **2021**, *12*, 858–876. [CrossRef] [PubMed]
66. Maltez, V.I.; Tubbs, A.L.; Cook, K.D.; Aachoui, Y.; Falcone, E.L.; Holland, S.M.; Whitmire, J.K.; Miao, E.A. Inflammasomes Coordinate Pyroptosis and Natural Killer Cell Cytotoxicity to Clear Infection by a Ubiquitous Environmental Bacterium. *Immunity* **2015**, *43*, 987–997. [CrossRef]
67. Rogers, C.; Fernandes-Alnemri, T.; Mayes, L.; Alnemri, D.; Cingolani, G.; Alnemri, E.S. Cleavage of DFNA5 by caspase-3 during apoptosis mediates progression to secondary necrotic/pyroptotic cell death. *Nat. Commun.* **2017**, *8*, 14128. [CrossRef]
68. Wang, Y.; Gao, W.; Shi, X.; Ding, J.; Liu, W.; He, H.; Wang, K.; Shao, F. Chemotherapy drugs induce pyroptosis through caspase-3 cleavage of a gasdermin. *Nature* **2017**, *547*, 99–103. [CrossRef]
69. Schneider, K.S.; Groß, C.J.; Dreier, R.F.; Saller, B.S.; Mishra, R.; Gorka, O.; Heilig, R.; Meunier, E.; Dick, M.S.; Ćiković, T. The Inflammasome Drives GSDMD-Independent Secondary Pyroptosis and IL-1 Release in the Absence of Caspase-1 Protease Activity. *Cell Rep.* **2017**, *21*, 3846–3859. [CrossRef]
70. Aizawa, E.; Karasawa, T.; Watanabe, S.; Komada, T.; Kimura, H.; Kamata, R.; Ito, H.; Hishida, E.; Yamada, N.; Kasahara, T. GSDME-Dependent Incomplete Pyroptosis Permits Selective IL-1α Release under Caspase-1 Inhibition. *iScience* **2020**, *23*, 101070. [CrossRef]
71. Maroju, P.K.; Grazioli, C.; Di Fraia, M.; Moioli, M.; Ertel, D.; Ahmadi, H.; Plekan, O.; Finetti, P.; Allaria, E.; Giannessi, L. Gasdermin E suppresses tumour growth by activating anti-tumour immunity. *Nature* **2020**, *579*, 415–420.

72. Wang, L.; Qin, X.; Liang, J.; Ge, P. Induction of Pyroptosis: A Promising Strategy for Cancer Treatment. *Front. Oncol.* **2021**, *11*, 635774. [CrossRef] [PubMed]
73. Li, X.; Zhang, P.; Yin, Z.; Xu, F.; Yang, Z.-H.; Jin, J.; Qu, J.; Liu, Z.; Qi, H.; Yao, C. Caspase-1 and Gasdermin D Afford the Optimal Targets with Distinct Switching Strategies in NLRP1b Inflammasome-Induced Cell Death. *Research* **2022**, *2022*, 9838341. [CrossRef] [PubMed]
74. Dixon, S.J.; Lemberg, K.M.; Lamprecht, M.R.; Skouta, R.; Zaitsev, E.M.; Gleason, C.E.; Patel, D.N.; Bauer, A.J.; Cantley, A.M.; Yang, W.S. Ferroptosis: An iron-dependent form of nonapoptotic cell death. *Cell* **2012**, *149*, 1060–1072. [CrossRef]
75. Stockwell, B.R.; Angeli, J.P.F.; Bayir, H.; Bush, A.I.; Conrad, M.; Dixon, S.J.; Fulda, S.; Gascón, S.; Hatzios, S.K.; Kagan, V.E. Ferroptosis: A Regulated Cell Death Nexus Linking Metabolism, Redox Biology, and Disease. *Cell* **2017**, *171*, 273–285. [CrossRef]
76. Yang, W.S.; SriRamaratnam, R.; Welsch, M.E.; Shimada, K.; Skouta, R.; Viswanathan, V.S.; Cheah, J.H.; Clemons, P.A.; Shamji, A.F.; Clish, C.B. Regulation of ferroptotic cancer cell death by GPX4. *Cell* **2014**, *156*, 317–331. [CrossRef]
77. Doll, S.; Proneth, B.; Tyurina, Y.Y.; Panzilius, E.; Kobayashi, S.; Ingold, I.; Irmler, M.; Beckers, J.; Aichler, M.; Walch, A. ACSL4 dictates ferroptosis sensitivity by shaping cellular lipid composition. *Nat. Chem. Biol.* **2017**, *13*, 91–98. [CrossRef]
78. Yang, W.S.; Stockwell, B.R. Synthetic lethal screening identifies compounds activating iron-dependent, nonapoptotic cell death in oncogenic-RAS-harboring cancer cells. *Chem. Biol.* **2008**, *15*, 234–245. [CrossRef]
79. Gao, M.; Monian, P.; Quadri, N.; Ramasamy, R.; Jiang, X. Glutaminolysis and Transferrin Regulate Ferroptosis. *Mol. Cell* **2015**, *59*, 298–308. [CrossRef]
80. Torti, S.V.; Manz, D.H.; Paul, B.T.; Blanchette-Farra, N.; Torti, F.M. Iron and Cancer. *Annu. Rev. Nutr.* **2018**, *38*, 97–125. [CrossRef]
81. Paton, C.M.; Ntambi, J.M. Biochemical and physiological function of stearoyl-CoA desaturase. *Am. J. Physiol. Endocrinol. Metab.* **2009**, *297*, E28–E37. [CrossRef] [PubMed]
82. Wan, C.; Li, S.; Wen, L.; Kong, J.; Wang, K.; Zhu, Y. Damage of oxidative stress on mitochondria during microspores development in Honglian CMS line of rice. *Plant Cell Rep.* **2007**, *26*, 373–382. [CrossRef] [PubMed]
83. Ou, Y.; Wang, S.-J.; Li, D.; Chu, B.; Gu, W. Activation of SAT1 engages polyamine metabolism with p53-mediated ferroptotic responses. *Proc. Natl. Acad. Sci. USA* **2016**, *113*, E6806–E6812. [CrossRef]
84. Lei, G.; Zhuang, L.; Gan, B. Targeting ferroptosis as a vulnerability in cancer. *Nat. Rev. Cancer* **2022**, *22*, 381–396. [CrossRef]
85. Galluzzi, L.; Buqué, A.; Kepp, O.; Zitvogel, L.; Kroemer, G. Immunogenic cell death in cancer and infectious disease. *Nat. Rev. Immunol.* **2017**, *17*, 97–111. [CrossRef] [PubMed]
86. Fuchs, Y.; Steller, H. Live to die another way: Modes of programmed cell death and the signals emanating from dying cells. *Nat. Rev. Mol. Cell Biol.* **2015**, *16*, 329–344. [CrossRef]
87. Matzinger, P. The danger model: A renewed sense of self. *Science* **2002**, *296*, 301–305. [CrossRef]
88. Garg, A.D.; Dudek-Peric, A.M.; Romano, E.; Agostinis, P. Immunogenic cell death. *Int. J. Dev. Biol.* **2015**, *59*, 131–140. [CrossRef]
89. Serrano-del Valle, A.; Anel, A.; Naval, J.; Marzo, I. Immunogenic Cell Death and Immunotherapy of Multiple Myeloma. *Front. Cell Dev. Biol.* **2019**, *7*, 50. [CrossRef]
90. Stoll, G.; Viara, E.; Barillot, E.; Calzone, L. Continuous time boolean modeling for biological signaling: Application of Gillespie algorithm. *BMC Syst. Biol.* **2012**, *6*, 116. [CrossRef]
91. Wang, J.; Xu, L.; Wang, E. Potential landscape and flux framework of nonequilibrium networks: Robustness, dissipation, and coherence of biochemical oscillations. *Proc. Natl. Acad. Sci. USA* **2008**, *105*, 12271–12276. [CrossRef] [PubMed]
92. Demongeot, J.; Jelassi, M.; Hazgui, H.; Ben Miled, S.; Bellamine Ben Saoud, N.; Taramasco, C. Biological Networks Entropies: Examples in Neural Memory Networks, Genetic Regulation Networks and Social Epidemic Networks. *Entropy* **2018**, *20*, 36. [CrossRef]
93. Li, Y.; Yi, M.; Zou, X. Identification of the molecular mechanisms for cell-fate selection in budding yeast through mathematical modeling. *Biophys. J.* **2013**, *104*, 2282–2294. [CrossRef] [PubMed]
94. Kraikivski, P. Do cells make decisions based on uncertainty in their biochemical networks? *Biophys. J.* **2013**, *104*, 2121–2122. [CrossRef] [PubMed]

entropy

MDPI

Article

On the Emergence of the Deviation from a Poisson Law in Stochastic Mathematical Models for Radiation-Induced DNA Damage: A System Size Expansion

Francesco Giuseppe Cordoni

Department of Civil, Environmental and Mechanical Engineering, University of Trento, 38123 Trento, Italy; francesco.cordoni@unitn.it

Abstract: In this paper, we study the system size expansion of a stochastic model for radiation-induced DNA damage kinetics and repair. In particular, we characterize both the macroscopic deterministic limit and the fluctuation around it. We further show that such fluctuations are Gaussian-distributed. In deriving such results, we provide further insights into the relationship between stochastic and deterministic mathematical models for radiation-induced DNA damage repair. Specifically, we demonstrate how the governing deterministic equations commonly employed in the field arise naturally within the stochastic framework as a macroscopic limit. Additionally, by examining the fluctuations around this macroscopic limit, we uncover deviations from a Poissonian behavior driven by interactions and clustering among DNA damages. Although such behaviors have been empirically observed, our derived results represent the first rigorous derivation that incorporates these deviations from a Poissonian distribution within a mathematical model, eliminating the need for specific ad hoc corrections.

Keywords: biophysical modeling; radiation-induced DNA damage; system size expansion; DNA damage repair

Citation: Cordoni, F.G. On the Emergence of the Deviation from a Poisson Law in Stochastic Mathematical Models for Radiation-Induced DNA Damage: A System Size Expansion. *Entropy* **2023**, 25, 1322. https://doi.org/10.3390/e25091322

Academic Editor: Pavel Kraikivski

Received: 18 July 2023
Revised: 2 September 2023
Accepted: 5 September 2023
Published: 11 September 2023

1. Introduction

Radiotherapy is one of the most effective and used cancer treatment modalities [1]. Traditionally, radiotherapy relies on photons; however, in recent decades, there has been a growing interest in advanced radiotherapy using ion beams. Ion beams offer several advantages over photons [2], particularly their ability to release energy in a highly localized manner within tissues, potentially leading to a more effective biological response with reduced collateral effects in healthy tissues. Extensive research of the scientific community has focused on studying the effects of radiation on biological tissue, with DNA being identified as the most vulnerable target for radiation-induced damage leading to cell death [3]. Despite the theoretical advantages of using ion beams, further research is necessary to integrate this treatment modality into clinical practice fully. One significant challenge in the widespread adoption of ion beams lies in accurately determining their biological effects, as this is crucial for prescribing the most suitable treatment. Over the years, mathematical models have been developed to understand and predict the biological impact of ions on tissue, particularly in relation to DNA *Double Strand Breaks* (DSB) [4–10]. These mathematical approaches aim to describe the formation, progression, and clustering of DSBs, ultimately striving to predict the cell survival probability following radiation exposure.

Despite the inherently stochastic nature of biological pathways, most existing mathematical models, until now, have relied on deterministic frameworks with a priori assumptions about the Poisson distribution and disregarded the stochastic fluctuations in energy deposition. These fluctuations occur from cell to cell, particularly in complex radiation environments. In particular, the *Microdosimetric Kinetic Model* (MKM) [5,11], together with the *Local Effect Model* (LEM) [12,13], is one of the only two mechanistic models used in clinical treatment planning, which describes the temporal evolution of the average number of DNA damages in a single cell nucleus to obtain a linear–quadratic survival probability starting from energy deposition at the micron scale, in the order of the cell nucleus. A funding assumption of the linear–quadratic relation between dose and the corresponding survival probability is the Poissonian distribution for DNA damage. Evidence has emerged that in certain situations, such as for heavy ions or at high doses, the Poissonian assumption does not hold. Therefore, in recent years, the community tried to overcome this assumption with some *ad hoc* correction terms related to intercellular damage interaction [4,11,14–17]. Though attempts have been made to address the limitations of the Poissonian assumption, a stochastic representation that encompasses the spatial and temporal aspects of dose deposition was lacking in the description of radiation-induced DNA damage formation and dynamics.

Recently, a series of papers addressed these issues, and the *Generalized Stochastic Microdosimetric Model* (GSM^2) [10,18–21] has been introduced. GSM^2 is a probabilistic model able to describe the time evolution of the DNA damage in a cell nucleus based on a differential equation governing the time evolution of the probability distribution of the number of DNA damages. Among the most relevant GSM^2 strengths, there is the capability to efficiently treat the several levels of spatiotemporal stochasticity happening during protracted irradiation without relying on the typically used Poissonian assumption on the number of DNA damages induced by radiation [19]. It is further described in [10,18] how different parameters, initial DNA damage distribution, or irradiation conditions can lead naturally to several possible probability distributions that can be significantly different from the typically assumed Poissonian law.

The main *master equation* governing the time evolution of the probability distribution of the number of DNA damages derived in [10] is non-linear due to the presence of a quadratic term that accounts for DNA clustering, which has been recognized as one of the main factors that leads to cell inactivation in radiobiology [7]. On the one side, this quadratic term plays a crucial role in the emergence of non-Poissonian behaviors; on the other side, it makes it difficult to obtain an explicit solution for the probability distribution of the number of DNA damages.

In this study, we present a system size expansion of the GSM^2 master equation based on the pioneering work [22]. The approximation that will be carried out in the present work is usually referred to in the literature as *system size expansion* [22,23], and it is widely used in the physics community to provide an appropriate macroscopic approximation of *microscopic* systems. As for all formal expansions, a suitable parameter is needed, around which the approximation is performed. In concrete applications, the domain size usually provides a suitable parameter to carry out a formal approximation, so that an asymptotic expansion of the main GSM^2 *master equation* will be carried out as the system size increases. It must be stressed that the approximation is generally valid as the number of lesions increases [23], so that the approximation derived in the present work provides a relevant description for high-dose irradiation, where the number of lesions even in small domains is high. Such a case is of particular relevance, since most of the existing models fail to give a precise description of the cell survival probability at high doses [4], and suitable correction terms are needed to match experimental data.

We will derive an asymptotic expansion for the GSM^2 master equation computing both the macroscopic limit and the fluctuation around such a macroscopic limit. Besides allowing us to calculate an approximate distribution for the number of DNA damage, the expansion derived in the present work provides further insights into the relationship between stochastic and deterministic mathematical models, already highlighted in previous works [18,19]. Having in mind the above-mentioned regimes of validity for the proposed expansion, we will further strengthen the connection between GSM^2 and the MKM, showing that as the system size increases, the *master equation* derived in [10] converges toward the main deterministic kinetic equations of the MKM [4,5]. This emphasizes how deterministic macroscopic behavior emerges from stochastic microscopic fluctuations. We will go a step further so that we will also prove a suitable central limit theorem, in the sense that we will characterize the stochastic fluctuations around the macroscopic average value. These fluctuations are usually assumed to be Poissonian. By contrast, we will show that such fluctuation described by the macroscopic approximation of GSM^2 are Gaussian-distributed, as described by a *linear Fokker–Planck equation* (FPE) [23]. Recalling that for large mean value λ, a Poisson random variable of mean λ is approximated in a probabilistic sense by a Gaussian random variable with mean and variance λ, we will show that the derived limiting model can be seen as a correction around a Poisson distribution due to the clustering of lesions. In this sense, the present work shows how typical non-Poissonian correction terms of the MKM that have been proposed over the years naturally emerge in the fully probabilistic description of the GSM^2. Lastly, it is worth stressing that the present paper sheds light on another possible future connection to existing radiobiological models. In fact, in the literature, some models have been derived that attempt to describe DNA damage at high doses using a Gaussian formulation of a multi-hit model (MHM) [24,25]. It has already been shown in [19] that GSM^2 is closely connected to some multi-hit models [26,27], so the results derived in the present paper further connect GSM^2 to the multi-hit models derived in the literature.

The main contributions of the present paper are:

(i) to derive a system size expansion for the master equation governing GSM^2 studying both the macroscopic limit and the fluctuations around the average;
(ii) to show how the nonlinear terms accounting for DNA clustering give rise to a non-Poissonian behavior;
(iii) to shed light on another insightful connection between existing radiobiological models.

The present work is structured as follows: Section 2.1 recalls the basic facts of GSM^2. Section 3.1 contains the main result of the present research, with the formal asymptotic expansion and a rigorous description of the stochastic fluctuations around the macroscopic average value. Section 4 provides some numerical results on the derived approximation.

2. Material and Methods

2.1. The Generalized Stochastic Microdosimetric Model and the Microdosimetric Kinetic Model

GSM^2 [10] is a stochastic model that provides a probabilistic description of DNA damage formation and evolution, with particular attention to the link to DNA damage formation and energy deposition. The final goal of the model is to overcome existing models mainly based on the Poissonian assumption of energy deposition to provide a better characterization of some relevant biological endpoints such as the cell survival fraction.

GSM^2 considers two types of DNA damage, called sub-lethal and lethal lesions. Lethal lesions Y represent damage that cannot be repaired, leading to cell inactivation. By contrast, sublethal lesions X can be repaired at rate r or become a lethal lesion either by direct death at rate a or interacting with another sublethal lesion at rate b. This latter term b accounts for the clustering of DNA damage and gives rise to a nonlinearity in the governing master equation.

Denoted by $(Y(t), X(t))$ is the state of the system at time t, where X and Y are two $\mathbb{N}-$valued random variables counting the number of the lethal and sub-lethal lesion, respectively. In the following, we will consider a standard complete filtered probability space $(\Omega, \mathcal{F}, (\mathcal{F}_t)_{t\geq 0}, \mathbb{P})$ satisfying usual assumptions, namely right-continuity and saturation by $\mathbb{P}$-null sets.

The above reasoning can be represented by the following pathways:

$$
\begin{aligned}
&X \xrightarrow{r} \emptyset\,,\\
&X \xrightarrow{a} Y\,,\\
&X + X \xrightarrow{b} Y\,,
\end{aligned}
$$

for three positive constant rates, r, a, and b.

In what follows, we denote by:

$$p_{t_0,y_0,x_0}(t,y,x) := p(t,y,x|t_0,y_0,x_0) = \mathbb{P}(\,(Y(t),X(t)) = (y,x)|(Y(t_0),X(t_0)) = (y_0,x_0))\,.$$

If no confusion is possible, we will avoid stating the initial time and state, writing for short $p(t,y,x)$.

Following [10], the *microdosimetric master equation* (MME) can be derived:

$$\begin{aligned}\frac{\partial}{\partial t}p(t,y,x) &= -[(a+r)x + bx(x-1)]p(t,y,x) + (x+1)rp(t,y,x+1)+ \\ &+ (x+1)ap(t,y-1,x+1) + (x+2)(x+1)bp(t,y-1,x+2)\,.\end{aligned} \tag{1}$$

The MME (1) can be written for short as:

$$\begin{aligned}\frac{\partial}{\partial t}p(t,y,x) &= \left(E^{-1,2}-1\right)[x(x-1)bp(t,y,x)] + \left(E^{-1,1}-1\right)[xap(t,y,x)]+ \\ &+ \left(E^{0,1}-1\right)[xrp(t,y,x)] = \\ &= \mathcal{E}^{-1,2}[x(x-1)bp(t,y,x)] + \mathcal{E}^{-1,1}[xap(t,y,x)] + \mathcal{E}^{0,1}[xrp(t,y,x)]\,,\end{aligned} \tag{2}$$

where above we have denoted the creation operators as:

$$\mathcal{E}^{i,j}[f(y,x)] := \left(E^{i,j}-1\right)[f(y,x)] := f(y+i,x+j) - f(y,x)\,.$$

The above MME (2) is coupled with an initial distribution:

$$p_0(y,x) := p(0,y,x)\,,$$

as described in [10,18].

Remark 1. *The above choice is made to closely follow the existing literature on the topic. However, other choices for the pathways r, a, and b can be made. For instance, it could be assumed that the death rate a is logistic, including an increment in the death as the number of lesions becomes bigger. In such a case, the MME would become:*

$$\frac{\partial}{\partial t}p(t,y,x) = \mathcal{E}^{-1,2}[x(x-1)bp(t,y,x)] + \mathcal{E}^{-1,1}[x(a+\bar{a}x)p(t,y,x)] + \mathcal{E}^{0,1}[xrp(t,y,x)]\,. \tag{3}$$

It has been shown in [10] that GSM^2 is closely connected to the MKM, where, in fact, the former represents a stochastic reformulation of the latter. The MKM postulates the same assumptions of GSM^2 with two key additions. First, the MKM considers the time evolution for the average number of lethal lesions $\bar{y}$ and sublethal lesions $\bar{x}$, and second, $\bar{y}$ is assumed to be Poissonian-distributed.

In particular, the MKM assumes the $\bar{y}$ and $\bar{x}$ follows the set of coupled ODE:

$$\begin{cases}\frac{d}{dt}\bar{y}(t) = a\bar{x}(t) + b\bar{x}^2(t)\,, \\ \frac{d}{dt}\bar{x}(t) = -(a+r)\bar{x}(t) - 2b\bar{x}^2(t)\,.\end{cases} \tag{4}$$

Typically, it is further assumed that $(a+r)x >> 2bx^2$, so that Equation (4) reduces to:

$$\begin{cases}\frac{d}{dt}\bar{y}(t) = a\bar{x}(t) + b\bar{x}^2(t)\,, \\ \frac{d}{dt}\bar{x}(t) = -(a+r)\bar{x}(t)\,.\end{cases} \tag{5}$$

The connection between the MME (2) and the system of ODE (4) has already been shown in [10,18], and this connection will be further deepened in the present paper.

3. Theory and Calculations

3.1. Macroscopic Description for the GSM^2

In the present Section, we will derive a rigorous expansion that provides a macroscopic linear approximation counter-part of the master equation derived in [10]. The following expansion is at the basis of a macroscopic deterministic description of a microscopic stochastic system.

In the following, we will assume that the coefficients of the master Equation (2) depend on a parameter K in a suitable manner; namely, they are of order $\mathcal{O}(1)$ with respect to the parameter K; under this assumption, we are able to characterize the limit for the master Equation (2) as $\mathrm{K} \to \infty$. As mentioned in the introduction, the typical approach is to consider K as the system size, from which the name *system size expansion* is derived.

We will prove both the convergence of the microscopic system towards a macroscopic mean value, which corresponds to *the law of large numbers*, and also provide a description for the fluctuations of the system around such a mean value. This description of the fluctuations allows us to describe the system in terms of an FPE so that we will show that the order of the fluctuation is not Poissonian, as typically assumed in most of the existing literature on the subject. In this sense, the current research highlights how GSM^2 provides a rigorous non-Poissonian correction to the MKM.

In the following, in order to carry out the expansion, we assume that the parameter b depends on the K as $\tilde{b}_{\mathrm{K}} := \frac{b}{\mathrm{K}}$. Therefore, the MME (2) now becomes:

$$\frac{\partial}{\partial t}p(t,y,x) = \mathcal{E}^{-1,2}\left[x(x-1)\tilde{b}_{\mathrm{K}}p(t,y,x)\right] + \mathcal{E}^{-1,1}\left[xap(t,y,x)\right] + \mathcal{E}^{0,1}\left[xrp(t,y,x)\right]. \tag{6}$$

The main idea that will be carried out in the current section is to let $\mathrm{K} \to \infty$ in order to approximate the master equation by a continuous equation. The first-order approximation will satisfy a linear FPE, whose marginals, under some specific initial distribution to be better specified in later sections, can be shown to be Gaussian-distributed. The derived Gaussian approximation for the lesion distribution will be shown to provide better insights than the classical Poissonian hypothesis regarding lethal damage.

In order to prove the expansion, we set:

$$\begin{aligned} X(t) &= \mathrm{K}\bar{x}(t) + \sqrt{\mathrm{K}}\xi(t), \\ Y(t) &= \mathrm{K}\bar{y}(t) + \sqrt{\mathrm{K}}v(t), \end{aligned} \tag{7}$$

with $(\xi(t))_{t\geq 0}$ and $(v(t))_{t\geq 0}$ being two stochastic processes, so that for any $t \geq 0$, $\xi(t)$ and $v(t)$ are two centered random variables, i.e., with null mean value, whereas $\bar{x}(t)$ and $\bar{y}(t)$ are two suitable deterministic functions to be derived later. Heuristically speaking, $\bar{x}$ and $\bar{y}$ will play the role of the macroscopic deterministic behavior, which we will show to agree with the differential equations governing the MKM, as given in (4). Therefore, the above assumptions can be interpreted intuitively as an expansion of variables x and y around the macroscopic behavior, whereas the terms ξ and v represent the fluctuations around the mean value.

It is worth noting that the following holds true:

$$\begin{aligned} \mathbb{E}[X(t)] &= \mathrm{K}\bar{x}(t) + \sqrt{\mathrm{K}}\mathbb{E}[\xi(t)], \\ \mathbb{E}[Y(t)] &= \mathrm{K}\bar{y}(t) + \sqrt{\mathrm{K}}\mathbb{E}[v(t)], \\ \mathbb{V}ar[X(t)] &= \mathrm{K}\mathbb{V}ar[\xi(t)], \\ \mathbb{V}ar[Y(t)] &= \mathrm{K}\mathbb{V}ar[v(t)]. \end{aligned} \tag{8}$$

Define the new distribution with respect to the new variables as:

$$p(t,y,x) = p\left(t, \mathrm{K}\bar{x} + \sqrt{\mathrm{K}}\xi, \mathrm{K}\bar{y} + \sqrt{\mathrm{K}}v\right) = P(t,v,\xi).$$

The standard chain rule applied to $P(t, v, \xi)$ yields:

$$\frac{\partial}{\partial t}p(t,y,x) = \frac{\partial}{\partial t}P(t,v,\xi)\left(1+\frac{\partial}{\partial t}v+\frac{\partial}{\partial t}\xi\right),$$

so that inverting transformation (7) for v and ξ gives:

$$\frac{\partial}{\partial t}p(t,y,x) = \frac{\partial}{\partial t}P(t,v,\xi) - \sqrt{K}\frac{d}{dt}\bar{y}\frac{\partial}{\partial v}P(t,v,\xi) - \sqrt{K}\frac{d}{dt}\bar{x}\frac{\partial}{\partial \xi}P(t,v,\xi).$$

Regarding the step operators appearing in the MME (6), it can be shown that the following holds true:

$$\mathcal{E}^{i,j} = \frac{1}{\sqrt{K}}\left(i\frac{\partial}{\partial v}+j\frac{\partial}{\partial \xi}\right) + \frac{1}{2}\frac{1}{K}\left(i\frac{\partial}{\partial v}+j\frac{\partial}{\partial \xi}\right)^2.$$

The above computations substituted into Equation (6) yields:

$$\begin{aligned}
\frac{\partial}{\partial t}P(t,v,\xi) &= \sqrt{K}\frac{d}{dt}\bar{y}\frac{\partial}{\partial_v}P(t,v,\xi) + \sqrt{K}\frac{d}{dt}\bar{x}\frac{\partial}{\partial_\xi}P(t,v,\xi) + \\
&+ \frac{b}{K}\left[\frac{1}{\sqrt{K}}\left(2\frac{\partial}{\partial_\xi}-\frac{\partial}{\partial_v}\right)\right]\left(K\bar{x}+\sqrt{K}\xi\right)\left(K\bar{x}+\sqrt{K}\xi-1\right)P(t,v,\xi) + \\
&+ \frac{b}{K}\left[\frac{1}{2}\frac{1}{K}\left(2\frac{\partial}{\partial_\xi}-\frac{\partial}{\partial_v}\right)^2\right]\left(K\bar{x}+\sqrt{K}\xi\right)\left(K\bar{x}+\sqrt{K}\xi-1\right)P(t,v,\xi) + \\
&+ a\left[\frac{1}{\sqrt{K}}\left(\frac{\partial}{\partial_\xi}-\frac{\partial}{\partial_v}\right) + \frac{1}{2}\frac{1}{K}\left(\frac{\partial}{\partial_\xi}-\frac{\partial}{\partial_v}\right)^2\right]\left(K\bar{x}+\sqrt{K}\xi\right)P(t,v,\xi) + \\
&+ r\left[\frac{1}{\sqrt{K}}\frac{\partial}{\partial_\xi} + \frac{1}{2}\frac{1}{K}\frac{\partial^2}{\partial_\xi^2}\right]\left(K\bar{x}+\sqrt{K}\xi\right)P(t,v,\xi).
\end{aligned} \tag{9}$$

Grouping the terms of order $\sqrt{K}$, we obtain:

$$\begin{aligned}
\sim \sqrt{K} : \; &\frac{d}{dt}\bar{y}\frac{\partial}{\partial_v}P(t,v,\xi) + \frac{d}{dt}\bar{x}\frac{\partial}{\partial_\xi}P(t,v,\xi) + a\left(\frac{\partial}{\partial_\xi}-\frac{\partial}{\partial_v}\right)\bar{x}P(t,v,\xi) + \\
&+ r\frac{\partial}{\partial_\xi}xP(t,v,\xi) + 2bx^2\frac{\partial}{\partial_\xi}P(t,v,\xi) - b\bar{x}^2\frac{\partial}{\partial_v}P(t,v,\xi).
\end{aligned} \tag{10}$$

In order to compensate for the terms of order $\sqrt{K}$, we set the macroscopic system as:

$$\begin{cases} \frac{d}{dt}\bar{y} = a\bar{x} + b\bar{x}^2 \\ \frac{d}{dt}\bar{x} = -(a+r)\bar{x} - 2b\bar{x}^2, \end{cases} \tag{11}$$

so that all terms or order $\sqrt{K}$ in Equation (10) vanishes. Therefore, we have shown that the macroscopic limit of GSM^2 MME coincides with the main deterministic governing equation of the MKM (4).

Explicit solutions to system (11) can be derived with the further property that they are globally stable and converging to a stationary solution [19]. Consider first:

$$\frac{d}{dt}\bar{x}(t) = -(a+r)\bar{x}(t) - 2b\bar{x}^2(t), \quad \bar{x}(0) = x_0. \tag{12}$$

Such an equation (12) is known as the Bernoulli equation. Applying the transformation $u = \frac{1}{\bar{x}}$ leads to the following differential equation:

$$\frac{d}{dt}u(t) = (a+r)u(t) + 2b.$$

This last equation is a linear equation in u, so the explicit solution is given by:

$$u(t) = ce^{(a+r)t} - \frac{2b}{a+r}.$$

Coming back to the original equation, we obtain:

$$\bar{x}(t) = \frac{a+r}{ce^{(a+r)t} - 2b},$$

with

$$c := \frac{a+r}{x_0} + 2b.$$

We can, therefore, substitute $\bar{x}$ into Equation (11) to obtain:

$$\bar{y}(t) = y_0 + \frac{a+r}{4b - 2ce^{t(a+r)}} + \frac{r}{2b}\left((a+r)t - \log\left[\left|2b - ce^{t(a+r)}\right|\right]\right).$$

We can eventually calculate the long-time convergence toward the stationary solution of the above equations to be:

$$\lim_{t\to\infty} \bar{x}(t) =: \bar{x}_\infty = 0,$$
$$\lim_{t\to\infty} \bar{y}(t) =: \bar{y}_\infty = y_0 - \frac{r}{2b}\log\left(\frac{a+r}{x_0} + 2b\right).$$

Remark 2. *It is worth remarking that, for low-dose and sparsely ionizing radiation, such as X-rays or high-energy protons, the following assumption typically holds true, $(a+r)x >> 2bx^2$; therefore, the above calculations simplify so that the explicit solution to Equation* (11) *is given by [5,28]:*

$$\begin{cases} \bar{y}(t) = y_0 + ax_0\left(\frac{1-e^{-(a+r)t}}{a+r}\right) + bx_0^2\left(\frac{1-e^{-2(a+r)t}}{a+r}\right) \\ \bar{x}(t) = x_0 e^{-(a+r)t}. \end{cases} \tag{13}$$

In particular, Equation (13) *converges as $t \to \infty$ towards:*

$$\lim_{t\to\infty} \bar{y}(t) = y_0 + x_0\left(\frac{a}{a+r} + x_0\frac{b}{a+r}\right), \quad \lim_{t\to\infty} \bar{x}(t) = 0.$$

3.2. The Linear Noise Approximation and Moments Estimates

Having cancelled out terms of order $\sqrt{K}$, taking the limit as $K \to \infty$, all terms containing K vanish and only the zero$-th$ order terms remain, yielding:

$$\begin{aligned}
\frac{\partial}{\partial t}P(t,v,\xi) &= 2b\left(2\frac{\partial}{\partial_\xi} - \frac{\partial}{\partial_v}\right)[\bar{x}\xi P(t,v,\xi)] + \frac{1}{2}b\left(2\frac{\partial}{\partial_\xi} - \frac{\partial}{\partial_v}\right)^2\left[\bar{x}^2 P(t,v,\xi)\right] + \\
&+ a\left(\frac{\partial}{\partial_\xi} - \frac{\partial}{\partial_v}\right)[\xi P(t,v,\xi)] + r\frac{\partial}{\partial_\xi}[\xi P(t,v,\xi)] + \\
&+ \frac{1}{2}a\left(\frac{\partial}{\partial_\xi} - \frac{\partial}{\partial_v}\right)^2[\bar{x}P(t,v,\xi)] + \frac{1}{2}r\frac{\partial^2}{\partial_\xi^2}[\bar{x}P(t,v,\xi)] + \\
&= \frac{\partial}{\partial_\xi}[\xi P(t,v,\xi)](4b\bar{x} + a + r) - \frac{\partial}{\partial_v}[\xi P(t,v,\xi)](2b\bar{x} + a) + \\
&- \partial_{\xi v}P(t,v,\xi)\left(2b\bar{x}^2 + a\bar{x}\right) + \\
&+ \frac{1}{2}\frac{\partial^2}{\partial_\xi^2}P(t,v,\xi)\left((a+r)\bar{x} + 4b\bar{x}^2\right) + \frac{1}{2}\frac{\partial^2}{\partial_v^2}P(t,v,\xi)\left(a\bar{x} + b\bar{x}^2\right).
\end{aligned} \tag{14}$$

Equation (14) is a linear *Fokker–Planck* equation of dimension 2 that describes the fluctuations of the system around the average values $\bar{x}(t)$ and $\bar{y}(t)$. The solution to the linear FP Equation (14), under suitable initial conditions that will be specified later, can be shown to be the bi-dimensional Gaussian density.

Until now, we have avoided explicitly considering the initial condition both for the original MME (6) and for the approximating linear FPE (14).

As shown in [18], much of the stochasticity regarding lesion formation lies in the initial condition, in the sense that the distribution of initial lethal and sub-lethal damage deeply affects the subsequent time evolution of the probability density function. We will avoid an extensive treatment of such a topic in the present paper and focus more on the stochasticities inherent to the kinetics of the interaction of DNA damages, considering instead two simple and yet relevant cases for the initial damage distribution.

Let us start by assuming that the initial number of lesions is deterministic and is given by (y_0, x_0). We, therefore, equip the MME (6) with a deterministic initial condition given by:

$$p(0,y,x) = \delta(x - x_0)\delta(y - y_0),$$

with $\delta(x - x_0)$ and $\delta(y - y_0)$ being the Dirac delta centered at x_0 and y_0, respectively. It can be shown that [23,29] the solution to the linear FPE (14) is given by a bivariate Gaussian distribution:

$$P(t,v,\xi) = \frac{1}{2\pi}\frac{1}{\sqrt{detC}}\exp\left\{-\frac{1}{2}(v - \bar{v}, \xi - \bar{\xi})^T C^{-1}(v - \bar{v}, \xi - \bar{\xi})\right\},$$

where $\bar{v}$ and $\bar{\xi}$ are the mean values and C is the covariance matrix with entries:

$$C = \begin{pmatrix} c_{vv} & c_{\xi v} \\ c_{\xi v} & c_{\xi\xi} \end{pmatrix},$$

where c_{vv}, resp. $c_{\xi v}$, and resp. $c_{\xi\xi}$ are the variance of v, resp. covariance of ξ and v, and resp. variance of ξ. It is worth stressing that, given the properties of the multivariate Gaussian distributions, ξ and v are univariate Gaussian random variables.

Upon the multiplication of Equation (14) by ξ and v, it follows after integrating by parts that the first moment of ξ and v satisfies:

$$\begin{aligned} \frac{d}{dt}\bar{v} &= (2b\bar{x} + a)\bar{\xi}, \quad \bar{v}(0) = 0, \\ \frac{d}{dt}\bar{\xi} &= -(4b\bar{x} + a + r)\bar{\xi}, \quad \bar{\xi}(0) = 0. \end{aligned} \tag{15}$$

It immediately follows from Equation (15) that:

$$\bar{\xi}(t) = \bar{v}(t) \equiv 0.$$

This result is in agreement with the fact that ξ and v are centered random variables.

Multiplying Equation (14) by ξ^2, ξv, and v^2, we obtain, again after integration by parts, that the variance and covariance satisfy the following set of coupled ODEs:

$$\begin{cases} \frac{d}{dt}c_{vv} &= 2(2b\bar{x} + a)c_{\xi v} + a\bar{x} + b\bar{x}^2, \\ \frac{d}{dt}c_{\xi v} &= (2b\bar{x} + a)c_{\xi\xi} - (4b\bar{x} + a + r)c_{\xi v} - (2b\bar{x}^2 + a\bar{x}), \\ \frac{d}{dt}c_{\xi\xi} &= -2(4b\bar{x} + a + r)c_{\xi\xi} + (a + r)\bar{x} + 4b\bar{x}^2, \end{cases} \tag{16}$$

with the initial condition $c_{vv}(0) = c_{\xi v}(0) = c_{\xi\xi}(0) = 0$. The last two equations in (16) can be computed to be:

$$\begin{cases} c_{\xi\xi}(t) &= \dfrac{e^{2t(a+r)}\left((a+r)\left(-4b^2e^{t(-a-r)}-4bct(a+r)+c^2e^{t(a+r)}\right)+c(a+r)\left(4b^2-c^2\right)+ac\right)}{\left(ce^{t(a+r)}-2b\right)^4}, \\ c_{\xi v}(t) &= \dfrac{e^{t(a+r)}\left(-\dfrac{ce^{t(a+r)}\left(2a^2(2brt+b)+ar\left(-4b(b-2rt-1)+c^2-1\right)+r^2\left(2b(-2b+2rt+1)+c^2\right)\right)}{2b(a+r)\left(ce^{t(a+r)}-2b\right)}\right)}{\left(ce^{t(a+r)}-2b\right)^2}+ \\ & + \dfrac{e^{t(a+r)}\left(\dfrac{c^2e^{2t(a+r)}\left(-4ab^2+4bt(a+r)^2+ac^2-a-4b^2r+c^2r\right)}{4b\left(ce^{t(a+r)}-2b\right)^2}+2bre^{t(-a-r)}-t(a+r)(ac-r)+c_{\xi v}\right)}{\left(ce^{t(a+r)}-2b\right)^2}, \end{cases} \tag{17}$$

with $c_{\xi v}$ a suitable constant to ensure the initial condition. It can be seen that it holds:

$$\lim_{t\to\infty} c_{\xi\xi}(t) = \lim_{t\to\infty} c_{\xi v}(t) = 0\,.$$

In particular, we are mostly concerned with the term c_{vv} and with its stationary solution, as we aim to show that the distribution of lethal lesions differs from a Poisson distribution, as it is typically assumed in radiobiological models. It can, thus, be noticed that, integrating the third equation in (16), we obtain:

$$\begin{aligned} c_{vv}(t) &= \int_0^t \left[2(2b\bar{x}(s)+a)c_{\xi v}(s) + a\bar{x}(s) + b\bar{x}^2(s)\right]ds = \\ &= \bar{y}(t) + \int_0^t 2(2b\bar{x}(s)+a)c_{\xi v}(s)ds = \bar{y}(t) - \delta(t)\,, \end{aligned} \tag{18}$$

with $\bar{y}(t)$ being the mean value for lethal lesions, as computed in Equation (11), and:

$$\delta(t) := -\int_0^t 2(2b\bar{x}(s)+a)c_{\xi v}(s)ds\,.$$

The negative sign in δ is used to emphasize that the covariance is, in fact, negative, since a decrease in sublethal lesions correlates with an increase in lethal lesions.

The long time behaviour for c_{vv} can be explicitly computed using Equation (18) to be:

$$\lim_{t\to\infty} c_{vv}(t) =: \bar{y}_\infty - \delta_\infty\,, \tag{19}$$

with $\bar{y}_\infty$ being the long-time solution to the macroscopic average value $\bar{y}(t)$.

Recalling that for a large mean value, a Poisson distribution can be approximated by a Gaussian random variable with equal mean and variance, in order to infer that the lethal lesion distribution obeys a Poisson random variable, we must obtain $\lim_{t\to\infty} c_{vv}(t) = \bar{y}_\infty$. By contrast, the above calculations show that the variance is given by the average value corrected by a term given by the covariance of two types of lesions. In particular, as there is a negative correlation between the two variables, we can infer that the lethal lesion distribution is almost a Poisson random variable, where the variance is adjusted by subtracting a term due to pairwise interactions.

Moments Estimates for a Stochastic Initial Condition

In general, we cannot expect the initial number of lesions to be deterministic, so previous arguments must be slightly modified.

To explicitly compute the marginal distribution for the solution to the linear Fokker–Planck Equation (14), we assume the initial distribution for the MME to be normally distributed with mean (y_0, x_0) and variance Σ. It is worth remarking that such an assumption is not restrictive, as the standard assumption for the initial condition is to be a Poisson random variable, which, as mentioned above, under certain assumptions, can be approximated by a Gaussian random variable.

In particular, we assume that the initial number of lethal and sublethal lesions follows a Gaussian random variable with mean and variance given by x_0 and y_0:

$$p(0, y, x) = \frac{1}{2\pi} \frac{1}{\sqrt{det\Sigma}} \exp\left\{ -\frac{1}{2} (y - y_0, x - x_0)^T \Sigma^{-1} (y - y_0, x - x_0) \right\}, \tag{20}$$

with

$$\Sigma = \begin{pmatrix} x_0 & 0 \\ 0 & y_0 \end{pmatrix}.$$

Similar arguments as above imply that the initial condition for the linear FPE (14), under Equation (20), becomes a centered Gaussian random variable:

$$P(0, v, \xi) = \frac{1}{2\pi} \frac{1}{\sqrt{det\Sigma}} \exp\left\{ -\frac{1}{2} (v, \xi)^T \Sigma^{-1} (v, \xi) \right\}.$$

Therefore, the initial fluctuations around the mean value are Gaussian-distributed with a null average.

Therefore, all calculations above follow alike, implying that, again, the solution to the linear Fokker-Planck Equation (14) is given by:

$$P(t, v, \xi) = \frac{1}{2\pi} \frac{1}{\sqrt{detC}} \exp\left\{ -\frac{1}{2} (v, \xi)^T C^{-1} (v, \xi) \right\},$$

where now the covariance matrix incorporates the initial stochastic condition so that its entries satisfy the following set of differential equations:

$$\begin{cases} \frac{\partial}{\partial t} c_{vv} &= 2(2b\bar{x} + a)c_{\xi v} + a\bar{x} + b\bar{x}^2, \quad c_{vv}(0) = y_0, \\ \frac{\partial}{\partial t} c_{\xi v} &= (2b\bar{x} + a)c_{\xi\xi} - (4b\bar{x} + a + r)c_{\xi v} - 2\left(4b\bar{x}^2 + 2a\bar{x}\right), \quad c_{\xi v}(0) = 0, \\ \frac{\partial}{\partial t} c_{\xi\xi} &= -2(4b\bar{x} + a + r)c_{\xi\xi} + (a + r)\bar{x} + 2b\bar{x}^2, \quad c_{\xi\xi}(0) = x_0. \end{cases} \tag{21}$$

Analogously to what is shown at the end of Section 3.2, the variance of lethal lesion obeys:

$$c_{vv}(t) = \bar{y}(t) - \delta(t),$$

with $\bar{y}(t)$ the average deterministic value and:

$$\delta(t) = -\int_0^t 2(2b\bar{x}(s) + a)c_{\xi v}(s)ds,$$

so that, again, the variance for the lethal lesion is given by the macroscopic mean corrected by a covariance term.

Remark 3. *The solution to the linear FPE* (14) *can be shown [29] to be the probability density function of the time-dependent Ornstein–Uhlenbeck (OU) process, defined as:*

$$dZ(t) = -A(t)Z(t)dt + Q(t)dW(t), \quad Z(0) = z_0, \tag{22}$$

with W a bidimensional standard Brownian motion, $Z = (v, \xi)$, and z_0 a bivariate Gaussian random variable with mean (x_0, y_0) and variance:

$$\Sigma = \begin{pmatrix} x_0 & 0 \\ 0 & y_0 \end{pmatrix}.$$

Additionally:

$$A(t) = \begin{pmatrix} 0 & -2b\bar{x}(t) - a \\ 0 & 4b\bar{x}(t) + a + r \end{pmatrix},$$
$$Q(t) = \begin{pmatrix} \sqrt{a\bar{x}(t) + b\bar{x}^2(t) + \frac{(a\bar{x}(t)+b\bar{x}^2(t))^2}{(a+r)\bar{x}(t)+4b\bar{x}^2(t)}} & -\frac{a\bar{x}(t)+b\bar{x}^2(t)}{\sqrt{(a+r)\bar{x}(t)+4b\bar{x}^2(t)}} \\ 0 & \sqrt{(a+r)\bar{x}(t) + 4b\bar{x}^2(t)}. \end{pmatrix}$$

By simulating various trajectories of the OU process described in Equation (22)*, we can effectively estimate the solution to the FPE given by Equation* (14)*. Furthermore, it is important to note that the boundary $x = 0$ acts as an absorbing boundary in the original GSM^2. This implies that once the number of sublethal lesions reaches zero, it remains at zero. Therefore, to guarantee the positivity of the solution to the FPE defined in Equation* (14)*, it is necessary to impose a similar boundary condition on the OU process in Equation* (22)*. This ensures that the number of lethal lesions remains positive and is absorbed at zero upon reaching the boundary.*

Remark 4. *It has been shown in [23,30] that a different and yet related Fokker–Planck equation can be obtained without any truncation at first order. In fact, if the above assumption on b holds true, then the master Equation* (2)*, following [23] (Chapter 7.5), can be expanded as:*

$$\frac{\partial}{\partial t} p(t, y, x) = -\sum_{n} \sum_{(i,j)} \frac{((i,j) \cdot \nabla)^n}{n!} \left[C^{(i,j)} p(t, y, x) \right],$$

for a suitable term $C^{(i,j)}$.

Truncating at the second order, we obtain the following Fokker–Planck equation:

$$\begin{aligned} \frac{\partial}{\partial t} p(t, y, x) &= -\sum_{w=x,y} \frac{\partial}{\partial_w} \left[\sum_{i,j} (i,j) \cdot C^{(i,j)} p(t, y, x) \right] + \frac{1}{2} \sum_{w,q=x,y} \frac{\partial}{\partial_w} \frac{\partial}{\partial_q} \left[ijC^{(i,j)} p(t, y, x) \right] = \\ &= -\sum_{w=x,y} \frac{\partial}{\partial_w} [B(x) p(t, y, x)] + \frac{1}{2} \sum_{q=x,y} \sum_{w=x,y} \frac{\partial}{\partial_w} \frac{\partial}{\partial_q} [Q(x) p(t, y, x)], \end{aligned} \tag{23}$$

where the above coefficients in Equation (23) *are given explicitly by:*

$$B(x) = \begin{pmatrix} x(x-1)b + xa \\ -2x(x-1)b - x(a+r) \end{pmatrix},$$
$$Q(x) = \begin{pmatrix} x(x-1)b + xa & -2x(x-1)b - xa \\ -2x(x-1)b - xa & 4x(x-1)b + x(a+r) \end{pmatrix}.$$

The connection between Equations (14) *and* (23) *can be made rigorous by introducing the new variables:*

$$\bar{x} := \frac{x}{K}, \quad \bar{y} := \frac{y}{K},$$

into Equation (23)*, to obtain:*

$$\frac{\partial}{\partial t} p(t, \bar{y}, \bar{x}) = -\sum_{w=\bar{y},\bar{x}} \frac{\partial}{\partial_w} \left[\tilde{B}(\bar{x}) p(t, \bar{y}, \bar{x}) \right] + \frac{1}{2K} \sum_{w,q=\bar{y},\bar{x}} \frac{\partial}{\partial_w} \frac{\partial}{\partial_q} \left[\tilde{Q}(\bar{x}) p(t, \bar{y}, \bar{x}) \right]. \tag{24}$$

Performing a small-noise expansion [23,31,32], we can expand Equation (24) *around $\epsilon := \frac{1}{K}$, so that, using the new variables:*

$$\xi = \sqrt{K}(\bar{x} - \bar{x}(t)), \quad v = \sqrt{K}(\bar{y} - \bar{y}(t)),$$

we can see that the small noise expansion to the lowest order does coincide with the linear FP Equation (14).

The two expansions have different advantages and disadvantages. In fact, on the one side, considering the full expansion as described above, the nonlinearity of the system is preserved. However, on the other side, given the nonlinear diffusion term, the simulation of Equation (23) *is more complicated. Further, since Equation* (14) *is linear, its solution can be computed analytically, showing that the process follows a Gaussian distribution.*

4. Results

The present Section reports the implementation of the results derived in Section 3. Figure 1 shows a comparison between the distribution of sublethal lesions (top row) and lethal lesions (bottom row) for the solution to the MME (2) (histogram) and the solution to the FPE (14) (yellow line). Different columns refer to different times: the left column reports time 0.5 [a.u.], the central column reports time 0.7 [a.u], and the right column reports time 0.9 [a.u.]. The solution of the FPE has been centered around the average values $\bar{x}$ and $\bar{y}$, whereas the MME (2) is solved via the *stochastic simulation algorithms* (SSA) [33] (Chapter 13). A deterministic initial value of $x_0 = 100$ and $y_0 = 0$ has been prescribed. Further, GSM^2 parameters have been chosen to be $r = 4$, $a = 0.1$, $\tilde{b}_K = 0.01$; these parameters are in agreement with the parameters typically used [20]. A good agreement can be seen between the system size approximation and the original solution to the MME, particularly for lethal lesions. The approximation, as expected, shows a small discrepancy in the case of sublethal lesions at higher times, since the solution is closer to 0.

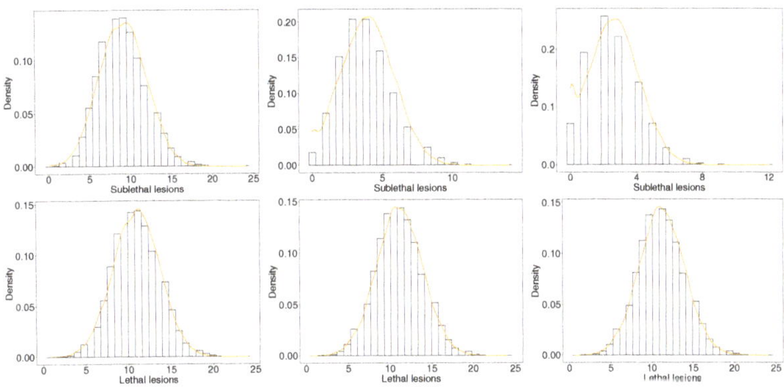

Figure 1. Sublethal lesion density evolution (**top row**) and lethal lesion density evolution (**bottom row**) as predicted by GSM^2 (histogram) and the linear noise approximation (yellow line). The left column reports time equal to 0.5 [a.u.], the central column reports time equal to 0.7 [a.u], and the right column reports time equal to 0.9 [a.u.].

Figure 2 shows the time evolution for the moments Equations (11)–(16): in yellow is the solution to the average number of sublethal lesions $\bar{x}$, whereas in blue is the average number of lethal lesions $\bar{y}$. In black is depicted the covariance between lethal and sublethal lesions $c_{\xi v}$, in purple the variance of lethal lesions $c_{\xi\xi}$, and in red the variance of lethal lesions c_{vv}. Both the average and variance of lethal lesions converge to 0 for a long time. By contrast, the average and the variance of lethal lesions converge toward a strictly positive value, with the latter being strictly lower than the former. Additionally, the covariance is strictly negative and converges to 0 at long times.

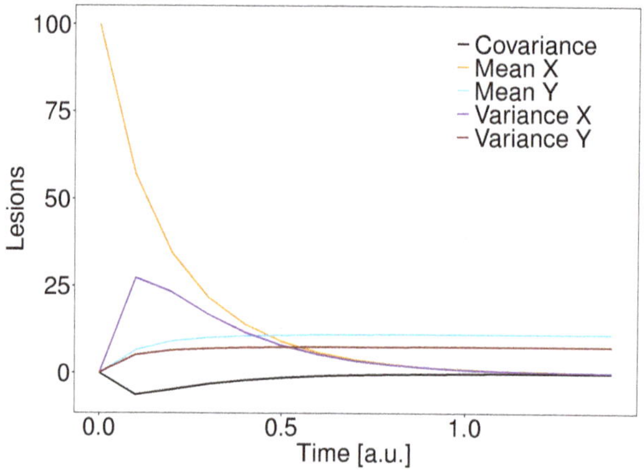

Figure 2. Time evolution for the average number and variance of sublethal lesions (yellow and purple), the average number and variance of lethal lesions (blue and red), and the covariance of lethal and sublethal lesions (black).

Figure 3 shows a comparison of 10 path solutions of the average values (black), original GSM^2 formulation (yellow), and linear noise approximation (blue) for sublethal lesions (left panel) and lethal lesions (right panel). It can be seen how the approximation and the original GSM^2 formulation produce similar patterns, with the average values being in the middle of the stochastic solutions.

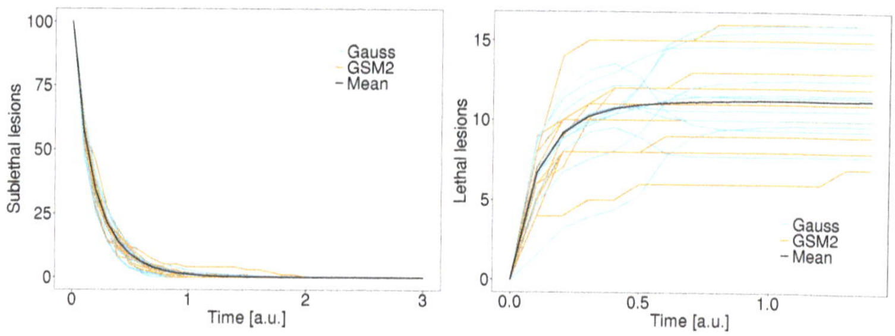

Figure 3. Comparison between path solutions to the GSM^2 (yellow), the linear noise approximation (blue), and the average value (black) of sublethal lesions (**left panel**) and lethal lesions (**right panel**).

5. Discussion

In the present paper, we presented a linear noise approximation of a stochastic model for radiation-induced DNA damage repair and kinetics [10]. Such approximation is carried out by expanding around the system size so that that it holds true for a high number of particles, which can be approximated as a continuum. The fluctuations, for the number of particles sufficiently far from the origin, are predicted to be Gaussian-distributed. The importance of the result is twofold: (i) it allows for the fast computation and simulation of GSM^2 as certain, and (ii) it theoretically shows that the number of lethal lesions deviates from a Poisson distribution, as typically assumed in the vast majority of radiobiological models.

The results show a good agreement between the solution to the MME (2) and the linear noise approximation (14). This is particularly true when the number of lesions is far from the origin. In fact, in this situation, the description of GSM^2 as a continuum of lesions is not valid and discrepancies between the two representations emerge. This is, however, mitigated by equipping the linear FPE with a suitable boundary condition, preserving the positivity of the solution. Further, the main interest in the long-time distribution lies in the distribution of lethal lesions, which could allow characterizing several relevant biological endpoints such as cell survival and cell killing. This implies that such approximation can be effectively used in several concrete applications even if it is typically difficult to estimate the experimental range in which the approximation proposed is valid.

The numerical solutions to the moments Equations (11)–(16) confirm the theoretical analysis performed in Section 3. In particular, the variance of the lethal lesions is strictly lower than the average; this, together with the Gaussianity of the distribution, implies a divergence from the Poissonianity of the number of lethal lesions. This is one of the first theoretically grounded results showing that a model can predict lethal lesions with non-Poisson distribution. Additionally, the covariance is strictly negative, since an increase in the number of lethal lesions can only be caused by a decrease in the number of sublethal lesions.

Finally, it is worth remarking that this paper furthers the investigation and comparison of diverse existing radiobiology models, revealing the underlying commonalities and shared perspectives among these approaches. The present paper shows the connections of GSM^2 to two other models proposed in the literature. Firstly, the main equations of the MKM arise formally within the context of GSM^2, with, however, an extremely relevant difference in the fluctuations around the average values. This has been already noted in previous research [10,19]. Secondly, the incorporation of a Gaussian distribution, previously employed in radiobiology studies [24,25], emerges as a deviation from a Poisson distribution. As a result, the proposed model establishes a remarkably insightful link between two seemingly different radiobiological models: the MKM and the Gaussian formulation of a multi-hit model. Recognizing the significance of this subject, future research will be dedicated to further exploring the interconnections among diverse radiobiological models.

6. Conclusions

The present research continues the investigation of how the stochastic nature of energy deposition affects DNA damage evolution and how this is, in turn, related to the overall probability distribution of the number of lethal and sublethal lesions. In [10], a *master equation* for the probability distribution of DNA damage has been derived. However, due to the non-linear terms, besides some cases such as the computation of the survival probability [18], its analytical solution is unfeasible. In the present work, we have shown how a proper expansion can be applied to the MME derived in [10]. Such expansion highlights, on one side, how the GSM^2 is connected to the MKM and, on the other side, how non-Poissonian effects naturally emerge with no need for *ad hoc* corrections.

Funding: This work was partially supported by INFN CSN5 projects MICROBE-IT and FRIDA.

Data Availability Statement: No new data have been created.

Acknowledgments: I would like to thank M. Missiaggia for several fruitful discussions that helped significantly improve the results presented in this research.

Conflicts of Interest: The authors declare no conflict of interest

References

1. Thariat, J.; Hannoun-Levi, J.M.; Sun Myint, A.; Vuong, T.; Gérard, J.P. Past, present, and future of radiotherapy for the benefit of patients. *Nat. Rev. Clin. Oncol.* **2013**, *10*, 52–60. [CrossRef] [PubMed]
2. Durante, M.; Paganetti, H. Nuclear physics in particle therapy: A review. *Rep. Prog. Phys.* **2016**, *79*, 096702. [CrossRef]
3. Durante, M.; Loeffler, J.S. Charged particles in radiation oncology. *Nat. Rev. Clin. Oncol.* **2010**, *7*, 37–43. [CrossRef] [PubMed]
4. Bellinzona, V.; Cordoni, F.; Missiaggia, M.; Tommasino, F.; Scifoni, E.; La Tessa, C.; Attili, A. Linking Microdosimetric Measurements to Biological Effectiveness in Ion Beam Therapy: A review of theoretical aspects of MKM and other models. *Front. Phys.* **2021**, *8*, 578492. [CrossRef]
5. Hawkins, R.B. A statistical theory of cell killing by radiation of varying linear energy transfer. *Radiat. Res.* **1994**, *140*, 366–374. [CrossRef]
6. Hawkins, R.B.; Inaniwa, T. A microdosimetric-kinetic model for cell killing by protracted continuous irradiation including dependence on LET I: Repair in cultured mammalian cells. *Radiat. Res.* **2013**, *180*, 584–594. [CrossRef]

7. Kellerer, A.M.; Rossi, H.H. The theory of dual radiation action. In *Current Topics in Radiation Research Quarterly*; North-Holland Publishing Company: Amsterdam, The Netherlands, 1974; pp. 85–158. Available online: https://www.osti.gov/biblio/4611340 (accessed on 13 June 2023).
8. Herr, L.; Friedrich, T.; Durante, M.; Scholz, M. A comparison of kinetic photon cell survival models. *Radiat. Res.* **2015**, *184*, 494–508. [CrossRef]
9. Pfuhl, T.; Friedrich, T.; Scholz, M. Prediction of cell survival after exposure to mixed radiation fields with the local effect model. *Radiat. Res.* **2020**, *193*, 130–142. [CrossRef]
10. Cordoni, F.; Missiaggia, M.; Attili, A.; Welford, S.; Scifoni, E.; La Tessa, C. Generalized stochastic microdosimetric model: The main formulation. *Phys. Rev. E* **2021**, *103*, 012412. [CrossRef]
11. Hawkins, R.B. A microdosimetric-kinetic model for the effect of non-Poisson distribution of lethal lesions on the variation of RBE with LET. *Radiat. Res.* **2003**, *160*, 61–69. [CrossRef]
12. Scholz, M.; Kellerer, A.; Kraft-Weyrather, W.; Kraft, G. Computation of cell survival in heavy ion beams for therapy. *Radiat. Environ. Biophys.* **1997**, *36*, 59–66. [CrossRef]
13. Friedrich, T.; Scholz, U.; Elsässer, T.; Durante, M.; Scholz, M. Calculation of the biological effects of ion beams based on the microscopic spatial damage distribution pattern. *Int. J. Radiat. Biol.* **2012**, *88*, 103–107. [CrossRef]
14. Hawkins, R.B. A microdosimetric-kinetic model of cell killing by irradiation from permanently incorporated radionuclides. *Radiat. Res.* **2018**, *189*, 104–116. [CrossRef]
15. Inaniwa, T.; Suzuki, M.; Furukawa, T.; Kase, Y.; Kanematsu, N.; Shirai, T.; Hawkins, R.B. Effects of dose-delivery time structure on biological effectiveness for therapeutic carbon-ion beams evaluated with microdosimetric kinetic model. *Radiat. Res.* **2013**, *180*, 44–59. [CrossRef]
16. Sato, T.; Furusawa, Y. Cell survival fraction estimation based on the probability densities of domain and cell nucleus specific energies using improved microdosimetric kinetic models. *Radiat. Res.* **2012**, *178*, 341–356. [CrossRef]
17. Kase, Y.; Kanai, T.; Matsumoto, Y.; Furusawa, Y.; Okamoto, H.; Asaba, T.; Sakama, M.; Shinoda, H. Microdosimetric measurements and estimation of human cell survival for heavy-ion beams. *Radiat. Res.* **2006**, *166*, 629–638. [CrossRef]
18. Cordoni, F.G.; Missiaggia, M.; Scifoni, E.; La Tessa, C. Cell Survival Computation via the Generalized Stochastic Microdosimetric Model (GSM2); Part I: The Theoretical Framework. *Radiat. Res.* **2022**, *197*, 218–232. [CrossRef]
19. Cordoni, F.G.; Missiaggia, M.; La Tessa, C.; Scifoni, E. Multiple levels of stochasticity accounted for in different radiation biophysical models: From physics to biology. *Int. J. Radiat. Biol.* **2022**, *99*, 807–822. [CrossRef] [PubMed]
20. Missiaggia, M.; Cordoni, F.G.; Scifoni, E.; La Tessa, C. Cell Survival Computation via the Generalized Stochastic Microdosimetric Model (GSM2)-Part II: Numerical results. *Radiat. Res.* 2022, *submitted*.
21. Cordoni, F.G. A spatial measure-valued model for radiation-induced DNA damage kinetics and repair under protracted irradiation condition. *arXiv* **2023**, arXiv:2303.14784.
22. Van Kampen, N.G. *Stochastic Processes in Physics and Chemistry*; Elsevier: Amsterdam, The Netherlands, 1992; Volume 1.
23. Gardiner, C.W. *Handbook of Stochastic Methods*; Springer: Berlin/Heidelberg, Germany, 1985; Volume 3.
24. Zhao, L.; Mi, D.; Sun, Y. A novel multitarget model of radiation-induced cell killing based on the Gaussian distribution. *J. Theor. Biol.* **2017**, *420*, 135–143. [CrossRef] [PubMed]
25. Rossi, H.H.; Zaider, M. Saturation in dual radiation action. In *Quantitative Mathematical Models in Radiation Biology*; Springer: Berlin/Heidelberg, Germany, 1988; pp. 111–118.
26. Rossi, H.H.; Zaider, M. *Microdosimetry and its Applications*; Springer: Berlin/Heidelberg, Germany, 1996.
27. Vassiliev, O.N. Formulation of the multi-hit model with a non-Poisson distribution of hits. *Int. J. Radiat. Oncol. Biol. Phys.* **2012**, *83*, 1311–1316. [CrossRef]
28. Manganaro, L.; Russo, G.; Cirio, R.; Dalmasso, F.; Giordanengo, S.; Monaco, V.; Muraro, S.; Sacchi, R.; Vignati, A.; Attili, A. A Monte Carlo approach to the microdosimetric kinetic model to account for dose rate time structure effects in ion beam therapy with application in treatment planning simulations. *Med. Phys.* **2017**, *44*, 1577–1589. [CrossRef] [PubMed]
29. Karatzas, I.; Shreve, S.E. Brownian motion. In *Brownian Motion and Stochastic Calculus*; Springer: Berlin/Heidelberg, Germany, 1998; pp. 47–127.
30. Gillespie, D.T. The chemical Langevin equation. *J. Chem. Phys.* **2000**, *113*, 297–306. [CrossRef]
31. Albeverio, S.; Cordoni, F.; Di Persio, L.; Pellegrini, G. Asymptotic expansion for some local volatility models arising in finance. *Decis. Econ. Financ.* **2019**, *42*, 527–573. [CrossRef]
32. Cordoni, F.; Di Persio, L. Small noise expansion for the Lévy perturbed Vasicek model. *Int. J. Pure Appl. Math.* **2015**, *98*, 291–301. [CrossRef]
33. Weinan, E.; Li, T.; Vanden-Eijnden, E. *Applied Stochastic Analysis*; American Mathematical Society: Providence, RI, USA, 2019; Volume 199.

entropy

Article

DNA Code from Cyclic and Skew Cyclic Codes over $\mathbb{F}_4[v]/\langle v^3\rangle$

Om Prakash [1], Ashutosh Singh [1], Ram Krishna Verma [2], Patrick Solé [3,*] and Wei Cheng [4,5]

1 Department of Mathematics, Indian Institute of Technology Patna, Bihar 801106, India
2 Department of Mathematics, SRM Institute of Science and Technology, Delhi-NCR Campus, Ghaziabad 201204, India
3 I2M, (CNRS, Aix-Marseille University, Centrale Marseille), 13009 Marseilles, France
4 LTCI, Télécom Paris (IP Paris), 19 Place Marguerite Perey, 91120 Palaiseau, France
5 Secure-IC S.A.S., 104 Boulevard du Montparnasse, 75014 Paris, France
* Correspondence: patrick.sole@telecom-paris.fr

Abstract: The main motivation of this work is to study and obtain some reversible and DNA codes of length n with better parameters. Here, we first investigate the structure of cyclic and skew cyclic codes over the chain ring $\mathcal{R} := \mathbb{F}_4[v]/\langle v^3\rangle$. We show an association between the codons and the elements of $\mathcal{R}$ using a Gray map. Under this Gray map, we study reversible and DNA codes of length n. Finally, several new DNA codes are obtained that have improved parameters than previously known codes. We also determine the Hamming and the Edit distances of these codes.

Keywords: reversible code; gray map; DNA codes

check for updates

Citation: Prakash, O.; Singh, A.; Verma, R.K.; Solé, P.; Cheng, W. DNA Code from Cyclic and Skew Cyclic Codes over $\mathbb{F}_4[v]/\langle v^3\rangle$. *Entropy* **2023**, 25, 239. https://doi.org/10.3390/e25020239

Academic Editor: Pavel Kraikivski

Received: 31 December 2022
Revised: 16 January 2023
Accepted: 19 January 2023
Published: 28 January 2023

1. Introduction

DNA is a nucleic acid used for carrying genetic information in living organisms. It is a double-strand molecule formed from two possible nitrogenous bases—Purines (Adenine and Guanine) and Pyrimidines (Cytosine—and Thymine) and two chemically polar ends, namely, $5'$ and $3'$. The Watson–Crick complementary (WCC) relation, which is characterized as $A^c = T$, $G^c = C$, and vice versa, is used to bind the bases of DNA. In 1994, Adleman [1] discussed the Hamiltonian path problem using DNA molecules. This (NP-complete) problem is solved by encoding a small graph in DNA molecules where all the operations were carried out using standard protocols such as the WCC relation. Due to massive parallelism, DNA computing emerged as a powerful tool among researchers to solve computationally difficult problems. Further, the experiments are performed on synthesized DNA and RNA molecules to control their combinatorial constraints such as constant GC-content and Hamming distance.

Linear codes over finite fields have been explored for almost three decades, but this research area experienced an astonishing rate after the remarkable work of Hammons et al. [2] when they established a relation between linear codes over $\mathbb{Z}_4$ with other non-linear binary codes. Afterward, many authors [3–6] considered alphabets endowed with a ring structure and found many good linear codes over finite fields via specific Gray maps. Within the class of linear codes, cyclic codes are the pivotal and the most studied codes due to their theoretical richness and practical implementation. Recently, many authors [7–13] constructed DNA codes using cyclic codes over rings. For instance, Bayram et al. [7] and Yildiz and Siap [13] explored DNA codes over the rings $\mathbb{F}_4 + v\mathbb{F}_4, v^2 = v$ and $\mathbb{F}_2[v]/\langle v^4 - 1\rangle$, respectively. In 2019, Mostafanasab and Darani [12] discussed the structure of cyclic DNA codes over the chain ring $\mathbb{F}_2 + u\mathbb{F}_2 + u^2\mathbb{F}_2$. Liu et al. [14] worked on cyclic DNA codes of an odd length over $\mathbb{F}_4[u]/\langle u^3\rangle$. On the other hand, Boucher et al. [15] introduced skew cyclic codes and discovered many new linear codes. Further, in [16,17], more properties of these codes over chain rings have been established. Recently, Gursoy et al. [18] studied reversible DNA codes by using skew cyclic codes. Later on, Cengellenmis et al. [19] studied DNA codes from skew cyclic codes over the rings $F_2[u, v, w]$, where $u^2 = v^2 + v = w^2 + w =$

$uv + vu = uw + wu = vw + wv = 0$. Motivated by the above works, we consider cyclic as well as skew cyclic codes over the finite chain ring $\mathcal{R} = \mathbb{F}_4[v]/\langle v^3 \rangle$ to construct DNA codes of arbitrary lengths. Hamming and edit distances are also calculated for the obtained codes. Interestingly, we obtain several new codes with better parameters than known codes [14].

The article is structured as follows: The Gray map, together with the correspondence of the codons and the other basic results of cyclic codes, are in Section 2. Reversible cyclic codes over the ring $\mathcal{R}$ are covered in Section 3, whereas the reversible skew cyclic codes are studied in Section 4. Some results related to the complement and reverse complement of obtained codes are presented in Section 5. Based on our established results from the previous Sections and magma computer algebra system [20], we provide a few examples of DNA codes of arbitrary lengths in Section 6. In the end, we conclude our work in Section 7.

2. Preliminaries

Let $\mathbb{F}_4 = \{0, 1, \mathfrak{t}, \mathfrak{t}^2\}$, where $\mathfrak{t}^2 = \mathfrak{t} + 1$ be a finite field. Then $\mathcal{R} := \mathbb{F}_4[v]/\langle v^3 \rangle$ is a finite chain ring with characteristic 2 and every element r of $\mathcal{R}$ can be represented as $r = \mathfrak{b}_1 + \mathfrak{b}_2 v + \mathfrak{b}_3 v^2$ where $\mathfrak{b}_i \in \mathbb{F}_4$, for $i = 0, 1, 2$ and $v^3 = 0$. It is easy to show that $\mathcal{R}$ is a principal ideal ring with unique maximal ideal $\langle vs. \rangle$ and $\mathcal{R}/\langle vs. \rangle$ is isomorphic to $\mathbb{F}_4$. Recall that the ring $\mathcal{R}$ has 48 invertible elements of the form $r = \mathfrak{b}_1 + \mathfrak{b}_2 v + \mathfrak{b}_3 v^2$, where $\mathfrak{b}_1$ is invertible in $\mathbb{F}_4$.

A linear code $\mathcal{C}$ of length n and alphabets from $\mathcal{R}$ is a submodule of an $\mathcal{R}$-module $\mathcal{R}^n$. The elements of $\mathcal{C}$ are called the codewords. The Hamming weight of an element $\mathfrak{b} = (\mathfrak{b}_0, \mathfrak{b}_1, \ldots, \mathfrak{b}_n) \in \mathcal{C}$ is defined as $w_H(\mathfrak{b}) = |\{i \mid \mathfrak{b}_i \neq 0\}|$ and Hamming distance $d_H(\mathfrak{b}, \mathfrak{k})$ between any two elements $\mathfrak{b} = (\mathfrak{b}_0, \mathfrak{b}_1, \ldots, \mathfrak{b}_n)$ and $\mathfrak{k} = (\mathfrak{k}_0, \mathfrak{k}_1, \ldots, \mathfrak{k}_n)$ in $\mathcal{C}$ is defined as $d_H(\mathfrak{b}, \mathfrak{k}) = w_H(\mathfrak{b} - \mathfrak{k})$. Additionally, the lowest value in the set $\{d_H(\mathfrak{b}, \mathfrak{k}) \mid \mathfrak{b} \neq \mathfrak{k}, \forall\, \mathfrak{b}, \mathfrak{k} \in \mathcal{C}\}$ is considered as the the Hamming distance $d_H(\mathcal{C})$ of the code $\mathcal{C}$.

Now, we describe a Gray map $\Phi : \mathcal{R} \longrightarrow \mathbb{F}_4^3$ as:

$$\Phi(\mathfrak{b}_0 + \mathfrak{b}_1 v + \mathfrak{b}_2 v^2) = (\mathfrak{b}_0 + \mathfrak{b}_1 + \mathfrak{b}_2, \mathfrak{b}_1 + \mathfrak{b}_2, \mathfrak{b}_2), \tag{1}$$

where $\mathfrak{b}_i \in \mathbb{F}_4$ for $i = 0, 1, 2$. It is easy to see that the function Φ is a distance-preserving map and is extendable to $\mathcal{R}^n$ component-wise. In Table 1, we establish the connection between the ring elements and the codons by using the Gray map (1).

Definition 1. *For a given polynomial $\mathfrak{g}(z) = \mathfrak{g}_0 + \mathfrak{g}_1 z + \ldots + \mathfrak{g}_m z^m \in \mathbb{F}_4[z]$, the reciprocal polynomial is denoted by $\mathfrak{g}^*(z)$ and defined as $\mathfrak{g}^*(z) = \sum_{i=0}^{m} \mathfrak{g}_{m-i} z^i$. A polynomial $\mathfrak{g}(z)$ is said to be self-reciprocal if and only if $\mathfrak{g}^*(z) = b\mathfrak{g}(z)$ for some non-zero element b in $\mathbb{F}_4$.*

Now, we present some useful lemmas that appeared in [8,14].

Lemma 1. *Let $\mathfrak{g}(z)$ and $\mathfrak{h}(z)$ be polynomials over $\mathcal{R}$ of degrees r and s, respectively, with $r \geq s$. Then:*

1. $[\mathfrak{g}(z)\mathfrak{h}(z)]^* = \mathfrak{g}^*(z)\mathfrak{h}^*(z)$
2. $[\mathfrak{g}(z) + \mathfrak{h}(z)]^* = \mathfrak{g}^*(z) + z^{(r-s)}\mathfrak{h}^*(z)$.

Lemma 2. *Let $\mathfrak{f}(z)$, $\mathfrak{g}(z)$, and $\mathfrak{h}(z)$ be polynomials over $\mathcal{R}$ of degrees r, s, and t, respectively, where $r \geq s, t$. Then:*

1. $[\mathfrak{f}(z)\mathfrak{g}(z)\mathfrak{h}(z)]^* = \mathfrak{f}^*(z)\mathfrak{g}^*(z)\mathfrak{h}^*(z)$
2. $[\mathfrak{f}(z) + \mathfrak{g}(z) + \mathfrak{h}(z)]^* = \mathfrak{f}^*(z) + z^{(r-s)}\mathfrak{g}^*(z) + z^{(r-t)}\mathfrak{h}^*(z)$.

Using the Watson–Crick complementary relation, we define the reverse (**R**) and the reverse complement (**RC**) of a DNA codeword $\mathfrak{b} = (\mathfrak{b}_0, \mathfrak{b}_1, \ldots, \mathfrak{b}_{n-1})$ by $\mathfrak{b}^r = (\mathfrak{b}_{n-1}, \ldots, \mathfrak{b}_1, \mathfrak{b}_0)$ and $\mathfrak{b}^{rc} = (\mathfrak{b}^c_{n-1}, \ldots, \mathfrak{b}^c_1, \mathfrak{b}^c_0)$, respectively. For example, given $\mathfrak{b} = ATCCGT$, we obtain $\mathfrak{b}^r = TGCCTA$ and $\mathfrak{b}^{rc} = ACGGAT$.

We have the following observations based on the Gray map provided in Equation (1).

Table 1. Codons correspondence with the elements of $\mathcal{R}$.

0	AAA	v^2	TTT	tv^2	GGG	t^2v^2	CCC
1	TAA	v^2+1	ATT	tv^2+1	CGG	t^2v^2+1	GCC
t	GAA	v^2+t	CTT	tv^2+t	AGG	t^2v^2+t	TCC
t^2	CAA	v^2+t^2	GTT	tv^2+t^2	TGG	$t^2v^2+t^2$	ACC
v	TTA	v^2+v	AAT	tv^2+v	CCG	t^2v^2+v	GGC
$v+1$	ATA	v^2+v+1	TAT	tv^2+v+1	GCG	t^2v^2+v+t	AGC
$v+t$	CTA	v^2+v+t	GAT	tv^2+v+t	TCG	t^2v^2+v+1	CGC
$v+t^2$	GTA	v^2+v+t^2	CAT	tv^2+v+t^2	ACG	$t^2v^2+v+t^2$	TGC
tv	GGA	v^2+tv	CCT	tv^2+tv	AAG	t^2v^2+tv	TTC
$tv+1$	CGA	v^2+tv+1	GCT	tv^2+tv+1	TAG	t^2v^2+tv+1	ATC
$tv+t$	AGA	v^2+tv+t	TCT	tv^2+tv+t	GAG	t^2v^2+tv+t	CTC
$tv+t^2$	TGA	v^2+tv+t^2	ACT	tv^2+tv+t^2	CAG	$t^2v^2+tv+t^2$	GTC
t^2v	CCA	v^2+t^2v	GGT	tv^2+t^2v	TTG	$t^2v^2+t^2v$	AAC
t^2v+1	GCA	v^2+t^2v+1	CGT	tv^2+t^2v+1	ATG	$t^2v^2+t^2v+1$	TAC
t^2v+t	TCA	v^2+t^2v+t	AGT	tv^2+t^2v+t	CTG	$t^2v^2+t^2v+t$	GAC
t^2v+t^2	ACA	$v^2+t^2v+t^2$	TGT	$tv^2+t^2v+t^2$	GTG	$t^2v^2+t^2v+t^2$	CAC

Lemma 3. 1. *For any* $a=(\mathfrak{b}_0+\mathfrak{b}_1v+\mathfrak{b}_2v^2)\in\mathcal{R}$, *we have* $\Phi(\mathfrak{b}_0+\mathfrak{b}_1v+\mathfrak{b}_2v^2)^r=\mathfrak{b}_1+\mathfrak{b}_0v+(\mathfrak{b}_0+\mathfrak{b}_1+\mathfrak{b}_2)v^2$, *where* $\mathfrak{b}_0,\mathfrak{b}_1,\mathfrak{b}_2\in\mathbb{F}_4$.
2. $\Phi(\mathfrak{b}_0+\mathfrak{b}_1)^r=\Phi(\mathfrak{b}_0)^r+\Phi(\mathfrak{b}_1)^r$, *where* $\mathfrak{b}_0,\mathfrak{b}_1\in\mathbb{F}_4$.

3. Reversible Cyclic Codes over $\mathcal{R}$

In the present section, we investigate the structure of cyclic codes and prove reversible conditions on these codes. The cyclic codes of odd lengths are provided in [14] and a detailed discussion on cyclic codes of arbitrary length with alphabets from $\mathbb{Z}_2[u]/\langle v^3\rangle$ is explored in [6]. Now, in the subsequent theorems, we describe the structure of the cyclic code. We omit the proof due to its similarity to the proof provided in [6].

Theorem 1. *Let C be a cyclic code of length n over $\mathcal{R}$. Then the code C is provided by:*

$$C=\langle\mathfrak{g}_0(z)+v\mathfrak{g}_1(z)+v^2\mathfrak{g}_2(z),va_1(z)+v^2p(z),v^2a_2(z)\rangle$$

where $a_2(z)|a_1(z)|\mathfrak{g}_0(z)|(z^n-1)$ *over* $\mathbb{F}_4$, $a_1(z)|\mathfrak{g}_1(z)(\frac{z^n-1}{\mathfrak{g}_0(z)})$, $a_2(z)|p(z)(\frac{z^n-1}{a_1(z)})$, *and* $a_2(z)|\mathfrak{g}_2(z)$ $(\frac{z^n-1}{\mathfrak{g}_0(z)})(\frac{z^n-1}{a_1(z)})$ *over* $\mathbb{F}_4$. *Moreover,* $deg(\mathfrak{g}_2(z))<deg(a_2(z))$, $deg(p(z))<deg(a_2(z))$, *and* $deg(\mathfrak{g}_1(z))<deg(a_1(z))$.

Corollary 1. *If the length of a cyclic code C is odd and* $\mathfrak{g}_1(z)=\mathfrak{g}_2(z)=p(z)=0$, *then* $C=\langle\mathfrak{g}_0(z),va_1(z),v^2a_2(z)\rangle=\langle\mathfrak{g}_0(z)+va_1(z)+v^2a_2(z)\rangle$.

A similar result is also possible when n is not odd. In this case, we assume that $\gcd(\frac{z^n-1}{a_2(z)},\mathfrak{g}_0(z))=1$ and consequently obtain the following result.

Corollary 2. *If a cyclic code C is of even length n and* $\gcd(\frac{z^n-1}{a_2(z)},\mathfrak{g}_0(z))=1$, *then* $\mathfrak{g}_1(z)=\mathfrak{g}_2(z)=p(z)=0$.

When $a_2(z) = \mathfrak{g}_0(z)$, then $a_2(z) = a_1(z) = \mathfrak{g}_0(z)$ and $\mathcal{C}$ as a subset of $\langle \mathfrak{g}_0(z) + v\mathfrak{g}_1(z) + v^2\mathfrak{g}_2(z)\rangle$. Since the other containment is true by the definition of $\mathcal{C}$, we, therefore, obtain the following corollary.

Corollary 3. *For a cyclic code* $\mathcal{C} = \langle \mathfrak{g}_0(z) + v\mathfrak{g}_1(z) + v^2\mathfrak{g}_2(z), va_1 + v^2p(z), v^2a_2(z)\rangle$, *if* $a_2(z) = \mathfrak{g}_0(z)$, *then* $\mathcal{C} = \langle \mathfrak{g}_0(z) + v\mathfrak{g}_1(z) + v^2\mathfrak{g}_2(z)\rangle$.

Definition 2. *Given a code* $\mathcal{C} = \langle \mathfrak{g}_0(z) + v\mathfrak{g}_1(z) + v^2\mathfrak{g}_2(z), va_1(z) + v^2p(z), v^2a_2(z)\rangle$ *over* $\mathcal{R}$, *we define* $\mathcal{C}_{v^2}$ *by* $\{q(z) \in \mathbb{F}_4[z] \mid v^2q(z) \in \mathcal{C}\}$. *Particularly, since* $a_2(z)|a_1(z)|\mathfrak{g}_0(z)$, $\mathcal{C}_{v^2} = \langle a_2(z)\rangle$.

In the next result, we determine the Hamming distance of the code $\mathcal{C}$ by using the above definition in terms of the Hamming distance of $\mathcal{C}_{v^2}$.

Theorem 2. *Let* $\mathcal{C}$ *be a code provided by* $\mathcal{C} = \langle \mathfrak{g}(z) + v\mathfrak{g}_1(z) + v^2\mathfrak{g}_2(z), va_1(z) + v^2p(z), v^2a_2(z)\rangle$. *Then Hamming distance of* $\mathcal{C}$ *and* $\mathcal{C}_{v^2}$ *are equal, i.e.,* $d_H(\mathcal{C}) = d_H(\mathcal{C}_{v^2})$.

Proof. It can be obtained from [4]. □

Remark 1. *For the sake of brevity, we use b for polynomial* $b(z)$ *whenever* $b(z)$ *belongs to the field* $\mathbb{F}_4$.

Lemma 4. *Let* $\mathfrak{g}_0(z), \mathfrak{g}_1(z)$ *and* $\mathfrak{g}_2(z) \in \mathbb{F}_4[z]$ *of degrees r, s and t, respectively. Then* $(\mathfrak{g}_0(z) + v\mathfrak{g}_1(z) + v^2\mathfrak{g}_2(z))^* = \mathfrak{g}_0^*(z) + vz^{r-s}\mathfrak{g}_2^*(z) + v^2z^{r-t}\mathfrak{g}_2^*(z)$.

Theorem 3. *Let* $\mathcal{C} = \langle \mathfrak{g}_0(z) + v\mathfrak{g}_1(z) + v^2\mathfrak{g}_2(z)\rangle$ *be a cyclic code of even length over* $\mathcal{R}$ *with monic polynomials* $\mathfrak{g}_0(z)$, $\mathfrak{g}_1(z)$ *and* $\mathfrak{g}_2(z)$ *of degrees r, s and t, respectively. Then the code* $\mathcal{C}$ *is reversible if and only if:*

(1) $\mathfrak{g}_0(z)$ *is a self-reciprocal polynomial;*
(2) $z^{r-s}\mathfrak{g}_1^*(z) = b_0\mathfrak{g}_1(z) + b_1\mathfrak{g}_0(z)$ *and* $z^{r-s}\mathfrak{g}_2^*(z) = b_0\mathfrak{g}_2(z) + b_1\mathfrak{g}_1(z) + b_2\mathfrak{g}_0(z)$, *where* $b_0 \in \mathbb{F}_4 \setminus \{0\}$ *and* $b_1, b_2 \in \mathbb{F}_4$.

Proof. Let $\mathcal{C}$ be a reversible cyclic code. Then

$$
\begin{aligned}
(\mathfrak{g}_0(z) + v\mathfrak{g}_1(z) + v^2\mathfrak{g}_2(z))^* &= \mathfrak{g}_0^*(z) + vz^{r-s}\mathfrak{g}_2^*(z) + v^2z^{r-t}\mathfrak{g}_2^*(z) \text{ and} \\
(\mathfrak{g}_0(z) + v\mathfrak{g}_1(z) + v^2\mathfrak{g}_2(z))^* &= b(z)(\mathfrak{g}_0(z) + v\mathfrak{g}_1(z) + v^2\mathfrak{g}_2(z)) \in \mathcal{C} \\
&= (b_0(z) + vb_1(z) + v^2b_2(z))(\mathfrak{g}_0(z) + v\mathfrak{g}_1(z) + v^2\mathfrak{g}_2(z)) \\
&= b_0(z)\mathfrak{g}_0(z) + v(b_0(z)\mathfrak{g}_1(z) + b_1(z)\mathfrak{g}_0(z)) \\
&\quad + v^2(b_0(z)\mathfrak{g}_2(z) + b_1(z)\mathfrak{g}_1(z) + b_2(z)\mathfrak{g}_0(z)).
\end{aligned}
$$

Comparing right side of the two equations, we obtain $\mathfrak{g}_0^*(z) = b_0(z)\mathfrak{g}_0(z)$, $z^{r-s}\mathfrak{g}_1^*(z) = b_0(z)\mathfrak{g}_1(z) + b_1(z)\mathfrak{g}_0(z)$ and $z^{r-t}\mathfrak{g}_2^*(z) = b_0(z)\mathfrak{g}_2(z) + b_1(z)\mathfrak{g}_1(z) + b_2(z)\mathfrak{g}_0(z)$. Now, using $\deg \mathfrak{f}^*(z) \leq \deg \mathfrak{f}(z)$, we obtain $b_0(z) \neq 0$ in $\mathbb{F}_4$ and this implies that the polynomial $\mathfrak{g}_0(z)$ is self-reciprocal. Therefore, $z^{r-s}\mathfrak{g}_1^*(z) = b_0\mathfrak{g}_1(z) + b_1(z)\mathfrak{g}_0(z)$ where $b_0 = b_0(z)$ is a non-zero element in $\mathbb{F}_4$. Now comparing the degrees of both sides, we obtain a constant polynomial $b_1(z) \in \mathbb{F}_4$, say, b_1. We have $z^{r-t}\mathfrak{g}_2^*(z) = b_0\mathfrak{g}_2(z) + b_1\mathfrak{g}_1(z) + b_2(z)\mathfrak{g}_0(z)$. Again, comparing the degrees of both sides, we obtain $b_2(z)$ in $\mathbb{F}_4$, say b_2. Thus, $z^{r-s}\mathfrak{g}_1^*(z) = b_0\mathfrak{g}_1(z) + b_1\mathfrak{g}_0(z)$ and $z^{r-t}\mathfrak{g}_2^*(z) = b_0\mathfrak{g}_2 + b_1\mathfrak{g}_1(z) + b_2\mathfrak{g}_0(z)$ where $b_0 \in \mathbb{F}_4 \setminus \{0\}$ and $b_1, b_2 \in \mathbb{F}_4$.

Conversely, assume (1) and (2) hold. Then

$$\begin{aligned}(\mathfrak{g}_0(z)+v\mathfrak{g}_1(z)+v^2\mathfrak{g}_2(z))^* =& \mathfrak{g}_0^*(z)+vz^{r-s}\mathfrak{g}_1^*(z)+v^2z^{r-t}\mathfrak{g}_2^*(z)\\ =& b_0\mathfrak{g}_0(z)+vb_0\mathfrak{g}_1(z)+vb_1\mathfrak{g}_0(z)+v^2b_0\mathfrak{g}_2(z)\\ &+v^2b_1\mathfrak{g}_1(z)+v^2b_2\mathfrak{g}_0(z)\\ =& b_0(\mathfrak{g}_0(z)+v\mathfrak{g}_1(z)+v^2\mathfrak{g}_2(z))+b_1(v\mathfrak{g}_0+v^2\mathfrak{g}_1)\\ &+b_2(v^2\mathfrak{g}_0(z))\in\mathcal{C}\end{aligned}$$

Thus, the code $\mathcal{C}$ is reversible. □

Theorem 4. *Let $\mathcal{C}=\langle \mathfrak{g}_0(z)+v\mathfrak{g}_1(z)+v^2\mathfrak{g}_2(z), v^2a_2(z)\rangle$ be a cyclic code of even length n over $\mathcal{R}$ with polynomials $\mathfrak{g}_0(z)$, $\mathfrak{g}_1(z)$, and $\mathfrak{g}_2(z)$ of degrees r,s, and t, respectively, and $r>\max\{s,t\}$. Furthermore, assume that $a_2(z)|\mathfrak{g}_0(z)|(z^n-1)$. Then the code $\mathcal{C}$ is reversible if and only if:*

(1) $\mathfrak{g}_0(z)$ and $a_2(z)$ are self-reversible;
(2) $z^{r-s}\mathfrak{g}_1^*(z)=b_0\mathfrak{g}_1(z)+b_1\mathfrak{g}_0(z)$, and $a_2(z)|(z^{r-t}\mathfrak{g}_2^*(z)+b_0\mathfrak{g}_2(z)+b_1\mathfrak{g}_1(z)$, where $b_0\in\mathbb{F}_4\setminus\{0\}$ and $b_1\in\mathbb{F}_4$.

Proof. Let $\mathcal{C}$ be a reversible code. Then

$$(\mathfrak{g}_0(z)+v\mathfrak{g}_1(z)+v^2\mathfrak{g}_2(z))^*=\mathfrak{g}_0^*(z)+vz^{r-s}\mathfrak{g}_1^*(z)+v^2z^{r-t}\mathfrak{g}_2^*(z).$$

Furthermore,

$$\begin{aligned}(\mathfrak{g}_0(z)+v\mathfrak{g}_1(z)+v^2\mathfrak{g}_2(z))^* &= b(z)(\mathfrak{g}_0(z)+v\mathfrak{g}_1(z)+v^2\mathfrak{g}_2(z))+v^2c(z)a_2(z)\\ &=(b_0(z)+vb_1(z)+v^2b_2(z))(\mathfrak{g}_0(z)+v\mathfrak{g}_1(z)+\\ &\quad v^2\mathfrak{g}_2(z))+v^2c(z)a_2(z) \text{ where } b_i(z),\ c(z)\in\mathbb{F}_4[z]\\ &=b_0(z)\mathfrak{g}_0(z)+v(b_0(z)\mathfrak{g}_1(z)+b_1(z)\mathfrak{g}_0(z))+v^2\\ &\quad(b_0(z)\mathfrak{g}_2(z)+b_1(z)\mathfrak{g}_1(z)+b_2(z)\mathfrak{g}_0(z)+c(z)a_2(z)).\end{aligned}$$

Comparing both equations, we obtain $b_0(z)\in\mathbb{F}_4\setminus\{0\}$, say b_0, this implies that $\mathfrak{g}_0(z)$ is self-reciprocal. Therefore, $z^{r-s}\mathfrak{g}_1^*(z)=b_0\mathfrak{g}_1(z)+b_1\mathfrak{g}_0(z)$ and $z^{r-t}\mathfrak{g}_2^*(z)=b_0\mathfrak{g}_2(z)+b_1\mathfrak{g}_1(z)+b_2(z)\mathfrak{g}_0(z)+c(z)a_2(z)$; this implies that $a_2(z)$ divides $z^{r-t}\mathfrak{g}_2^*(z)+b_0\mathfrak{g}_2(z)+b_1\mathfrak{g}_1(z)$. Again, $v^2a_2^*(z)\in\mathcal{C}$ and hence $a_2(z)|\mathfrak{g}_0(z)$ implies that $a_2(z)$ is self reversible.

Conversely, suppose conditions (1) and (2) hold. Then

$$\begin{aligned}(\mathfrak{g}_0(z)+v\mathfrak{g}_1(z)+v^2\mathfrak{g}_2(z))^* &= \mathfrak{g}_0^*(z)+vz^{r-s}\mathfrak{g}_1^*(z)+v^2z^{r-t}\mathfrak{g}_2^*(z)\\ &=b_0\mathfrak{g}_0(z)+v(b_0\mathfrak{g}_1(z)+b_1\mathfrak{g}_0(z))+v^2(b_0\mathfrak{g}_2(z)\\ &\quad+b_1\mathfrak{g}_1(z)+c(z)a_2(z)) \textit{ for some } c(z)\in\mathbb{F}_4[z]\\ &=b_0(\mathfrak{g}_0(z)+v\mathfrak{g}_1(z)+v^2\mathfrak{g}_2(z))+vb_1(\mathfrak{g}_0(z)\\ &\quad+v\mathfrak{g}_1(z)+v^2\mathfrak{g}_2(z))+c(z)v^2a_2(z)\in\mathcal{C}.\end{aligned}$$

Therefore, $\mathcal{C}$ is reversible. □

The following theorem states the reversible condition of odd length codes or a code satisfying Corollary 2.

Theorem 5. *Let $\mathcal{C}=\langle\mathfrak{g}_0(z), va_1(z), v^2a_2(z)\rangle$ be a cyclic code over $\mathcal{R}$ with $a_2(z)|a_1(z)|\mathfrak{g}_0(z)|(z^n-1)$. Then code $\mathcal{C}$ is reversible if and only if polynomials $\mathfrak{g}_0(z)$, $a_1(z)$ and $a_2(z)$ are self-reversible.*

Proof. Let $\mathcal{C}$ be a reversible code. Then for some polynomials $b_0(z)$, $b_1(z)$ and $b_2(z)$ in $\mathbb{F}_4[z]$, we have $(\mathfrak{g}_0(z))^*=b_0(z)\mathfrak{g}_0(z)+vb_1(z)a_1(z)+v^2b_2(z)a_2(z)$.

Comparing both sides, we obtain $b_0(z) \in \mathbb{F}_4 \setminus \{0\}$, say b_0, since $deg\mathfrak{f}^*(z) \leq deg\mathfrak{f}(z)$, then $\mathfrak{g}_0(z)$ is self-reciprocal. Similarly, $a_1(z)$ and $a_2(z)$ are self-reciprocal polynomials.

Conversely, let the polynomials $\mathfrak{g}_0(z)$, $a_1(z)$, and $a_2(z)$ be self-reciprocal. Then, elements of C are provided by the polynomial $b_0(z)\mathfrak{g}_0(z) + vb_1(z)a_1(z) + v^2b_2(z)a_2(z)$, therefore by Lemma 4, we have

$$\begin{aligned}(b_0(z)\mathfrak{g}_0(z) + vb_1(z)a_1(z) + v^2b_2(z)a_2(z))^* &= (b_0(z)\mathfrak{g}_0(z))^* + v(b_1(z)a_1(z))^*z^{r-s} \\ &+ v^2(b_2(z)a_2(z))^*z^{r-t}. \\ &= b_0^*(z)\mathfrak{g}_0^*(z) + vz^{r-s}b_1^*(z)a_1^*(z) \\ &+ v^2z^{r-t}b_2^*(z)a_2^*(z) \in C.\end{aligned}$$

Thus, C is reversible. □

Now, in the following result, we determine the rank of a code C. The proof is followed by similar arguments as in Theorem 3 of [6].

Theorem 6. *Let C be a cyclic code of length n over $\mathcal{R}$ such that*

$$C = \langle \mathfrak{g}_0(z) + v\mathfrak{g}_1(z) + v^2\mathfrak{g}_2(z), va_1(z) + vp(z), v^2a_2(z)\rangle,$$

where $\mathfrak{g}_0(z)$, $\mathfrak{g}_1(z)$, $\mathfrak{g}_2(z)$, and $a_2(z)$ are polynomials in $\mathbb{F}_4[z]$ and $deg(\mathfrak{g}_0(z) + v\mathfrak{g}_1(z) + v^2\mathfrak{g}_2(z)) = r$, $deg(a_1(z)) = s$ and $deg(a_2(z)) = t$. Then C is a free module and $rank(C) = n - t$. Moreover, the basis of C is provided by the set S, where

$$\begin{aligned}S =\{&(\mathfrak{g}_0(z) + v\mathfrak{g}_1(z) + v^2\mathfrak{g}_2(z)), x(\mathfrak{g}_0(z) + v\mathfrak{g}_1(z) + v^2\mathfrak{g}_2(z)), \ldots, z^{n-r-1}(\mathfrak{g}_0(z) + v\mathfrak{g}_1(z) \\ &+ v^2\mathfrak{g}_2(z)), (va_1(z) + v^2p(z)), x(va_1(z) + v^2p(z)), \ldots, z^{r-s-1}(va_1(z) + v^2p(z)), \\ &v^2a_2(z), v^2xa_2(z), \ldots, v^2z^{s-t-1}a_2(z))\}.\end{aligned}$$

4. Reversible Skew Cyclic Codes over $\mathcal{R}$

In this part, we focus on the structure of skew cyclic codes over $\mathcal{R}$ and establish a necessary and sufficient condition for these codes to be reversible. We first define the skew polynomial ring over $\mathcal{R}$ and provide some definitions that will be used in this section.

Let $\theta \in Aut(\mathbb{F}_4)$ be defined by $\theta(a) = a^2$. Now, consider a map $\sigma : \mathcal{R} \longrightarrow \mathcal{R}$ defined by:

$$\sigma(a_0 + a_1v + a_2v^2) = \theta(a_0) + \theta(a_1)v + \theta(a_2)v^2,$$

where $a_0, a_1, a_2 \in \mathbb{F}_4$. Since σ is an extension of θ, σ is an automorphism of $\mathcal{R}$. Let us consider the set:

$$\mathcal{R}[z;\sigma] = \{a_0 + a_1z + \ldots + a_nz^n \mid a_i \in \mathcal{R}\ \forall\, i, n \in \mathbb{N}\}.$$

Define the addition on $\mathcal{R}[z;\sigma]$ as the usual addition of polynomials and multiplication under the rule $(a_iz^i)(a_jz^j) = a_i\sigma^i(a_j)z^{i+j}$. Then, it is easy to show that $\mathcal{R}[z;\sigma]$ forms a ring under the above binary operations, known as a skew polynomial ring. Here, $(a_iz^i)(a_jz^j) \neq (a_jz^j)(a_iz^i)$ unless σ is identity automorphism.

Definition 3. *Let $\tau_\sigma : \mathcal{R}^n \longrightarrow \mathcal{R}^n$ be a skew cyclic shift operator defined by:*

$$\tau_\sigma(a_0, a_1, \ldots, a_{n-1}) = (\sigma(a_{n-1}), \sigma(a_0), \ldots, \sigma(a_{n-2})), \forall\ (a_0, a_1, \ldots, a_{n-1}) \in \mathcal{R}^n.$$

, a linear code C of length n over $\mathcal{R}$ is said to be skew cyclic code if for any codeword $c \in C$, their skew cyclic shift $\tau_\sigma(c)$ belongs to C, that is, $\tau_\sigma(C) = C$.

Definition 4. *For skew polynomials, $a(z)$ and $b(z) \neq 0$, the polynomial $b(z)$ is said to be rightly divided by $a(z)$ if and only if there exists a skew polynomial $q(z)$ such that $a(z) = q(z)b(z)$ and we denote it by $b(z)|_r a(z)$.*

Using similar arguments as in the commutative case, we provide the structure of the skew cyclic codes over $\mathcal{R}$ for automorphism σ.

Theorem 7. *Let C be a skew cyclic code in $\frac{\mathcal{R}[z;\sigma]}{\langle z^n-1\rangle}$. Then, $C = \langle \mathfrak{g}_0(z) + v\mathfrak{g}_1(z) + v^2\mathfrak{g}_2(z), va_1(z) + v^2p(z), v^2a_2(z)\rangle$ with $a_2(z)|_r a_1(z)|_r \mathfrak{g}_0(z)|_r(z^n-1)$ in $\mathbb{F}_4[z;\theta]$, $a_1(z)|_r \mathfrak{g}_1(z)(\frac{z^n-1}{\mathfrak{g}_0(z)})$ and $a_2(z)$ right divides $p(z)(\frac{z^n-1}{a_1(z)})$, and $\mathfrak{g}_2(z)(\frac{z^n-1}{\mathfrak{g}_0(z)})(\frac{z^n-1}{a_1(z)})$.*

Proof. Consider the ring $\mathcal{R}' = \frac{\mathbb{F}_4[v]}{\langle v^2\rangle}$ and $\sigma' \in Aut(\mathcal{R}')$. For a skew cyclic code C over $\mathcal{R}$, define a map $\psi_1 : \mathcal{R} \to \mathcal{R}'$ by $\psi_1(a + bv + cv^2) = a + bv$ where $a, b, c \in \mathbb{F}$. Then, ψ_1 is a ring homomorphism that can be extended to a homomorphism $\phi : C \to \frac{\mathcal{R}'[z;\sigma']}{\langle z^n-1\rangle}$ defined by

$$\phi(c_0 + c_1z + \ldots + c_{n-1}z^{n-1}) = \psi_1(c_0) + \psi_1(c_1)z + \ldots + \psi_1(c_{n-1})z^{n-1}.$$

Then $ker(\phi) = \{v^2r(z) : r(z) \in \mathbb{F}_4[z;\theta]/\langle z^n-1\rangle\}$.

In order to determine the generators of cyclic code in $\mathcal{R}_n = \mathcal{R}[z,\sigma]/\langle z^n-1\rangle$, we need to know the image of ϕ which is a skew cyclic code in $\mathcal{R}'_n = \mathcal{R}'[z,\sigma_2]/\langle z^n-1\rangle$.

Let D be a cyclic code in $\mathcal{R}'_n$. Now, define a map $\psi_2 : \mathcal{R}' \to \mathbb{F}_4$ by $\psi_2(a + ub) = a^2$. Then ψ_2 is a ring homomorphism. We extend ψ_2 to a ring homomorphism $\varphi : D \to \mathbb{F}_4[z;\theta]/\langle z^n-1\rangle$ defined by

$$\varphi(d_0 + d_1z + \ldots + d_{n-1}z^{n-1}) = \psi_2(d_0) + \psi_2(d_1)z + \ldots + \psi_2(d_{n-1})z^{n-1}.$$

Then,

$$\begin{aligned} ker(\varphi) &= \{vr'(z) : r'(z) \text{ is a skew polynomial in } \mathbb{F}_4[z;\theta]/\langle z^n-1\rangle\} \\ &= \langle va_1(z)\rangle \textit{ with } a_1(z)|_r(z^n-1). \end{aligned}$$

Since the set image(φ) is also an ideal and hence a skew cyclic code generated by $\mathfrak{g}_0(z)$, where $\mathfrak{g}_0(z)$ right divides (z^n-1). Therefore, $D = \langle \mathfrak{g}_0(z) + v\mathfrak{g}_1(z), va_1(z)\rangle$ where $a_1(z)|_r\mathfrak{g}_0(z)$ and $a_1(z)|_r(\mathfrak{g}_1(z)\frac{z^n-1}{\mathfrak{g}_0(z)})$

Similarly, the set image(ϕ) is an ideal over $\mathcal{R}'$. Therefore, skew cyclic code C over $\mathcal{R}$ is provided by $C = \langle \mathfrak{g}_0(z) + v\mathfrak{g}_1(z) + v^2\mathfrak{g}_2(z), va_1(z) + v^2p(z), v^2a_2(z)\rangle$ with $a_2(z)|_r a_1(z)|_r \mathfrak{g}_0(z)|_r(z^n-1)$ and $a_1(z)|_r(\mathfrak{g}_1(z)\frac{z^n-1}{\mathfrak{g}_0(z)})$, $a_2(z)|_r(\mathfrak{g}_1(z)\frac{z^n-1}{\mathfrak{g}_0(z)})$. □

Definition 5. *Let $\mathfrak{g}(z) = \mathfrak{g}_0 + \mathfrak{g}_1z + \ldots + \mathfrak{g}_mz^m$ be a polynomial in $\mathbb{F}_4[z,\theta]$. Then, $\mathfrak{g}(z)$ is said to be a palindromic polynomial if $\mathfrak{g}_i = \mathfrak{g}_{m-i}$ and θ-palindromic if $\mathfrak{g}_i = \theta(\mathfrak{g}_{m-i})$ where $i \in \{1, 2, \ldots, m\}$.*

Note that if the length of the code C is odd, then the skew cyclic codes and cyclic codes are equivalent (Theorem 8 in [17]). Now, we provide two lemmas to check the reversibility of the even length skew cyclic codes over the field $\mathbb{F}_4$.

Lemma 5. *Let C be a skew cyclic code of even length generated by a monic polynomial $\mathfrak{f}(z) = 1 + \mathfrak{f}_1z + \ldots + \mathfrak{f}_{m-1}z^{m-1} + z^m$ of even degree, where $\mathfrak{f}(z)|_r(z^n-1)$ in $\mathbb{F}_4[z,\theta]$. Then, the code C is reversible if and only if skew polynomial $\mathfrak{f}(z)$ is θ-palindromic.*

Proof. Let C be a skew cyclic code of even length generated by the θ-palindromic polynomial $\mathfrak{f}(z)$ of even degree m over the ring $\mathbb{F}_4$. Then, the elements of the generated code are provided by $\sum_{i=0}^{k-1} \alpha_i z^i \mathfrak{f}(z)$. From the repetitive use of Lemma 3, for $c = \phi(\sum_{i=0}^{k-1} \alpha_i z^i \mathfrak{f}(z)) \in \mathcal{C}$, we obtain:

$$(\phi(\sum_{i=0}^{k-1} \alpha_i z^i \mathfrak{f}(z)))^r = \phi(\sum_{i=0}^{k-1} \alpha_i z^{k-i-1} \mathfrak{f}(z)) \in \mathcal{C}.$$

where $\alpha \in \mathbb{F}_4$ and $k = n - m$. Since c^r belongs to the code $\mathcal{C}$, $\mathcal{C}$ is a reversible code.

Conversely, let $\mathcal{C}$ be a reversible code generated by $\mathfrak{f}(z) = 1 + \mathfrak{f}_1 z + \ldots + \mathfrak{f}_{m-1} z^{m-1} + z^m$. Then, because $n - m - 1$ is odd:

$$z^{n-m-1}\mathfrak{f}(z) = z^{n-m-1} + \theta(\mathfrak{f}_1) z^{n-m} + \ldots + \theta(\mathfrak{f}_{m-1}) z^{n-2} + z^{n-1}.$$

Since $\mathcal{C}$ is a skew cyclic and reversible code,

$$[z^{n-m-1}\mathfrak{f}(z)]^r = 1 + \theta(\mathfrak{f}_{m-1}) z + \theta(\mathfrak{f}_{m-2}) z^2 + \ldots + \theta(\mathfrak{f}_1) z^{m-1} + z^m \in \mathcal{C}.$$

Further, we obtain $deg(\mathfrak{f}(z) - [z^{n-m-1}\mathfrak{f}(z)]^r) < m$, which contradicts the fact that $\mathfrak{f}(z)$ is a minimal degree polynomial in $\mathcal{C}$ implies $\mathfrak{f}(z) - [z^{n-m-1}\mathfrak{f}(z)]^r = 0$. Comparing coefficients, we obtain:

$$[\mathfrak{f}_i - \theta(\mathfrak{f}_{m-i})] = 0$$

for $i = 1, \ldots, m-1$. Thus, $\mathfrak{f}_i = \theta(\mathfrak{f}_{m-i})$ and the polynomial $\mathfrak{f}(z)$ is θ-palindromic. □

Lemma 6. *Let $\mathcal{C}$ be a skew cyclic code of even length generated by a monic polynomial $\mathfrak{f}(z) = 1 + \mathfrak{f}_1 z + \ldots + \mathfrak{f}_{m-1} z^{m-1} + z^m$ of odd degree, where $\mathfrak{f}(z)|_r(z^n - 1)$ in $\mathbb{F}_4[z, \theta]$. Then, the code $\mathcal{C}$ is reversible if and only if the skew polynomial $\mathfrak{f}(z)$ is palindromic.*

Proof. Let $\mathcal{C}$ be a skew cyclic code of even length generated by a palindromic polynomial $\mathfrak{f}(z)$ of odd degree m over the ring $\mathbb{F}_4$. Then, elements of the generated code are provided by $\sum_{j=0}^{k-1} \alpha_j z^j \mathfrak{f}(z)$. From the repetitive use of Lemma 3 and using the property of the palindromic polynomial, for $C = \phi(\sum_{j=0}^{k-1} \alpha_j z^j \mathfrak{f}(z)) \in \mathcal{C}$, we obtain:

$$(\phi(\sum_{j=0}^{k-1} \alpha_j z^j \mathfrak{f}(z)))^r = \phi(\sum_{j=0}^{k-1} \alpha_j z^{k-j-1} \mathfrak{f}(z)) \in \mathcal{C}.$$

where $\alpha \in \mathbb{F}_4$ and $k = n - m$. Since the reverse of $\mathcal{C}$ belongs to $\mathcal{C}$, the code $\mathcal{C}$ is reversible. Conversely, let $\mathcal{C}$ be a reversible code generated by $\mathfrak{f}(z) = 1 + \mathfrak{f}_1 z + \ldots + \mathfrak{f}_{m-1} z^{m-1} + z^m$. Since $n - m - 1$ is even:

$$z^{n-m-1}\mathfrak{f}(z) = z^{n-m-1} + \mathfrak{f}_1 z^{n-m} + \ldots + \mathfrak{f}_{m-1} z^{n-2} + z^{n-1}.$$

Furthermore, the code $\mathcal{C}$ is a skew cyclic as well as reversible code; therefore, $[z^{n-m-1}\mathfrak{f}(z)]^r \in \mathcal{C}$ and:

$$[z^{n-m-1}\mathfrak{f}(z)]^r = [1 + \mathfrak{f}_{m-1} z + \mathfrak{f}_{m-2} z^2 + \ldots + \mathfrak{f}_1 z^{m-1} + z^m] \in \mathcal{C}.$$

This implies that $deg(\mathfrak{f}(z) - [z^{n-m-1}\mathfrak{f}(z)]^r) < m$, which contradicts the fact that $\mathfrak{f}(z)$ is a minimal degree polynomial in $\mathcal{C}$. Hence, $\mathfrak{f}(z) - [z^{n-m-1}\mathfrak{f}(z)]^r = 0$. By comparing the coefficients, we obtain

$$[\mathfrak{f}_i - \mathfrak{f}_{m-i}] = 0 \text{ and } \mathfrak{f}_i = \mathfrak{f}_{m-i},$$

for $i = 1, \ldots, m-1$. Thus, the given polynomial $\mathfrak{f}(z)$ is palindromic. □

Now, in the next theorem, we provide necessary and sufficient conditions for a skew cyclic code $\mathcal{C}$ to be reversible in terms of palindromic and θ-palindromic polynomials. These conditions depend on the degree of generator polynomials of $\mathcal{C}$.

Theorem 8. *Let $\mathcal{C} = \langle \mathfrak{g}_0(z), v\mathfrak{g}_1(z), v^2\mathfrak{g}_2(z)\rangle$ be a skew cyclic code of even length, where $\mathfrak{g}_i(z)$ right divides $(z^n - 1)$ in $\mathbb{F}_4[z,\theta]$ and $deg(\mathfrak{g}_i(z))$ is even (odd), for $i = 0,1,2$. Then, the code $\mathcal{C}$ is reversible if and only if skew polynomials $\mathfrak{g}_i(z)$ are θ-palindromic (palindromic) for $i = 0,1,2$.*

5. DNA Codes over $\mathcal{R}$

In this section, we discuss the complementary condition of the codes obtained from previous sections to obtain DNA codes. For a DNA code, the reversible and complement conditions are essential [21].

Definition 6. *Let $\mathcal{C}$ be a code of length n over $\mathcal{R}$. If $\Phi(\mathcal{C})^{rc} \in \Phi(\mathcal{C})$ for all $c \in \mathcal{C}$, then $\mathcal{C}$ or equivalently $\Phi(\mathcal{C})$ is called a DNA code.*

In the following lemma, we provide some relations on ring elements and their complement using the Gray map provided in Equation (1).

Lemma 7. *For the given cyclic code in Section 3, the following conditions hold:*

(1) For any $r \in \mathcal{R}$, $r + r^c = v^2$.
(2) For any $r_1, r_2 \in \mathcal{R}$, $r_1^c + r_2^c = (r_1 + r_2)^c + v^2$.

Proof. This lemma can easily be proved by observing Table 1. □

Remark 2. *We identify $\mathfrak{i}(z)$ by the polynomial $1 + z + z^2 + \cdots + z^{n-1}$.*

Theorem 9. *Given a polynomial $\mathfrak{a}(z)$ in $\mathcal{R}[z]$. We have $\mathfrak{a}(z)^{rc} = \mathfrak{a}(z)^r + v^2\mathfrak{i}(z)$.*

Proof. Let $\mathcal{C}$ be a reversible-complement code. Then, by definition, $\mathcal{C}$ is reversible and $0 \in \mathcal{C}$ implies that $(0 + 0z + \ldots + 0z^{n-1})^c \in \mathcal{C}$. That is, $\mathcal{C}$ is reversible and $v^2 + v^2z + \ldots + v^2z^{n-1} \in \mathcal{C}$.

Conversely, let $\mathfrak{a}(z) = \mathfrak{a}_0 + \mathfrak{a}_1 z + \ldots + \mathfrak{a}_{n-1}z^{n-1} + \mathfrak{a}_n z^n$ be a polynomial in $\mathcal{R}[z]$. Then:

$$\begin{aligned}\mathfrak{a}(z)^{rc} &= \mathfrak{a}_n^c + \mathfrak{a}_{n-1}^c z + \ldots + \mathfrak{a}_1^c z^{n-1} + \mathfrak{a}_0^c z^n \\ &= \mathfrak{a}_n + v^2 + (\mathfrak{a}_{n-1} + v^2)z + (\mathfrak{a}_{n-2} + v^2)z^2 + \ldots \\ &+ (\mathfrak{a}_1 + v^2)z^{n-1} + (\mathfrak{a}_0 + v^2)z^n \\ &= v^2\mathfrak{i}(z) + \mathfrak{a}(z)^r \in \mathcal{C}.\end{aligned}$$

Thus, cyclic code $\mathcal{C}$ is a reversible-complement code. □

Corollary 4. *Let $\mathcal{C}$ be a cyclic code of even length over $\mathcal{R}$. Then, $\mathcal{C}$ is a DNA code if and only if $\mathcal{C}$ is reversible and $v^2\mathfrak{i}(z)$ is in $\mathcal{C}$.*

Proof. It is obvious from above theorem. □

6. Computational Results

Now, we provide some examples of DNA codes satisfying the above-mentioned constraints. We consider DNA code of any length (even or odd). All the computational works are performed by using Magma software [20].

Example 1. *In $\mathbb{F}_4[z]$, we have:*

$$z^6 - 1 = (z+1)^2(z+\mathfrak{t})^2(z+\mathfrak{t}^2)^2.$$

Let $\mathcal{C}$ be a cyclic code of length $n = 6$ over $\mathcal{R}$ provided by:

$$\mathcal{C} = \langle z^4 + z^2 + 1, v(z^4 + z^2 + 1), v^2(z^4 + z^2 + 1)\rangle.$$

Then, using Theorem 2, we obtain $d(C) = 3$. *Furthermore,* $(x-1)$ *does not divide* (z^4+z^2+1) *and polynomial* (z^4+z^2+1) *is self reciprocal. Thus, we obtain a DNA code* C *of parameters* $(18, 4^6, 3)$.

In the next example, we provide some DNA codes of arbitrary lengths that are generated from cyclic codes over $\mathcal{R}$.

Example 2. *Suppose* C *is a cyclic code of the form* $C = \langle \mathfrak{g}_0(z) + v\mathfrak{g}_1(z) + v^2\mathfrak{g}_2(z), va_1(z) + v^2p(z), v^2a_2(z)\rangle$, *where* $\gcd(\frac{z^n-1}{a_2(z)}, \mathfrak{g}_0(z)) = 1$. *If* $\mathfrak{g}_0(z) = a_1(z) = a_2(z)$, *then we list several DNA codes in Table 2 that are obtained from cyclic code* C. *Since* $\mathfrak{g}_0(z)$, $a_1(z)$, *and* $a_2(z)$ *are equal, therefore, in Table 2, we mention only* $\mathfrak{g}_0(z)$. *For brevity, polynomial* $z^2 + \mathfrak{b}_1 z + \mathfrak{b}_0$ *is represented as* $\mathfrak{b}_0\mathfrak{b}_1 1$.

Table 2. DNA codes of different lengths.

Length	$\mathfrak{g}_0(z)$	Type of Code	Gray Image
5	$1\mathfrak{t}1$	$(5,3,3)$	$(15,4^9,3)$
5	11111	$(5,1,5)$	$(15,4^3,5)$
6	10101	$(6,2,3)$	$(18,4^6,3)$
10	101010101	$(10,2,5)$	$(30,4^6,5)$
13	$1\mathfrak{t}0(1+\mathfrak{t})0\mathfrak{t}1$	$(13,7,5)$	$(39,4^{21},5)$
14	1010101010101	$(14,2,7)$	$(42,4^6,7)$
17	$11\mathfrak{t}11$	$(17,13,4)$	$(51,4^{39},4)$
17	$1(1+\mathfrak{t})11\mathfrak{t}11(1+\mathfrak{t})1$	$(17,9,7)$	$(51,4^{27},7)$
29	$1\mathfrak{t}0\mathfrak{t}(1+\mathfrak{t})1(1+\mathfrak{t})\mathfrak{t}(1+\mathfrak{t})1(1+\mathfrak{t})\mathfrak{t}0\mathfrak{t}1$	$(10,1,5)$	$(30,4^3,5)$

Example 3. *Consider a cyclic code* C *of length* $n = 9$ *over ring* $\mathcal{R}$. *In* $\mathbb{F}_4[z]$, *we have:*

$$z^9 - 1 = (z+1)(z+\mathfrak{t})(z+\mathfrak{t}^2)(z^3+\mathfrak{t})(z^3+\mathfrak{t}^2).$$

To write briefly, we identify factors by $\mathfrak{g}_1, \mathfrak{g}_2, \mathfrak{g}_3, \mathfrak{g}_4$, *and* $\mathfrak{g}_5$, *respectively. The codes for* $n = 9$ *are provided in Table 3. All the codes are better than the codes that appeared in [14].*

Example 4. *Consider a cyclic code* C *of length* $n = 15$ *over ring* $\mathcal{R}$. *In* $\mathbb{F}_4[z]$, *we have*

$$\begin{aligned} z^{15} - 1 =& (z+1)(z+\mathfrak{t})(z+\mathfrak{t}^2)(z^2+z+\mathfrak{t})(z^2+z+\mathfrak{t}^2)(z^2+\mathfrak{t}z+1) \\ & (z^2+\mathfrak{t}z+\mathfrak{t})(z^2+\mathfrak{t}^2z+1)(z^2+\mathfrak{t}^2z+\mathfrak{t}^2). \end{aligned}$$

For brevity, we identify the factors by $\mathfrak{g}_1, \mathfrak{g}_2, \mathfrak{g}_3, \mathfrak{g}_5, \mathfrak{g}_6, \mathfrak{g}_7, \mathfrak{g}_8$, *and* $\mathfrak{g}_9$, *respectively. DNA codes for* $n = 15$ *are provided in Table 4. All the obtained DNA codes are better than the codes provided in [14].*

In particular, if $C = \langle \mathfrak{g}_2\mathfrak{g}_3\mathfrak{g}_4\mathfrak{g}_5\mathfrak{g}_6\mathfrak{g}_7\mathfrak{g}_8\mathfrak{g}_9, v\mathfrak{g}_2\mathfrak{g}_3\mathfrak{g}_4\mathfrak{g}_5\mathfrak{g}_6\mathfrak{g}_7\mathfrak{g}_8\mathfrak{g}_9, v^2\mathfrak{g}_2\mathfrak{g}_3\mathfrak{g}_4\mathfrak{g}_5\mathfrak{g}_6\mathfrak{g}_7\mathfrak{g}_8\mathfrak{g}_9\rangle$, *then we obtain a DNA code with parameters* $[45, 4^3, 15]$. *Further, we list all the DNA codewords of the obtained DNA code in Table 5. Furthermore, the edit distance of the obtained DNA code is 2, given by the codewords "TCCTCCTCCTCCTCCTCCTCCTCCTCC" and "CTCCTCCTCCTCCTCCTCCTCCTCCTC".*

Table 3. Codes of length 27.

Sr No	Generator of Code	Type of Code	Gray Image	DNA Code [14]
1	$\langle \mathfrak{g}_2\mathfrak{g}_3, v\mathfrak{g}_2\mathfrak{g}_3, v^2\mathfrak{g}_2\mathfrak{g}_3\rangle$	$(9,7,2)$	$(27,4^{21},2)$	$(27,4^{14},2)$
2	$\langle \mathfrak{g}_4\mathfrak{g}_5, v\mathfrak{g}_4\mathfrak{g}_5, v^2\mathfrak{g}_4\mathfrak{g}_5\rangle$	$(9,3,3)$	$(27,4^9,3)$	$(27,4^6,3)$
3	$\langle \mathfrak{g}_2\mathfrak{g}_3\mathfrak{g}_4\mathfrak{g}_5, v\mathfrak{g}_2\mathfrak{g}_3\mathfrak{g}_4\mathfrak{g}_5, v^2\mathfrak{g}_2\mathfrak{g}_3\mathfrak{g}_4\mathfrak{g}_5\rangle$	$(9,1,9)$	$(27,4^3,9)$	$(27,4^2,9)$

Table 4. Codes of length 45.

Code	Type of Code	Gray Image	DNA Code [14]
$\langle g_2g_3, vg_2g_3, v^2g_2g_3 \rangle$	$(15, 13, 2)$	$(45, 4^{39}, 2)$	$(45, 4^{26}, 2)$
$\langle g_2g_3g_6, vg_2g_3g_6, v^2g_2g_3g_6 \rangle$	$(15, 11, 4)$	$(45, 4^{33}, 4)$	$(45, 4^{24}, 3)$
$\langle g_4g_8g_9, vg_4g_8g_9, v^2g_4g_8g_9 \rangle$	$(15, 9, 5)$	$(45, 4^{27}, 5)$	$(45, 4^{18}, 5)$
$\langle g_2g_3g_5g_6g_7, vg_2g_3g_5g_6g_7, v^2g_2g_3g_5g_6g_7 \rangle$	$(15, 7, 7)$	$(45, 4^{21}, 7)$	$(45, 4^{14}, 7)$
$\langle g_2g_3g_4g_5g_6g_7g_9, vg_2g_3g_4g_5g_6g_7g_9, v^2g_2g_3g_4g_5g_6g_7g_9 \rangle$	$(15, 3, 9)$	$(45, 4^{9}, 9)$	$(45, 4^{6}, 9)$

Table 5. Codewords of length 45 and dimension 3.

AAAAAAAAAAAAAAAAAAAAAAAAAAA	TAATAATAATAATAATAATAATAATAA
GAAGAAGAAGAAGAAGAAGAAGAAGAA	CAACAACAACAACAACAACAACAACAA
TTATTATTATTATTATTATTATTATTA	ATAATAATAATAATAATAATAATAATA
CTACTACTACTACTACTACTACTACTA	GTAGTAGTAGTAGTAGTAGTAGTAGTA
GGAGGAGGAGGAGGAGGAGGAGGAGGA	CGACGACGACGACGACGACGACGACGA
AGAAGAAGAAGAAGAAGAAGAAGAAGA	TGATGATGATGATGATGATGATGATGA
CCACCACCACCACCACCACCACCACCA	GCAGCAGCAGCAGCAGCAGCAGCAGCA
TCATCATCATCATCATCATCATCATCA	ACAACAACAACAACAACAACAACAACA
TTTTTTTTTTTTTTTTTTTTTTTTTTT	ATTATTATTATTATTATTATTATTATT
CTTCTTCTTCTTCTTCTTCTTCTTCTT	GTTGTTGTTGTTGTTGTTGTTGTTGTT
AATAATAATAATAATAATAATAATAAT	TATTATTATTATTATTATTATTATTAT
GATGATGATGATGATGATGATGATGAT	CATCATCATCATCATCATCATCATCAT
CCTCCTCCTCCTCCTCCTCCTCCTCCT	GCTGCTGCTGCTGCTGCTGCTGCTGCT
TCTTCTTCTTCTTCTTCTTCTTCTTCT	ACTACTACTACTACTACTACTACTACT
GGTGGTGGTGGTGGTGGTGGTGGTGGT	CGTCGTCGTCGTCGTCGTCGTCGTCGT
AGTAGTAGTAGTAGTAGTAGTAGTAGT	TGTTGTTGTTGTTGTTGTTGTTGTTGT
GGGGGGGGGGGGGGGGGGGGGGGGGGG	CGGCGGCGGCGGCGGCGGCGGCGGCGG
AGGAGGAGGAGGAGGAGGAGGAGGAGG	TGGTGGTGGTGGTGGTGGTGGTGGTGG
CCGCCGCCGCCGCCGCCGCCGCCGCCG	GCGGCGGCGGCGGCGGCGGCGGCGGCG
TCGTCGTCGTCGTCGTCGTCGTCGTCG	ACGACGACGACGACGACGACGACGACG
AAGAAGAAGAAGAAGAAGAAGAAGAAG	TAGTAGTAGTAGTAGTAGTAGTAGTAG
GAGGAGGAGGAGGAGGAGGAGGAGGAG	CAGCAGCAGCAGCAGCAGCAGCAGCAG
TTGTTGTTGTTGTTGTTGTTGTTGTTG	ATGATGATGATGATGATGATGATGATG
CTGCTGCTGCTGCTGCTGCTGCTGCTG	GTGGTGGTGGTGGTGGTGGTGGTGGTG
CCCCCCCCCCCCCCCCCCCCCCCCCCC	GCCGCCGCCGCCGCCGCCGCCGCCGCC
TCCTCCTCCTCCTCCTCCTCCTCCTCC	ACCACCACCACCACCACCACCACCACC
GGCGGCGGCGGCGGCGGCGGCGGCGGC	AGCAGCAGCAGCAGCAGCAGCAGCAGC
CGCCGCCGCCGCCGCCGCCGCCGCCGC	TGCTGCTGCTGCTGCTGCTGCTGCTGC
TTCTTCTTCTTCTTCTTCTTCTTCTTC	ATCATCATCATCATCATCATCATCATC
CTCCTCCTCCTCCTCCTCCTCCTCCTC	GTCGTCGTCGTCGTCGTCGTCGTCGTC
AACAACAACAACAACAACAACAACAAC	CACCACCACCACCACCACCACCACCAC
TACTACTACTACTACTACTACTACTAC	GACGACGACGACGACGACGACGACGAC

7. Conclusions

In this paper, we have studied reversible and DNA codes using the chain ring $\mathcal{R} = \mathbb{F}_4[v]/\langle v^3 \rangle$. We have defined a Gray map on $\mathcal{R}$ and found codons corresponding to the elements of $\mathcal{R}$. In this way, we have obtained good DNA and reversible codes with the Hamming distances. In the future, one can work on DNA codes over a generalized structure of $\mathcal{R}$ as well as DNA codes by using skew polynomial rings.

Author Contributions: This work is initiated by A.S. and R.K.V. under the supervision of O.P. Then we discussed it with P.S. and W.C. to reach the final version. All authors have read and agreed to the published version of the manuscript.

Funding: This research received no external funding for APC. Financial support for this work is properly acknowledged in the acknowledgement section.

Data Availability Statement: The data that support the findings of this study are available from the corresponding author upon reasonable request.

Acknowledgments: The first and second authors are thankful to the DST and the CSIR, Govt. of India, for financial support under CRG/2020/005927, vide Diary No. SERB/F/6780/ 2020-2021 dated 31 December 2020 and under File No. 09/1023(0027)/2019- EMR-1, respectively. Furthermore, we thank the Indian Institute of Technology Patna for providing research facilities.

Conflicts of Interest: The authors declare that there is no conflict of interest regarding the publication of this manuscript.

References

1. Adleman, L.M. Molecular computation of solutions to combinatorial problems. *Science* **1994**, *266*, 1021–1024. [CrossRef] [PubMed]
2. Hammons, A.R.; Kumar, P.V.; Calderbank, A.R.; Sloane, N.J.; Solé, P. The $\mathbb{Z}_4-$linearity of Kerdock, Preparata, Goethals, and related codes. *IEEE Trans. Inf. Theory* **1994**, *40*, 301–319. [CrossRef]
3. Ling, S.; Xing, C. *Coding Theory: A First Course*; Cambridge University Press: Cambridge, UK, 2004.
4. Norton, G.H.; Sălăgean, A. On the Hamming distance of linear codes over a finite chain ring. *IEEE Trans. Inform. Theory* **2000**, *46*, 1060–1067. [CrossRef]
5. Norton, G.H.; Sălăgean, A. On the structure of linear and cyclic codes over a finite chain ring. *Appl. Algebra Engrg. Comm. Comput.* **2000**, *10*, 489–506. [CrossRef]
6. Abualrub, T.; Siap, I. Cyclic codes over the rings $\mathbb{Z}_2 + u\mathbb{Z}_2$ and $\mathbb{Z}_2 + u\mathbb{Z}_2 + u^2\mathbb{Z}_2$. *Des. Codes Cryptogr.* **2007**, *42*, 273–287. [CrossRef]
7. Bayram, A.; Oztas, E.S.; Siap, I. Codes over $\mathbb{F}_4 + v\mathbb{F}_4$ and some DNA applications. *Des. Codes Cryptogr.* **2016**, *80*, 379–393. [CrossRef]
8. Guenda, K.; Gulliver, T.A. Construction of cyclic codes over $\mathbb{F}_2 + u\mathbb{F}_2$ for DNA computing. *Appl. Algebra Engrg. Comm. Comput.* **2013**, *24*, 445–459. [CrossRef]
9. Prakash, O.; Patel, S.; Yadav, S. Reversible cyclic codes over some finite rings and their application to DNA codes. *Comput. Appl. Math.* **2021**, *40*, 1–17. [CrossRef]
10. Prakash, O.; Yadav, S.; Sharma, P. Reversible cyclic codes over a class of chain rings and their application to DNA codes. *Int. J. Inf. Coding Theory* **2022**, *6*, 52–70. [CrossRef]
11. Bennenni, N.; Guenda, K.; Mesnager, S. DNA cyclic codes over rings. *Adv. Math. Commun.* **2017**, *11*, 83. [CrossRef]
12. Mostafanasab, H.; Darani, A.Y. On Cyclic DNA Codes Over $\mathbb{F}_2 + u\mathbb{F}_2 + u^2\mathbb{F}_2$. *Commun. Math. Stat.* **2021**, *9*, 39–52. [CrossRef]
13. Yildiz, B.; Siap, I. Cyclic codes over $\mathbb{F}_2[u]/(u^4 - 1)$ and applications to DNA codes. *Comput. Math. Appl.* **2012**, *63*, 1169–1176. [CrossRef]
14. Liu, J.; Liu, H. DNA Codes Over the Ring $\mathbb{F}_4[u]/\langle u^3 \rangle$. *IEEE Access* **2020**, *8*, 77528–77534. [CrossRef]
15. Boucher, D.; Ulmer, F. Coding with skew polynomial rings. *J. Symbolic Comput.* **2009**, *44*, 1644–1656. [CrossRef]
16. Jitman, S.; Ling, S.; Udomkavanich, P. Skew constacyclic codes over finite chain rings. *Adv. Math. Commun.* **2012**, *6*, 39–62.
17. Siap, I.; Abualrub, T.; Aydin, N.; Seneviratne, P. Skew cyclic codes of arbitrary length. *Int. J. Inf. Coding Theory* **2011**, *2*, 10–20. [CrossRef]
18. Gursoy, F.; Oztas, E.S.; Siap, I. Reversible DNA codes using skew polynomial rings. *Appl. Algebra Engrg. Comm. Comput.* **2017**, *28*, 311–320. [CrossRef]
19. Cengellenmis, Y.; Aydin, N.; Dertli, A. Reversible DNA codes from skew cyclic codes over a ring of order 256. *J. Algebra Comb. Discret. Struct. Appl.* **2021**, *8*, 1–8.
20. Bosma, W.; Cannon, W.J.; Playoust, C. The Magma algebra system I: The user language. *J. Symbolic Comput.* **1997**, *24*, 235–265. [CrossRef]
21. Limbachiya, D.; Rao, B.; Gupta, M.K. The art of DNA strings: Sixteen years of DNA coding theory. *arXiv* **2016**, arXiv:1607.00266.

entropy

MDPI

Article

A Well-Posed Fractional Order Cholera Model with Saturated Incidence Rate

Isa Abdullahi Baba [1,2], Usa Wannasingha Humphries [2,*] and Fathalla A. Rihan [3,4]

1 Department of Mathematics, Bayero University, Kano 700241, Nigeria
2 Department of Mathematics, Faculty of Science, King Mongkuts University of Science and Technology Thonburi (KMUTT), Bangkok 10140, Thailand
3 Department of Mathematical Sciences, College of Science, United Arab Emirates University, Al Ain 15551, United Arab Emirates
4 Department of Mathematics, Faculty of Science, Helwan University, Cairo 11795, Egypt
* Correspondence: usa.wan@kmutt.ac.th

Abstract: A fractional-order cholera model in the Caputo sense is constructed. The model is an extension of the Susceptible–Infected–Recovered (SIR) epidemic model. The transmission dynamics of the disease are studied by incorporating the saturated incidence rate into the model. This is particularly important since assuming that the increase in incidence for a large number of infected individualsis equivalent to a small number of infected individualsdoes not make much sense. The positivity, boundedness, existence, and uniqueness of the solution of the model are also studied. Equilibrium solutions are computed, and their stability analyses are shown to depend on a threshold quantity, the basic reproduction ratio (R_0). It is clearly shown that if $R_0 < 1$, the disease-free equilibrium is locally asymptotically stable, whereas if $R_0 > 1$, the endemic equilibrium exists and is locally asymptotically stable. Numerical simulations are carried out to support the analytic results and to show the significance of the fractional order from the biological point of view. Furthermore, the significance of awareness is studied in the numerical section.

Keywords: mathematical model; fractional order; Caputo; cholera; well-posedness; saturated incidence rate

check for updates

Citation: Baba, I.A.; Humphries, U.W.; Rihan, F.A. A Well-Posed Fractional Order Cholera Model with Saturated Incidence Rate. *Entropy* **2023**, 25, 360. https://doi.org/10.3390/e25020360

Academic Editors: Pavel Kraikivski and António M. Lopes

Received: 1 December 2022
Revised: 11 January 2023
Accepted: 16 January 2023
Published: 15 February 2023

1. Introduction

Cholera is a prolific diarrheal disease that leads to death in a short period of time if treatment measures are not taken. The disease is estimated to cause about 21,000 to 143,000 deaths from the 1,300,000 to 4,000,000 cholera cases annually. This represents about 1.62% to 3.58% of the reported cases [1]. People who live in slums and refugee camps due to natural disasters, social conflicts, climate change, and economic meltdowns are the most affected. The camps often possess poor drinking water, which serves as a cholera-causing factor [2]. Many of the symptoms of cholera include vomiting, profuse rice–water stool, sunken eyes, cramps, shock, and severe dehydration. *Vibrio cholerae* carriers are those people that are exposed to an incomplete cholera-causing dose, therefore, the disease may not manifest any symptoms in their body [3]. Acute cholera leads to death in a short period that varies from hours up to three days. Exposed individuals have only half a chance of being infected with the disease if the concentration of *Vibrio cholerae* is 105 cells per milliliter. A minimum of 1 liter per day is the least daily consumption of untreated water that may cause the disease [4].

Up to 75% of *Vibrio cholerae* carriers do not show any symptoms of the disease, potentially spreading it through their feces [5]. This causes a big risk of cholera disease and outbreaks. The cholera infection falls into one of three classes: asymptomatic, mild, or severe. In a population of infected individuals, 80% have mild or moderate symptoms and only 20% develop severe watery diarrhea [6].

Mathematical models help in studying the dynamics of a given infectious disease [7–9]. The investigation of cholera models using the SIR epidemic model introduced in 1927 [10] provides a cavernous understanding of the transmission mechanisms of the disease, while other models use time-dependent coefficients [11]. This is why many researchers put more effort into constructing and analyzing mathematical models of a cholera epidemic. Cholera dynamical models involve the study of the disease transmission among humans and V. cholerae concentration in contaminated water. Most of the time, the direct and indirect pathway transmission of cholera gives rise to a basic framework for investigating the dynamics of the disease [12–18].

Due to their hereditary properties and memory description abilities, many researchers prefer to use fractional order derivatives and integrals as tools in the study of mathematical modeling. Nowadays, a fractional differential equation is used to study biological phenomenaby developing related mathematical models [19,20]. This is due to the fact that fractional calculus can be used to explain the retention and heritage properties of many materials more accurately compared to its corresponding integer-order analog [21,22]. In this paper, we use the Caputo fractional order to model the dynamics of cholera in a homogeneous setting.

The mechanism of transmission of any transmissible disease is controlled by a certain function that depends on the subpopulations of infected individuals called the incidence rate. In epidemic models, the most frequently used incidence rates are the standard incidence rate $\frac{\beta SI}{N}$ and the bilinear incidence rate βSI, where β is the contact rate, S stands for the susceptible population, I stands for the infected population, and N is the total population. Assuming the increase in incidence for a large number of infected individualsis equivalent to a small number of infected individuals does not make much sense, hence, there is a need for a more realistic incidence rate of the form $g(I)S$ [23] where g tends to a saturation level, as it is a nonnegative function, such that $g(0) = 0$. To incorporate the effect of the behavioral changes of the susceptible individuals, a more general incidence rate of the form $\frac{\beta SI^m}{1+\delta I^n}$, where m and n are positive constants and δ is a nonnegative term, was proposed [24]. These types of incidence rates that allow for the possibility of the introduction of psychological effects are called saturated incidence rates; δ and $1 + \delta I^n$ determine the amount of psychological effect and the inhibition effect, respectively. As the number of infective individuals increases, the rate of infection spread may decrease due to public awareness, potentially leading to contact reduction [25].

Many mathematical models of cholera transmission exist in the literature, and they study the dynamics of the disease. For example, Leo [26] developed an ML reference cholera model that can be used to overcome the existing complexities of modeling the disease. His results indicate, at an average of 87%, that the developed measures can integrate a large number of datasets, including environmental factors, for the timely prediction of cholera epidemics in Tanzania. Daudel et al. [27] constructed and studied a compartmental malaria model. Their results show that the higher implementation of strategies combining awareness programs and therapeutic treatments increases the efficacy of control measures. In [28], a stochastic norovirus epidemic model with a time delay and random perturbations was explored. In [29], a mathematical model for cholera considering vaccination effects was proposed. In [30], Capasso and Paveri-Fontana suggested a mathematical model for the 1973 cholera epidemic in the European Mediterranean region. In 2017, the transmission dynamics of cholera in Yemen were investigated by Nishiura et al. [31]. Lastly, a model containing optimal intervention strategies for cholera control was formulated in [32].

Luchko and Yamamoto [33] proposed a new differential operator with a general kernel function. Due to the existence of flexibility in choosing the kernel, they provide a basis for a broad range of applications [34]. Changing the kernel in the general derivative leads to the discovery of various asymptotic behaviors. Hence, the hidden features of real-world systems are more accurately observed than in the classical sense. However, both properties and applications regarding this operator must be studied in practical situations. There is also the need to state and prove some theorems in order to study models using this operator.

For cholera, the saturated incidence rate is more reasonable than the bilinear incidence rate. This is because it includes the behavioral change and crowding effect of the cholera-infective individuals and prevents the unboundedness of the contact rate by choosing suitable parameters. Motivated by the above discussions, we construct a novel mathematical model that studies the transmission dynamics of cholera in the Caputosense. The model is novel because, to the best of our knowledge, no previous study has analyzed a mathematical model with a saturated incidence rate and fractional derivative for the cholera disease in detail. The main contributions of this research and the new achievements obtained within this manuscript are summarized as follows:

- This paper addresses a new mathematical model of cholera disease which involves the Caputo fractional derivative.
- The fundamental characteristics of the new model are discussed in detail.
- A numerical scheme is developed to carry out numerical simulations.
- The effect of awareness is studied.
- Comparative results in this research show an obvious linkage between the mathematical and biological mechanisms.

Hence, in this paper, we study a fractional-order cholera model with a saturated incidence rate. The main contribution of the paper is the introduction of a more reasonable incidence rate of the form $\frac{\beta S I^m}{1+\delta I^n}$, which makes more sense when considering that assuming the increase in incidence for a large number of infected individualsis equivalent to a small number of infected individualsdoes not make much sense. It is also the aim of the paper to consider the effect of awareness in the control of cholera and study all the properties of well-posedness.

The paper is arranged as follows: Section 1 gives an introduction, Section 2 gives important definitions and preliminaries, Section 3 gives model formulation, Section 4 gives the well-posedness properties of the model, and Section 5 gives the numerical simulation results to support the analytic results. In Section 6, a discussion and conclusion are given.

2. Preliminary Definitions and Theorems

Definition 1 [35]. *The Caputo fractional derivative of order $\alpha \in (n-1,\ n]$ of $f(x)$ is defined as,*

$$ {}_a^C D_x^\alpha f(x) = \frac{1}{\Gamma(n-\alpha)} \int_a^x (x-t)^{n-\alpha-1} f^n(t)dt, \qquad n = [\alpha] + 1. $$

Definition 2 [36]. *The linearity of the fractional derivative.*

Let $f,\ g$ be continuous and $b,\ c$ be scalars, then

$$ {}_a^{RL} D_x^\alpha [bf(x) + dg(x)] = b_a^{RL} D_x^\alpha f(x) + d_a^{RL} D_x^\alpha g(x), $$

$$ {}_a^C D_x^\alpha [bf(x) + dg(x)] = b_a^C D_x^\alpha f(x) + d_a^C D_x^\alpha g(x). $$

Definition 3 [37]. *Contraction.*

An operator $f : X \to X$ that maps a metric space onto itself is said to be contractive if for $0 < q < 1$.

$$ d(f(x),\ f(y)) = qd(x,\ y),\ \forall\ x,\ y \in X. $$

Theorem 1 [37]. *Picard–Banach fixed point.*

Any contractive operator that maps a metric space onto itself has a unique fixed point. Furthermore, if $f : X \to X$ is a contractive operator that maps a metric space onto itself and a is its fixed point: $f(a) = a$; then for any iterative sequence

$$x_0, \quad x_1 = f(x_0), \quad x_2 = f(x_1), \ldots, x_{n+1} = f(x_n), \ldots$$

converges to a.

In other words, a is a solution or equilibrium for a continuous dynamical system and a fixed point for a discrete dynamical system.

Theorem 2 [36]. *The equilibrium solutions x^* of a system of fractional order differential equation is locally asymptotically stable if all the eigenvalues λ_i of the Jacobian matrix $\frac{\partial f}{\partial x_i}$ evaluated at the equilibrium points satisfy*

$$|\arg(\lambda_i)| > \frac{\alpha\pi}{2}, \qquad 0 < \alpha < 1.$$

Theorem 3 [36]. *Let $x(t) \in \mathbb{R}^+$ be a continuous and derivable function. Then, for any time instant $t \geq t_0$ and $\alpha \in (0, 1)$*

$${}_0^C D_t^\alpha [x(t) - x^* - x^* \ln\left(\frac{x(t)}{x^*}\right)] \leq \left(1 - \frac{x(t)}{x^*}\right) {}_0^C D_t^\alpha x(t), \qquad x^* \in \mathbb{R}^+.$$

3. Model Formulation

The model consists of the following classes: susceptible S, latently infected individuals E, infectious individuals I, and those that recovered from the infection R. We also define $N(t) = S(t) + E(t) + I(t) + R(t)$. Assuming homogeneous mixing of the population, the model is given as:

$$\begin{aligned}
{}_0^C D_t^\alpha S(t) &= \Lambda^\alpha - \tfrac{\beta^\alpha (1-k^\alpha) S(t) I(t)}{f(I)} + \gamma^\alpha E(t) + \xi^\alpha I(t) - \mu^\alpha S(t), \\
{}_0^C D_t^\alpha E(t) &= p \tfrac{\beta^\alpha (1-k^\alpha) S(t) I(t)}{f(I)} - (\gamma^\alpha + \eta^\alpha + \phi^\alpha + \mu^\alpha) E(t), \\
{}_0^C D_t^\alpha I(t) &= (1-p) \tfrac{\beta^\alpha (1-k^\alpha) S(t) I(t)}{f(I)} + \eta^\alpha E(t) - (\xi^\alpha + \delta^\alpha + q^\alpha + \mu^\alpha) I(t), \\
{}_0^C D_t^\alpha R(t) &= \delta^\alpha I(t) - \mu^\alpha R(t),
\end{aligned} \tag{1}$$

subject to the initial conditions;

$$S(0) > 0, \ E(0) \geq 0, \ I(0) \geq 0, \ R(0) \geq 0.$$

where the function $f(I)$ is given as;

$$f(I) = 1 + I^2.$$

We then modify the fractional operator via an auxiliary parameter $Y > 0$ to avoid dimensional mismatching [38] to obtain

$$\begin{aligned}
Y^{\alpha-1} {}_0^C D_t^\alpha S(t) &= \Lambda^\alpha - \tfrac{\beta^\alpha (1-k^\alpha) S(t) I(t)}{f(I)} + \gamma^\alpha E(t) + \xi^\alpha I(t) - \mu^\alpha S(t), \\
Y^{\alpha-1} {}_0^C D_t^\alpha E(t) &= p \tfrac{\beta^\alpha (1-k^\alpha) S(t) I(t)}{f(I)} - (\gamma^\alpha + \eta^\alpha + \phi^\alpha + \mu^\alpha) E(t), \\
Y^{\alpha-1} {}_0^C D_t^\alpha I(t) &= (1-p) \tfrac{\beta^\alpha (1-k^\alpha) S(t) I(t)}{f(I)} + \eta^\alpha E(t) - (\xi^\alpha + \delta^\alpha + q^\alpha + \mu^\alpha) I(t), \\
Y^{\alpha-1} {}_0^C D_t^\alpha R(t) &= \delta^\alpha I(t) - \mu^\alpha R(t),
\end{aligned} \tag{2}$$

subject to the initial conditions;

$$S(0) > 0, \ E(0) \geq 0, \ I(0) \geq 0, \ R(0) \geq 0.$$

The meaning of the parameters in the model are given in Table 1 as follows.

Table 1. Meaning of parameters.

Parameter	Meaning
Λ	Birthrate
β	Disease contact rate
μ	Natural death rate
ϕ	Disease-induced death rate in the Exposed class
γ	Rate at which the Exposed become Susceptible
ξ	Rate at which the Infectious become Susceptible
η	Rate at which the Exposed become Infectious
d	Disease-induced death rate in the Infectious class
δ	Rate of recovery
k	Awareness parameter
p	Fraction of individuals joining the Exposed class

4. Well-Posednessof the Model

In this section, the mathematical properties of the model are explored. This consists of the positivity and boundedness, the existence and uniqueness, the computation of equilibrium solutions and basic reproduction ratio, and the stability analysis of the solutions.

4.1. Positivity and Boundedness

The positivity of solutions means that the population thrives, while boundedness means that the population growth is restricted naturally due to limited resources.

To show positivity, consider Equation (1),

$$\begin{aligned}
{}_0^C D_t^\alpha S(t)|_{S=0} &= \Lambda^\alpha + \gamma^\alpha E(t) + \xi^\alpha I(t) > 0,\\
{}_0^C D_t^\alpha E(t)|_{E=0} &= p\frac{\beta^\alpha(1-k^\alpha)S(t)I(t)}{f(I)} \geq 0,\\
{}_0^C D_t^\alpha I(t)|_{I=0} &= \eta^\alpha E(t) \geq 0,\\
{}_0^C D_t^\alpha R(t)|_{R=0} &= \delta^\alpha I(t) \geq 0.
\end{aligned}$$

Hence, we can observe that the solution of (1) is non-negative.

For the boundedness, observe that,

$$N(t) = S(t) + E(t) + I(t) + R(t).$$

Then, by Definition 3, we have

$${}_0^C D_t^\alpha N(t) = {}_0^C D_t^\alpha S(t) + {}_0^C D_t^\alpha E(t) + {}_0^C D_t^\alpha I(t) + {}_0^C D_t^\alpha R(t).$$

Then,

$$\begin{aligned}
{}_0^C D_t^\alpha N(t) &= \Lambda^\alpha - \mu^\alpha N(t) - (\phi^\alpha E(t) + q^\alpha I(t)),\\
{}_0^C D_t^\alpha N(t) &\leq \Lambda^\alpha - \mu^\alpha N(t).
\end{aligned}$$

We apply the Laplace transform method to solve the Gronwall's like inequality with initial condition $N(t_0) \geq 0$ to find

$$L\left\{{}_0^{CF} D_t^\alpha N(t) + \mu^\alpha N\right\} \leq L\{\Lambda^\alpha\}.$$

By the linearity of the Laplace transform, we obtain

$$L\left\{ {}_{0}^{CF}D_{t}^{\alpha}N(t) \right\} + \mu^{\alpha}L\{N(t)\Big\} \leq L\{\Lambda^{\alpha}\},$$

we find,

$$S^{\alpha}L\{N(t)\} - \sum_{k=0}^{n-1} S^{\alpha-k-1}N^{k}(t_0) + \mu^{\alpha}L\{N(t)\} \leq \frac{\Lambda^{\alpha}}{S}.$$

Simplifying, we obtain

$$L\{N(t)\} \leq \Lambda^{\alpha}\left(\frac{1}{S} - \frac{1}{S}\frac{1}{\left(1+\frac{\mu^{\alpha}}{S^{\alpha}}\right)}\right) + \sum_{k=0}^{n-1}\frac{1}{S^{k+1}}\frac{1}{(1+\frac{\mu^{\alpha}}{S^{\alpha}})}N^{k}(t_0).$$

Using the Taylor series expansion, we have

$$\frac{1}{\left(1+\frac{\mu^{\alpha}}{S^{\alpha}}\right)} = \sum_{n=0}^{\infty}\left(\frac{-\mu^{\alpha}}{S^{\alpha}}\right)^{n}.$$

Therefore,

$$L\{N(t)\} \leq \Lambda^{\alpha}\left(\frac{1}{S} - \frac{1}{S}\sum_{n=0}^{\infty}\left(\frac{-\mu^{\alpha}}{S^{\alpha}}\right)^{n}\right) + \sum_{k=0}^{n-1}\frac{1}{S^{k+1}}N^{k}(t_0)\sum_{n=0}^{\infty}\left(\frac{-\mu^{\alpha}}{S^{\alpha}}\right)^{n}.$$

Taking the Laplace inverse, we find

$$N(t) \leq Y^{\alpha} - \Lambda^{\alpha}\sum_{n=0}^{\infty}\frac{-(\mu^{\alpha}t^{\alpha})^{n}}{\Gamma(\alpha n+1)} + \sum_{k=0}^{n-1}\sum_{n=0}^{\infty}\frac{-(\mu^{\alpha}t^{\alpha})^{n}}{\Gamma(\alpha n+k+1)}t^{k}N^{k}(t_0).$$

Substituting the Mittag–Leffler function, we have

$$N(t) \leq \Lambda^{\alpha}[1 - E_1(-\mu^{\alpha}t^{\alpha})] + \sum_{k=0}^{n-1}E_{k+1}(-\mu^{\alpha}t^{\alpha})t^{k}N^{k}(t_0).$$

where $E_1(-\mu^{\alpha}t^{\alpha}),\ E_{k+1}(-\mu^{\alpha}t^{\alpha})$ are the series of the Mittag–Leffler function which converges for any argument, hence, we say that the solution to the model is bounded.

Thus, we define,

$$\omega = \left\{(S(t),\ E(t), I(t), R(t)) \in R_{+}^{4} : S(t),\ E(t), I(t), R(t) \leq \Lambda^{\alpha}[1 - E_1(-\mu^{\alpha}t^{\alpha})] + \sum_{k=0}^{n-1}E_{k+1}(-\mu^{\alpha}t^{\alpha})t^{k}N^{k}(t_0)\right\}$$

hence, all solutions of (1) commencing in ω stays in ω for all $t \geq 0$.

4.2. Existence and Uniqueness

In this section, we study the existence and uniqueness properties of the solution of Equation (1). First, we consider the following theorem to show Lipschitz continuity.

Theorem 4. *The kernels of Equation (1) satisfy the Lipschitz continuity for* $L_i \geq 0,\ i = 1, 2, \ldots, 4$.

Proof. Let the kernel be defined as,

$$\begin{array}{c} g_1(t,S) = \Lambda^\alpha - \frac{\beta^\alpha(1-k^\alpha)S(t)I(t)}{f(I)} + \gamma^\alpha E(t) + \xi^\alpha I(t) - \mu^\alpha S(t), \\ g_2(t,E) = p\frac{\beta^\alpha(1-k^\alpha)S(t)I(t)}{f(I)} - (\gamma^\alpha + \eta^\alpha + \phi^\alpha + \mu^\alpha)E(t), \\ g_3(t,I) = (1-p)\frac{\beta^\alpha(1-k^\alpha)S(t)I(t)}{f(I)} + \eta^\alpha E(t) - (\xi^\alpha + \delta^\alpha + q^\alpha + \mu^\alpha)I(t), \\ g_4(t,R) = \delta^\alpha I(t) - \mu^\alpha R(t). \end{array} \tag{3}$$

Now,

$$\begin{array}{c} |g_1(t,\ S) - g_1(t,\ S^*)| = \left|\left(\frac{\beta^\alpha(1-k^\alpha)I(t)}{f(I)} + \mu^\alpha\right)(S - S^*)\right| \\ \leq \left(|\mu^\alpha| + \left|\frac{\beta^\alpha(1-k^\alpha)I(t)}{f(I)}\right|\right)\|S - S^*\| \\ \leq \left(|\mu^\alpha| + \beta^\alpha(1-k^\alpha)\max\limits_{t\in[0,\ h^*]}\left|\frac{(1-k^\alpha)I(t)}{f(I)}\right|\right)\|S - S^*\| \\ \leq L_1\|S - S^*\|, \quad L_1 = |\mu^\alpha| + \beta^\alpha(1-k^\alpha)\max\limits_{t\in[0,\ h^*]}\left|\frac{I(t)}{f(I)}\right|. \end{array}$$

Hence,

$$|g_1(t,\ S) - g_1(t,\ S^*)| \leq L_1\|S - S^*\|. \tag{4}$$

In a similar way, we obtain

$$\begin{array}{c} |g_2(t,\ E) - g_2(t,\ E^*)| \leq L_2\|E - E^*\|, \\ |g_3(t,\ I) - g_3(t,\ I^*)| \leq L_3\|I - I^*\|, \\ |g_4(t,\ R) - g_4(t,\ R^*)| \leq L_4\|R - R^*\|. \end{array} \tag{5}$$

where

$$L_2 = |\gamma^\alpha + \eta^\alpha + \phi^\alpha + \mu^\alpha|,\ L_3 = |\xi^\alpha + \delta^\alpha + q^\alpha + \mu^\alpha| + \beta^\alpha(1-k^\alpha)(1-p)\max_{t\in[0,\ h^*]}\left|\frac{I(t)}{f(I)}\right|,$$

and $L_4 = |\mu^\alpha|$. □

The following Lemma converts the system to Volterra integral equations.

Lemma 1. *The continuous system (1) can be transformed to equivalent Volterra integral equations.*

Proof. Consider

$${}_0^C D_t^\alpha S(t) = g_1(t,\ S(t)).$$

On integrating fractionally, we find

$$\begin{array}{c} {}_0^C I_t^\alpha [{}_0^C D_t^\alpha S(t)] = {}_0^C I_t^\alpha [g_1(t,\ S(t))] \\ S(t) - S(0) = \frac{1}{\Gamma(\alpha)}\int_0^t (t-\tau)^{\alpha-1} g_1(\tau,\ S(\tau))dt \\ S(t) = S(0) + \frac{1}{\Gamma(\alpha)}\int_0^t (t-\tau)^{\alpha-1} g_1(\tau,\ S(\tau))dt. \end{array} \tag{6}$$

Similarly,

$$\begin{array}{c} E(t) = E(0) + \frac{1}{\Gamma(\alpha)}\int_0^t (t-\tau)^{\alpha-1} g_2(\tau,\ E(\tau))dt, \\ I(t) = I(0) + \frac{1}{\Gamma(\alpha)}\int_0^t (t-\tau)^{\alpha-1} g_3(\tau,\ I(\tau))dt, \\ R(t) = R(0) + \frac{1}{\Gamma(\alpha)}\int_0^t (t-\tau)^{\alpha-1} g_4(\tau,\ R(\tau))dt. \ \square \end{array} \tag{7}$$

The following theorem provides the existence of the unique solution.

Theorem 5. *Let* $0 < \alpha < 1,\ I = [0,\ \mathrm{h}^*] \subseteq \mathbb{R}$ *and* $\mathrm{J} = |\mathrm{S(t)} - \mathrm{S(0)}| \leq \mathrm{k}_1$.

Let $g_1 : I \times J \rightarrow \mathbb{R}$ be continuous bounded function, that is $\exists!M > 0$ such that $|g_1(t, S)| \leq M_1$.

Assume that g_1 satisfies the Lipschitz conditions. If $L_1 k_1 < M_1$, then there exist unique $S \in C[0, h^*]$, *where* $h^* = min[h, \left(\frac{k_1\Gamma(\alpha+1)}{M_1}\right)^{\frac{1}{\alpha}}]$ that holds the equation.

Proof.

$$Let\ T = \{S \in C[0, h^*] : \ \|S(t) - S(0)\| \leq k_1\}.$$

Since $T \subseteq \mathbb{R}$ and its closed set, then T is a complete metric space.
Recall that,

$$S(t) = S(0) + \frac{1}{\Gamma(\alpha)} \int_0^t (t-\tau)^{\alpha-1} g_1(\tau, S(\tau))dt. \tag{8}$$

Define operator F in T by,

$$F(S(t)) = S(0) + \frac{1}{\Gamma(\alpha)} \int_0^t (t-\tau)^{\alpha-1} g_1(\tau, S(\tau))dt. \tag{9}$$

To show that (6) satisfies Theorem 1, we have

$$\begin{aligned} |F(S(t)) - S(0)| &= \left|\tfrac{1}{\Gamma(\alpha)} \textstyle\int_0^t (t-\tau)^{\alpha-1} g_1(\tau, S(\tau))dt\right| \\ &\leq \tfrac{1}{\Gamma(\alpha)} \textstyle\int_0^t (t-\tau)^{\alpha-1} |g_1(\tau, S(\tau))|dt \\ &\leq \tfrac{1}{\Gamma(\alpha)} \textstyle\int_0^t (t-\tau)^{\alpha-1} M_1 dt \\ &= \tfrac{M_1}{\Gamma(\alpha+1)} t^\alpha \\ &= \tfrac{M_1}{\Gamma(\alpha+1)} (h^*)^\alpha \\ &\leq \tfrac{M_1}{\Gamma(\alpha+1)} \tfrac{k_1\Gamma(\alpha+1)}{M_1} \\ &= k_1. \end{aligned}$$

Hence,

$$|F(S(t)) - S(0)| \leq k_1. \tag{10}$$

Similarly,

$$\begin{aligned} |F(E(t)) - E(0)| \leq k_2, \\ |F(I(t)) - I(0)| \leq k_3, \\ |F(R(t)) - R(0)| \leq k_4. \end{aligned} \tag{11}$$

Therefore F maps T onto itself.
Secondly, to show that T is a contraction, we have

$$F(S) - F(S^*) = S(0) - S^*(0) + \frac{1}{\Gamma(\alpha)} \int_0^t (t-\tau)^{\alpha-1} [g_1(\tau, S(\tau)) - g_1(\tau, S^*(\tau))]d\tau.$$

Since $S(0) = S^*(0)$

$$\begin{aligned}
|F(S) - F(S^*)| &= \left|\tfrac{1}{\Gamma(\alpha)}\int_0^t (t-\tau)^{\alpha-1}[g_1(\tau,\ S(\tau)) - g_1(\tau,\ S^*(\tau))]d\tau\right| \\
&\leq \tfrac{1}{\Gamma(\alpha)}\int_0^t (t-\tau)^{\alpha-1}|g_1(\tau,\ S(\tau)) - g_1(\tau,\ S^*(\tau))|d\tau. \\
&\leq \tfrac{1}{\Gamma(\alpha)}\int_0^t (t-\tau)^{\alpha-1}L_1\|S - S\|d\tau \\
&= \tfrac{L_1}{\Gamma(\alpha)}\|S - S\|\int_0^t (t-\tau)^{\alpha-1}\tau^0 d\tau \\
&= \tfrac{L_1}{\Gamma(\alpha)}\|S - S^*\|\tfrac{\Gamma(\alpha)}{\Gamma(\alpha+1)}t^\alpha \\
&= \tfrac{L_1}{\Gamma(\alpha+1)}\|S - S^*\|t^\alpha \\
&\leq \tfrac{L_1}{\Gamma(\alpha+1)}\|S - S^*\|(h^*)^\alpha \\
&\leq \tfrac{L_1}{\Gamma(\alpha+1)}\|S - S^*\|\tfrac{k_1\Gamma(\alpha+1)}{M_1}.
\end{aligned}$$

Hence,

$$\|FS - FS^*\| \leq \frac{L_1 k_1}{M_1}\|S - S^*\|. \tag{12}$$

Since, by hypothesis $\frac{L_1 k_1}{M_1} < 1$, T is contractive and has a unique fixed point.
In a similar way, we obtain

$$\begin{aligned}
\|F(E) - F(E^*)\| &\leq \tfrac{L_2 k_2}{M_2}\|E - E^*\|, \\
\|F(I) - F(I^*)\| &\leq \tfrac{L_3 k_3}{M_3}\|I - I^*\|, \\
\|F(R) - F(R^*)\| &\leq \tfrac{L_4 k_4}{M_4}\|R - R^*\|.
\end{aligned} \tag{13}$$

Thus, Equation (1) has a unique solution. □

4.3. Existence of Equilibrium Solutions

Since $R = N - (S + E + I)$, we can consider the first three equations in Equation (1) for this analysis.

Setting the first three equations in Equation (1) to zero and solving simultaneously, we find the following equilibrium solutions;

I. Disease-free equilibrium (E_0) is given as;

$$E_0 = \left(S^0, 0, 0\right) = \left(\frac{\Lambda^\alpha}{\mu^\alpha}, 0, 0\right).$$

II. Endemic equilibrium (E_1) is given as;

$$E_1 = \left(S^1, E^1, I^1\right),$$

where

$$\begin{aligned}
S^1 &= \tfrac{A_1 A_2 f(I)}{\beta^\alpha(1-k^\alpha)[(1-p)A_1 + \eta^\alpha p]}, \\
E^1 &= \tfrac{p A_2 I^1}{\beta^\alpha(1-k^\alpha)[(1-p)A_1 + \eta^\alpha p]},
\end{aligned}$$

and I^1 can be obtained by solving,

$$\Lambda^\alpha - \frac{\beta^\alpha(1-k^\alpha)SI}{f(I)} - \mu^\alpha S + \gamma^\alpha E + \xi^\alpha I = 0,$$

where

$$A_1 = \gamma^\alpha + \eta^\alpha + \phi^\alpha + \mu^\alpha,\ and\ A_2 = \xi^\alpha + \delta^\alpha + q^\alpha + \mu^\alpha.$$

Simplifying, we find

$$G(I) = \Lambda^\alpha - I\left[\frac{A_1A_2}{(1-p)A_1+\eta^\alpha p} - \left(\frac{\gamma^\alpha pA_2}{(1-p)A_1+\eta^\alpha p} + \zeta^\alpha\right)\right] - \frac{A_1A_2f(I)}{\beta^\alpha(1-k^\alpha)[(1-p)A_1+\eta^\alpha p]} = 0.$$

Since,

$$\frac{A_1A_2}{(1-p)A_1+\eta^\alpha p} - \left(\frac{\gamma^\alpha pA_2}{(1-p)A_1+\eta^\alpha p} + \zeta^\alpha\right) > 0, \ and\ f' \geq 0,$$

then G is an increasing function.

In addition,

$$G(I) < \Lambda^\alpha - I\left[\frac{A_1A_2}{(1-p)A_1+\eta^\alpha p} - \left(\frac{\gamma^\alpha pA_2}{(1-p)A_1+\eta^\alpha p} + \zeta^\alpha\right)\right],$$

then,

$$\lim_{t\to\infty} G(I) = -\infty.$$

If $f(0) = 1$, clearly

$$\begin{aligned} G(0) &= \Lambda^\alpha - \frac{A_1A_2}{\beta^\alpha(1-k^\alpha)[(1-p)A_1+\eta^\alpha p]} \\ &= \Lambda^\alpha\left[1 - \left(\frac{\mu^\alpha A_1A_2}{\eta^\alpha p\beta^\alpha(1-k^\alpha)\Lambda^\alpha} + \frac{\mu^\alpha A_2}{(1-p)\beta^\alpha(1-k^\alpha)\Lambda^\alpha}\right)\right]. \end{aligned}$$

Thus, a unique positive solution for G exists $iff\ G(0) > 0$, i.e., if

$$\frac{\mu^\alpha A_1A_2}{\eta^\alpha p\beta^\alpha(1-k^\alpha)\Lambda^\alpha} + \frac{\mu^\alpha A_2}{(1-p)\beta^\alpha(1-k^\alpha)\Lambda^\alpha} > 1.$$

We shall establish,

$$\frac{\mu^\alpha A_1A_2}{\eta^\alpha p\beta^\alpha(1-k^\alpha)\Lambda^\alpha} + \frac{\mu^\alpha A_2}{(1-p)\beta^\alpha(1-k^\alpha)\Lambda^\alpha} = R_0,$$

in the subsequent section where R_0 is the basic reproduction ratio.

4.4. Basic Reproduction Ratio

This is defined as the number of secondary cases caused by a single infected individual in a population of Susceptible [37]. Here we use the idea of the next-generation matrix as in [39].

Let,

Y represents new infection and Z represents the remaining terms in Equation (1). Then,

$$Y = \begin{bmatrix} p\frac{\beta^\alpha(1-k^\alpha)S(t)I(t)}{f(I)} \\ (1-p)\frac{\beta^\alpha(1-k^\alpha)S(t)I(t)}{f(I)} \end{bmatrix}, \ and\ Z = \begin{bmatrix} A_1E \\ -\eta^\alpha E + A_2I \end{bmatrix}.$$

Then,

$$Y(E_0) = \begin{bmatrix} 0 & p\beta^\alpha(1-k^\alpha)S^0 \\ 0 & (1-p)\beta^\alpha(1-k^\alpha)S^0 \end{bmatrix}, \ and\ Z(E_0) = \begin{bmatrix} A_1 & 0 \\ -\eta^\alpha & A_2 \end{bmatrix},$$

where $Y(E_0)$ is the Jacobian of Y at E_0 and $Z(E_0)$ is the Jacobian at Z at E_0.

Therefore,

$$X = Y(E_0)(Z(E_0)^{-1}) = \begin{bmatrix} \eta^\alpha p\beta^\alpha(1-k^\alpha)S^0 & A_1p\beta^\alpha(1-k^\alpha)S^0 \\ \eta^\alpha(1-p)\beta^\alpha(1-k^\alpha)S^0 & A_1(1-p)\beta^\alpha(1-k^\alpha)S^0 \end{bmatrix}.$$

Note: X is the next generation matrix and the dominant eigenvalue of X is the basic reproduction ratio (R_0).

Here,

$$R_0 = \frac{\mu^\alpha A_1 A_2}{\eta^\alpha p\beta^\alpha(1-k^\alpha)\Lambda^\alpha} + \frac{\mu^\alpha A_2}{(1-p)\beta^\alpha(1-k^\alpha)\Lambda^\alpha}.$$

4.5. Stability Analysis of the Equilibria

In this section, we carry out a local stability analysis of both disease-free and endemic equilibrium points.

First, consider the following Jacobian matrix from Equation (1);

$$J = \begin{bmatrix} -\frac{\beta^\alpha(1-k^\alpha)I}{f(I)} - \mu^\alpha & \gamma^\alpha & -\frac{\beta^\alpha(1-k^\alpha)S}{f(I)}\left(1-\frac{If'(I)}{f(I)}\right)+\xi^\alpha \\ \frac{p\beta^\alpha(1-k^\alpha)I}{f(I)} & -A_1 & \frac{p\beta^\alpha(1-k^\alpha)S}{f(I)}\left(1-\frac{If'(I)}{f(I)}\right) \\ \frac{(1-p)\beta^\alpha(1-k^\alpha)I}{f(I)} & \eta^\alpha & \frac{(1-p)\beta^\alpha(1-k^\alpha)S}{f(I)}\left(1-\frac{If'(I)}{f(I)}\right) - A_2 \end{bmatrix}. \tag{14}$$

Theorem 6. *The disease-free equilibrium (E_0) is locally asymptotically stable if $R_0 < 1$.*

Proof. Consider Equation (14) at E_0, then we have

$$J(E_0) = \begin{bmatrix} -\mu^\alpha & \gamma^\alpha & \beta^\alpha(1-k^\alpha)S^0 + \xi^\alpha \\ 0 & -A_1 & p\beta^\alpha(1-k^\alpha)S^0 \\ 0 & \eta^\alpha & (1-p)\beta^\alpha(1-k^\alpha)S^0 - A_2 \end{bmatrix}.$$

In echelon form, we find

$$J(E_0) = \begin{bmatrix} -\mu^\alpha & \gamma^\alpha & \beta^\alpha(1-k^\alpha)S^0 + \xi^\alpha \\ 0 & -A_1 & p\beta^\alpha(1-k^\alpha)S^0 \\ 0 & \eta^\alpha & A_1(1-p)\beta^\alpha(1-k^\alpha)S^0 - A_1A_2 + \eta^\alpha p\beta^\alpha(1-k^\alpha)S^0 \end{bmatrix}. \tag{15}$$

The eigenvalues of Equation (15) are:

$$\lambda_1 = -\mu^\alpha < 0,\ \lambda_2 = -A_1,\ and\ \lambda_3 = -A_1A_2(R_0 - 1) < 0\, if\ R_0 < 1.$$

Hence, E_0 is locally asymptotically stable if $R_0 < 1$. □

Theorem 7. *The endemic equilibrium (E_1) is locally asymptotically stable if $R_0 > 1$.*

Proof. Consider Equation (14) at E_1, then we find

$$J(E_1) = \begin{bmatrix} -\frac{\beta^\alpha(1-k^\alpha)I^1}{f(I)} - \mu^\alpha & \gamma^\alpha & -\frac{\beta^\alpha(1-k^\alpha)S^1}{f(I)}\left(1-\frac{I^1f'(I)}{f(I)}\right)+\xi^\alpha \\ \frac{p\beta^\alpha(1-k^\alpha)I^1}{f(I)} & -A_1 & \frac{p\beta^\alpha(1-k^\alpha)S^1}{f(I)}\left(1-\frac{I^1f'(I)}{f(I)}\right) \\ \frac{(1-p)\beta^\alpha(1-k^\alpha)I^1}{f(I)} & \eta^\alpha & \frac{(1-p)\beta^\alpha(1-k^\alpha)S^1}{f(I)}\left(1-\frac{I^1f'(I)}{f(I)}\right) - A_2 \end{bmatrix}.$$

After computing the echelon form,

$$\begin{aligned} trace(J(E_1)) = \ & -(\mu^\alpha A_1 + p(\mu^\alpha + \phi^\alpha))\frac{\beta^\alpha(1-k^\alpha)I}{f(I)}I((1-p)A_1 + p\eta^\alpha) \\ & -\frac{(1-p)\beta^\alpha(1-k^\alpha)I}{f(I)}(\mu^\alpha + q^\alpha + \delta^\alpha)I \\ & -A_2\mu^\alpha\left[((1-p)A_1 + p\eta^\alpha) - (1-p)A_1A_2\left(1-\frac{If'(I)}{f(I)}\right)\right] \end{aligned}$$

and

$$\begin{aligned} det(J(E_1)) = \;& \left(\mu^\alpha A_1 + \tfrac{p\beta^\alpha(1-k^\alpha)(\mu^\alpha+\phi^\alpha)I}{f(I)}\right)\Big[(\mu^\alpha + q^\alpha + \delta^\alpha)(1-p)\tfrac{\beta^\alpha(1-k^\alpha)I}{f(I)} \\ & + A_2\mu^\alpha((1-p)A_1 + p\eta^\alpha) - \mu^\alpha(1-p)A_1A_2\left(1 - \tfrac{If'(I)}{f(I)}\right)\Big] \\ & + (\mu^\alpha(1-p) + \mu^\alpha)\left[\frac{\beta^\alpha\mu^\alpha A_1A_2\left(1-\frac{If'(I)}{f(I)}\right)(1-k^\alpha)I^1}{((1-p)A_1+p\eta^\alpha)+f(I)p(\mu^\alpha+q^\alpha+\delta^\alpha)}I\right]. \end{aligned}$$

Clearly, $trace(J(E_1)) < 0$ and $det(J(E_1)) > 0$ if $((1-p)A_1 + p\eta^\alpha) - \mu^\alpha(1-p)$ $A_1A_2\left(1 - \frac{If'(I)}{f(I)}\right) > 0$. This is equivalent to $R_0 > 1$. □

5. Numerical Simulations

This section is devoted to testing the performance of the proposed fractional-order model (1) under the Caputo differential operator while using a numerical explicit technique called the Adams–Bashforth–Moulton technique, also known as the fractional predictor-corrector method introduced and analyzed for its convergence and error bounds in [40,41]; $s = min(1+\alpha, 2)$ is the order of accuracy for the numerical technique. It is also worth mentioning that, unlike newly established numerical techniques for classical initial value problems, the present literature is not rich enough with numerical techniques for fractional order differential equations.

In this research, the parameter values used are from [28]; $\Lambda = 6$, $\beta = 10$, $\mu = 0.02$, $\gamma = 0.01$, $\xi = 0.02$, $\phi = 0.01, q = 0.02$, $\eta = 0.2$, $\delta = 0.1$, $p \in [0,1]$, $k = 0.2$, and $\alpha \in (0,1]$. Consider an equi-spaced mesh $t_i = i\Delta t$, $i = 0, 1, \ldots, M$, where M is a positive integer and $\Delta t = \frac{T}{M}$, where T is the upper limit of the closed interval of integration $[0, T]$. This setting leads to the following structure for the predictor part required by the numerical technique under consideration:

$$\begin{aligned} S^s_{i+1} &= S(0) + \sum_{i=1}^{n} b_{\alpha,i,n+1}g_1(t_i, S_i, E_i, I_i, R_i), \\ E^s_{i+1} &= E(0) + \sum_{i=1}^{n} b_{\alpha,i,n+1}g_2(t_i, S_i, E_i, I_i, R_i), \\ I^s_{i+1} &= I(0) + \sum_{i=1}^{n} b_{\alpha,i,n+1}g_3(t_i, S_i, E_i, I_i, R_i), \\ R^s_{i+1} &= R(0) + \sum_{i=1}^{n} b_{\alpha,i,n+1}g_4(t_i, S_i, E_i, I_i, R_i). \end{aligned} \tag{16}$$

Similarly, for the corrector, we have

$$\begin{aligned} S^u_{i+1} &= S(0) + a_{\alpha,n+1,n+1}g_1(t_i, S_i^s, E_i^s, I_i^s, R_i^s) + \sum_{i=1}^{n} a_{\alpha,i,n+1}g_1(t_i, S_i, E_i, I_i, R_i), \\ E^u_{i+1} &= E(0) + a_{\alpha,n+1,n+1}g_2(t_i, S_i^s, E_i^s, I_i^s, R_i^s) + \sum_{i=1}^{n} a_{\alpha,i,n+1}g_2(t_i, S_i, E_i, I_i, R_i), \\ I^u_{i+1} &= I(0) + a_{\alpha,n+1,n+1}g_3(t_i, S_i^s, E_i^s, I_i^s, R_i^s) + \sum_{i=1}^{n} a_{\alpha,i,n+1}g_3(t_i, S_i, E_i, I_i, R_i), \\ R^u_{i+1} &= R(0) + a_{\alpha,n+1,n+1}g_4(t_i, S_i^s, E_i^s, I_i^s, R_i^s) + \sum_{i=1}^{n} a_{\alpha,i,n+1}g_4(t_i, S_i, E_i, I_i, R_i). \end{aligned} \tag{17}$$

where

$$a_{\alpha,i,n+1} = \frac{(\Delta t)^\alpha}{\Gamma(\alpha+2)}, \text{ and } b_{\alpha,i,n+1} = \frac{1}{\Gamma(\alpha+2)}\left[(n-i+1)^\alpha - (n-i)^\alpha\right].$$

and s denote the predictor and u represent the corrector.

Additionally,

$$\begin{cases} n^{\alpha+1} - (n-\alpha)(n+1)^{\alpha}, & i = 0 \\ \quad (n-i+2)^{\alpha} - 2(n-i+1)^{\alpha+1} + (n-1)^{\alpha+1}, & 1 \leq i \leq n \\ 1, & i = n+1. \end{cases}$$

The behavior of the Susceptible $S(t)$, Exposed $E(t)$, Infected $I(t)$, and Recovered $R(t)$ population can be seen in Figure 1a–d where the required simulations have been carried out for T = 100 days while varying the values of the fractional order parameter α. For the Susceptible population, it is observed that the population decays at a faster rate for higher values of the fractional-order parameter. On the contrary, the higher values of fractional-order parameter α lead to an increase in the Exposed, Infected, and Recovered populations.

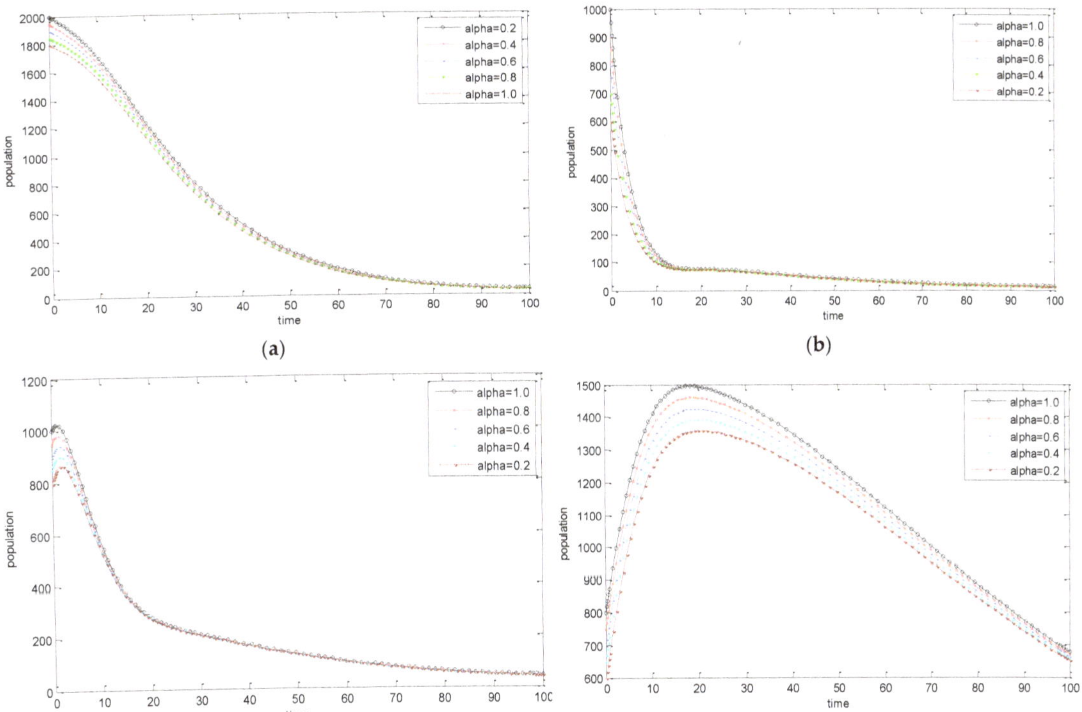

(**a**) (**b**) (**c**) (**d**)

Figure 1. (**a**) The dynamics of the Susceptible population for varying fractional order α. (**b**) The dynamics of the Exposed population for varying fractional order α. (**c**) The dynamics of the Infected population for varying fractional order α. (**d**) The dynamics of the Recovered population for varying fractional order α.

The behavior of the Exposed $E(t)$ and Infected $I(t)$ classes can be seen in Figure 2a,b where the required simulations have been carried out for T = 100 days while varying the values of the awareness parameter k. For both the Exposed and Infected classes, it can be observed that a higher value of awareness parameter k leads to a decrease in the Exposed and Infected populations.

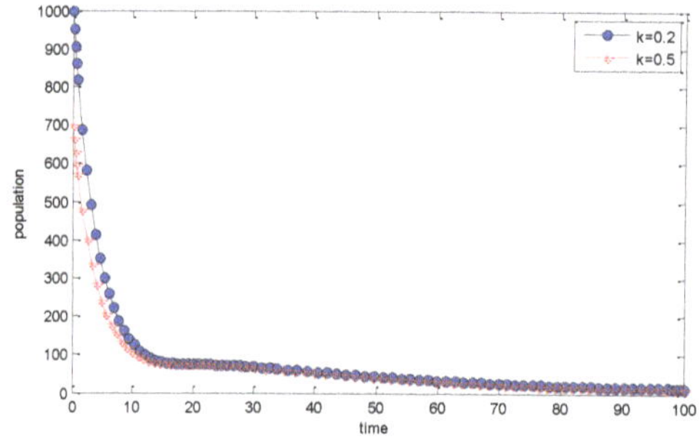

(a)

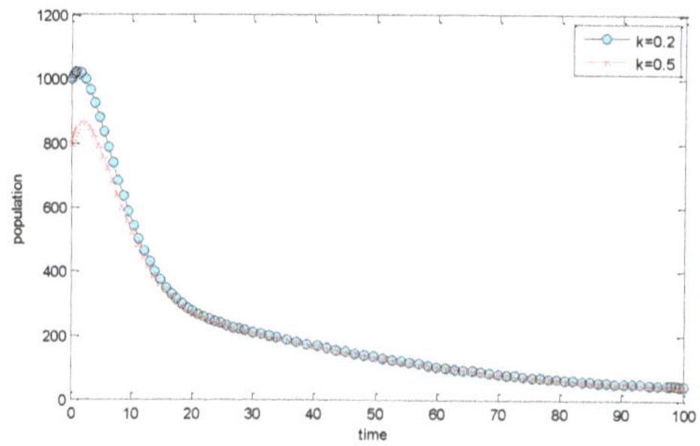

(b)

Figure 2. (**a**) The dynamics of the Exposed classes by increasing awareness parameter k. (**b**) The dynamics of the Infected classes by increasing awareness parameter k.

6. Summary and Conclusions

In this paper, a fractional order cholera model in the Caputo sense is constructed. The transmission dynamics of the disease are studied by incorporating the saturated incidence rate into the model. Various well-posedness properties such as positivity, boundedness, existence, and uniqueness of the solution are also studied. Equilibrium solutions were computed, and their stability analysis was shown to depend on a threshold quantity, the basic reproduction ratio (R_0). It was clearly shown that if $R_0 < 1$, the disease-free equilibrium is locally asymptotically stable, whereas if $R_0 > 1$, endemic equilibrium exists and is locally asymptotically stable. The model is found to be well-posed and more realistic than many related models in the literature due to the consideration of the saturated incidence rate.

Numerical simulations were carried out to support the analytic result and to show the significance of the fractional order from the biological point of view as well as show the significance of awareness programs in curtailing the spread of cholera in a population. From the result obtained, it is observed that the Susceptible population decay at a faster rate for higher values of the fractional-order parameter. On the contrary, the higher values of fractional-order parameter α lead to an increase in the Exposed, Infected, and Recovered populations. In addition, it is observed that for both the Exposed and Infected classes, the increase in the value of awareness parameter k leads to a decrease in the Exposed and Infected populations, respectively.

In conclusion, this paper studied a fractional order model in the Caputo sense with a saturated incidence rate. The entire well-posedness properties of the model were studied in detail. The awareness contribution in controlling cholera was studied numerically. From the findings of this research, one can see there is a need for the consideration of a saturated incidence rate in studying epidemic diseases such as cholera. Additionally, there is a need for studying well-posedness properties before making deductions on any mathematical model. From the findings of this model in particular, there is a need for relevant agencies to mount awareness programs in order to curtail the spread of cholera in a given population. Using a fractional-order to construct a mathematical model is also of paramount importance due to its hereditary properties and memory description ability. The limitation of this work is that there is a need to consider the agent-based approach of the model proposed in this paper which will help in providing more decision-making tools for the effective prevention and control of cholera and public health interventions.

Author Contributions: Conceptualization, I.A.B. and U.W.H.; methodology, I.A.B.; software, I.A.B. and F.A.R.; validation, I.A.B., U.W.H. and F.A.R.; formal analysis, I.A.B. and U.W.H.; investigation, F.A.R.; resources, I.A.B.; data curation, F.A.R.; writing—original draft preparation, I.A.B., U.W.H. and F.A.R.; writing—review and editing, I.A.B. and F.A.R.; visualization, F.A.R.; supervision, U.W.H.; project administration, U.W.H.; funding acquisition, U.W.H. and F.A.R. All authors have read and agreed to the published version of the manuscript.

Funding: This research received no external funding.

Data Availability Statement: Not applicable.

Acknowledgments: This research was supported by King Mongkut's University of Science and Technology Thonburi's Postdoctoral Fellowship.

Conflicts of Interest: The authors declare no conflict of interest.

References

1. Ganesan, D.; Gupta, S.S.; Legros, D. Cholera surveillance and estimation of burden of cholera. *Vaccine* **2020**, *38*, A13–A17. [CrossRef] [PubMed]
2. Idoga, P.E.; Toycan, M.; Zayyad, M.A. Analysis of factors contributing to the spread of cholera in developing countries. *Eurasian J. Med.* **2019**, *51*, 121–127. [CrossRef] [PubMed]
3. Llanes, R.; Somarriba, L.; Hern'andez, G.; Bardaj, Y.; Aguila, A.; Mazumder, R.N. Low detection of vibrio cholera carriage in healthcare workers returning to 12 Latin American countries from Haiti. *Epidemiol. Infect.* **2015**, *143*, 1016–1019. [CrossRef] [PubMed]
4. Momba, M.; El-Liethy, M.A. *Vibrio Cholerae and Cholera Biotypes*; Global Water Pathogen Project: Pretoria, South Africa, 2018.
5. Javidi, M.; Ahmad, B. A study of a fractional-order cholera model. *Appl. Math. Inf. Sci.* **2014**, *8*, 2195–2206. [CrossRef]
6. Lamond, E.; Kinyanjui, J. *Cholera Outbreak Guidelines: Preparedness, Prevention and Control*; Oxfam GB: Oxford, UK, 2012.
7. Eskandari, Z.; Avazzadeh, Z.; Khoshsiar, G.; Li, B. Dynamics and bifurcations of a discrete—Time Lotka–Volterra model using nonstandard finite difference discretization method. *Math. Methods Appl. Sci.* **2022**. [CrossRef]
8. Li, B.; Liang, H.; Shi, L.; He, Q. Complex dynamics of Kopel model with nonsymmetric response between oligopolists. *Chaos Solitons Fractals* **2022**, *156*, 111860. [CrossRef]
9. Liu, X.; Arfan, M.; Ur Rahman, M.; Fatima, B. Analysis of SIQR type mathematical model under Atangana-Baleanu fractional differential operator. *Comput. Methods Biomech. Biomed. Eng.* **2022**, *3*, 1–5. [CrossRef]
10. Kermack, W.O.; McKendrick, A.G. A contribution to the mathematical theory of epidemics. *Proc. R. Soc. Lond. Ser. A Contain. Pap. A Math. Phys. Character* **1927**, *115*, 700–721.
11. Marinov, T.; Marinova, R. Adaptive SIR model with vaccination: Simultaneous identification of rates and functions illustrated with COVID-19. *Sci. Rep.* **2022**, *12*, 1–13.
12. Alam, A.; LaRocque, R.C.; Harris, J.B.; Vanderspurt, C.; Ryan, E.T.; Qadri, F.; Calderwood, S.B. Hyperinfectivity of human-passaged Vibrio cholerae can be modelled by growth in the infant mouse. *Infect. Immun.* **2005**, *73*, 6674–6679. [CrossRef]
13. Codeço, C. Endemic and epidemic dynamics of cholera: The role of the aquatic reservoir. *BMC Infect. Dis.* **2001**, *1*, 1–14. [CrossRef]
14. Hartley, D.M., Jr.; Morris, J.; Smith, D. Hyperinfectivity: A critical element in the ability of *V. cholerae* to cause epidemics? *PLoS Med.* **2006**, *3*, 63–69. [CrossRef] [PubMed]
15. Mukandavire, Z.; Liao, S.; Wang, J.; Gaff, H.; Smith, D.; Morris, J. Estimating the reproductive numbers for the 2008–2009 cholera outbreaks in Zimbabwe. *Proc. Natl. Acad. Sci. USA* **2011**, *108*, 8767–8772. [CrossRef] [PubMed]
16. Nelson, E.; Harris, J.; Morris, J.; Calderwood, S.; Camilli, A. Cholera transmission: The host, pathogen and bacteriophage dynamics. *Nat. Rev. Microbiol.* **2009**, *7*, 693–702. [CrossRef] [PubMed]
17. Shuai, Z.; Tien, J.H.; Van den Driessche, P. Cholera models with hyper infectivity and temporary immunity. *Bull. Math. Biol.* **2012**, *74*, 2423–2445. [CrossRef]
18. Shuai, Z.; van den Driessche, P. Modeling and control of cholera on networks with a common water source. *J. Biol. Dyn.* **2015**, *9*, 90–103. [CrossRef]

19. Escalante-Martínez, J.E.; Gómez-Aguilar, J.F.; Calderón-Ramón, C.; Aguilar-Meléndez, A.; Padilla-Longoria, P. Synchronized bioluminescence behavior of a set of fireflies involving fractional operators of Liouville Caputo type. *Int. J. Biomath.* **2018**, *11*, 1–24. [CrossRef]
20. Escalante-Martínez, J.E.; Gómez-Aguilar, J.F.; Calderón-Ramón, C.; Aguilar-Meléndez, A.; Padilla-Longoria, P. A mathematical model of circadian rhythms synchronization using fractional differential equations system of coupled van der Pol oscillators. *Int. J. Biomath.* **2018**, *11*, 1850041. [CrossRef]
21. Ullah, S.; Khan, M.A.; Farooq, M. A fractional model for the dynamics of TB virus. *Chaos Solitons Fractals* **2018**, *116*, 63–71. [CrossRef]
22. Aguilar, J.F.G. Fundamental solutions to electrical circuits of non-integer order via fractional derivatives with and without singular kernels. *Eur. Phys. J. Plus* **2018**, *133*, 1–20.
23. Capasso, V.; Serio, G. A generalization of the Kermack-Mckendrick deterministic epidemic model. *Math. Biosci.* **1978**, *42*, 43–61. [CrossRef]
24. Liu, W.M.; Levin, S.A.; Iwasa, Y. Influence of nonlinear incidence rates upon the behavior of SIRS epidemiological models. *J. Math. Biol.* **1986**, *23*, 187–204. [CrossRef] [PubMed]
25. Xiao, D.; Ruan, S. Global analysis of an epidemic model with non monotone incidence rate. *Math. Biosci.* **2007**, *208*, 419–429. [CrossRef] [PubMed]
26. Leo, J. Complexity of Epidemics Models: A Case-Study of Cholera in Tanzania. In *Digital Transformation for Sustainability*; Marx Gómez, J., Lorini, M.R., Eds.; Springer: Cham, Switzerland, 2022. [CrossRef]
27. Tchatat, D.; Kolaye, G.; Bowong, S.; Temgoua, A. Theoretical assessment of the impact of awareness programs on cholera transmission dynamic. *Int. J. Nonlinear Sci. Numer. Simul.* **2022**. [CrossRef]
28. Wang, Y.; Abdeljawad, T.; Din, A. Modeling the dynamics of stochastic norovirus epidemic model with time-delay. *Fractals* **2022**, *30*, 1–13. [CrossRef]
29. Lemos-Paião, A.; Silva, C.; Torres, D. A cholera mathematical model with vaccination and the biggest outbreak of world's history. *AIMS Math.* **2018**, *3*, 448–463. [CrossRef]
30. Capasso, V.; Paveri-Fontana, S. A mathematical model for the 1973 cholera epidemic in the European Mediterranean region. *Rev. D'epidémiologie Et De St.* **1979**, *27*, 121–132.
31. Nishiura, H.; Tsuzuki, S.; Yuan, B.; Yamaguchi, T.; Asai, Y. Transmission dynamics of cholera in Yemen, 2017: A real time forecasting Theoret. *Biol. Med. Model.* **2017**, *14*, 14. [CrossRef]
32. Neilan, R.; Schaefer, E.; Gaff, H.; Fister, K.; Lenhart, S. Modeling optimal intervention strategies for cholera. *Bull. Math. Biol.* **2010**, *72*, 2004–2018.
33. Luchko, Y.; Yamamoto, M. General time-fractional diffusion equation: Some uniqueness and existence results for the initial-boundary-value problems. *Fract. Calculus Appl. Anal.* **2016**, *19*, 676–695. [CrossRef]
34. Jajarmi, A.; Baleanu, D.; Vahid, K.Z.; Pirouz, H.M.; Asad, J.H. A new and general fractional Lagrangian approach: A capacitor microphone case study. *Res. Phys.* **2021**, *31*, 104950. [CrossRef]
35. Podlubny, I. *Fractional Differential Equations*; Academic Press: San Diego, CA, USA, 1999.
36. Kilbas, A.A.; Srivastava, H.M.; Trujillo, J.J. *Theory and Application of Fractional Differential Equations*; Elsevier: Amsterdam, The Netherlands, 2006; Volume 204.
37. van den Driessche, P.; Watmough, J. Reproduction numbers and sub-threshold endemic equilibria for compartmental models of disease transmission. *Math. Biosci.* **2002**, *180*, 29–48. [CrossRef] [PubMed]
38. Ullah, M.; Baleanu, D. A new fractional SICA model and numerical method for the transmission of HIV/AIDS. *Math. Methods Appl. Sci.* **2021**, *44*, 8648–8659. [CrossRef]
39. Diethelm, K.; Ford, N.; Freed, A. A predictor-corrector approach for the numerical solution of fractional differential equations. *Nonlinear Dyn.* **2002**, *29*, 3–22. [CrossRef]
40. Diethelm, K.; Ford, N.; Freed, A. Detailed error analysis for a fractional Adams method. *Numer. Algorithms* **2004**, *36*, 31–52. [CrossRef]
41. Khan, M.; Parvez, M.; Islam, S.; Khan, I.; Shafie, S.; Gul, T. Mathematical Analysis of Typhoid Model with Saturated Incidence Rate. *Adv. Stud. Biol.* **2015**, *7*, 65–78. [CrossRef]

entropy

MDPI

Article

A Numerical Study of the Dynamics of Vector-Born Viral Plant Disorders Using a Hybrid Artificial Neural Network Approach

Hosam Alhakami [1,*], Muhammad Umar [2], Muhammad Sulaiman [2,*], Wajdi Alhakami [3] and Abdullah Baz [4]

1 Department of Computer Science, College of Computer and Information Systems, Umm Al-Qura University, Makkah 21955, Saudi Arabia
2 Department of Mathematics, Abdul Wali Khan University, Mardan 23200, Pakistan
3 Department of Information Technology, College of Computers and Information Technology, Taif University, Taif 21944, Saudi Arabia
4 Department of Computer Engineering, College of Computer and Information Systems, Umm Al-Qura University, Makkah 21955, Saudi Arabia
* Correspondence: hhhakam@uqu.edu.sa (H.A.); msulaiman@awkum.edu.pk (M.S.)

Abstract: Most plant viral infections are vector-borne. There is a latent period of disease inside the vector after obtaining the virus from the infected plant. Thus, after interacting with an infected vector, the plant demonstrates an incubation time before becoming diseased. This paper analyzes a mathematical model for persistent vector-borne viral plant disease dynamics. The backpropagated neural network based on the Levenberg—Marquardt algorithm (NN-BLMA) is used to study approximate solutions for fluctuations in natural plant mortality and vector mortality rates. A state-of-the-art numerical technique is utilized to generate reference data for obtaining surrogate solutions for multiple cases through NN-BLMA. Curve fitting, regression analysis, error histograms, and convergence analysis are used to assess accuracy of the calculated solutions. It is evident from our simulations that NN-BLMA is accurate and reliable.

Keywords: mathematical modeling; artificial neural networks; numerical solutions; delay differential equations; optimization techniques; machine learning; Levenberg—Marquardt algorithm

Citation: Alhakami, H.; Umar, M.; Sulaiman, M.; Alhakami, W.; Baz, A. A Numerical Study of the Dynamics of Vector-Born Viral Plant Disorders Using a Hybrid Artificial Neural Network Approach. *Entropy* **2021**, *24*, 1511. https://doi.org/10.3390/e24111511

Academic Editors: Pavel Kraikivski and Cristóbal López

Received: 14 August 2022
Accepted: 18 October 2022
Published: 22 October 2022

1. Introduction

Plant disease epidemiology studies how diseases affect plant populations and how to combat plant diseases. Using spatial and temporal plant epidemiology models can provide useful statistical and mathematical data about disease transmission. In the mid-20th century, plant epidemiological models became prominent [1]. Examples of actual uses of this type of model include cassava mosaic disease [2], pine wilt disease [3], and potato late blight [4]. Later, new methods for studying nonlinear dynamics and numerical simulations helped solve complex ecological problems [5,6]. This accelerated the creation of more realistic and complicated plant disease models.

An essential part of the plant epidemiological system is modeling the interactions between infected and healthy plant populations, either directly or via a vector. Infected vectors feed on healthy plants, infecting them. Similarly, non-infected vectors become infected by diseased plants. The vector-borne plant disease is classified as persistent, semipersistent, or non-persistent based on the infectious agent's residence period in the vector [7,8]. The vector ingests viruses while feeding on infected plant sap in persistent transmission. The salivary glands then release the viruses into the plant tissue as they penetrate the digestive system. The persistent mode of transmission differs from the other two because it takes a long time for a vector to become infected with the virus and become infectious [7,9]. In the case of vectors, this time lag is referred to as the latent phase of infection.

The latent period in plants is similar to the time it takes for a healthy plant to become infected following infection [10]. The incubation period (or incubation time) is the time it takes for symptoms to manifest following infection [1]. Depending on the plant species, the incubation period varies [11]. Incubation durations for beet mosaic virus (BMV), African cassava mosaic virus (ACMV), tobacco mosaic virus (TMV) and bean golden mosaic virus (BGMV) are 7–15 days [12], 3–5 weeks [13], 5 h [14], and 5–6 days [15], respectively. The incubation and latent periods in plants are distinct. However, the expression of disease symptoms correlates with disease transmission [16]. Furthermore, determining the latent period is challenging, whereas observing disease signs is straightforward. So our model development analysis considers the incubation period.

Among the most frequent vector-borne viral diseases affecting crops, leaf curl disease and mosaic disease are two of the most common. The whitefly (*Bemisia* sp.), which transmits several viral infections to Jatropha, cassava, tomato, tobacco, cotton, and other plants, is a hemipteran vector. Most of the disease is systematically spread by whiteflies, meaning that a latent period is frequently observed [17]. Unfortunately, information on the latent and incubation time of infection for various persistently transmitted diseases is lacking in the literature. Due to the variety of viral agents and host plant species, both delay methods have varying effects on disease severity. It also differs between whitefly species and host plants. These delays may vary due to genetic complexity, climate fluctuation, phenotypic heterogeneity, and plasticity [18]. The plant incubation period is usually longer than the latent period in vectors. For example, ACMV has a 6-hour latent period and a 3–5 week incubation period [13].

Ordinary differential equations (ODEs) models cannot account for the incubation or latent period. However, models based on delay differential equations (DDEs) allow system integration. It can represent a system's dynamics when its evolution depends on prior events. When time lag responses exist, delays are one of the most powerful mathematical modeling tools [19]. DDE models are more sophisticated than ODE models but more realistic. Prey–predator mathematical models with delay differential equations are commonly employed [20,21]. Delay can teach us dynamic phenomena, such as instability, oscillations, and bifurcation.

Van der Plank [1] used DDE to delay plant epidemics. Cooke [22] proposed a model with an incubation time state variable for vector-borne diseases. Wang et al. [23] discussed wheat starch and gluten's thermal characteristics and interactions. Zhang [24] added the plant incubation period to a Meng and Li [25] plant disease model, causing modifications in the model's dynamics. Munyasya et al. [26] proposed an integrated on-site and off-site rainwater-harvesting system that enhances rainfed maize output for better climate change adaption. Buonomo and Cerasuolo [27] presented and analyzed a soil-borne plant disease dynamics model. Miao [28] suggested an accuracy of space-for-time substitution for predicting vegetation status after shrub restoration.

An ODE model of the impact of replanting and roguing on eliminating plant disease latency comprises a compartment for latently diseased plant populations [29]. The model does not consider any vector compartment, but it includes classes of latently infected, healthy, post-infection, and infectious plants. Holt et al. [2] proposed a model with infected plants, healthy vectors, and susceptible vectors but no delays. The vector-borne plant disease model [30] was modified by Jackson and Chen [31] by delaying plant incubation and vector latent periods. The threshold value for delay-induced destabilization was determined by observing changes in system solution dynamics. Li et al. used an updated model [31] to analyze Hopf bifurcation, which included incubation and latent period characteristics [32].

Banerjee and Takeuchi [33] identified several critical elements of the dynamics that could lead to false findings. A long wait can stabilize or cure a system Buonomo, and Cerasuolo [27]. Transcritical bifurcations, periodic oscillations, and stability switches can be revealed if the vector-borne plant disease models' parameters change [2,27,34]. The undelayed model analysis cannot be ignored [31,32]. A mathematical model (1) with parameters given in Table 1 [2,35], which was previously analyzed by Basir et al. [35] for persistent vector-borne viral plant disease dynamics for the effect of both latent period and incubation delay of the dynamics of the deceased. This model is numerically analyzed using a gradient-based numerical technique. Numerous studies claimed that gradient-based techniques, such as RK-4, take up much more computer time than soft computing methods with comparable accuracy and that it is difficult to produce accurate global estimates of the truncation error [36,37]. For instance, at each step of the RK-4 method, the derivative must be evaluated n times. Here, 'n' is the order of accuracy of the RK-4 method, which is a significant drawback of gradient-based algorithms [38]. Moreover, RK-4 suffers from divergence for complex systems [39]. Failure in the case of singularity is another hurdle in using these gradient-based numerical techniques. Keeping these disadvantages in mind, the authors of this paper aimed to suggest an alternative gradient-free approach that can handle problems, such as model (1), with accuracy and reliability. The key features of this study are outlined as follows:

- In this paper, we analyzed an established mathematical model (1) for persistent vector-borne viral plant disease dynamics, which is presented in Section 2. The set of parameters substituted in the model is for the case of cassava mosaic disease.
- A gradient-free intelligent design of a two-layer artificial neural network architecture and the Levenberg—Marquardt algorithm is utilized to formulate surrogate solutions. A state-of-the-art numerical method is used to calculate reference solutions for establishing the accuracy, validity, and reliability of NN-BLMA; see Section 3.
- The impact of variations in parameters, such as plant mortality and vector mortality rate, on the model of persistent vector-borne viral plant disease dynamics is observed through the surrogate solutions formulated by the designed NN-BLMA; see Section 4. Graphical analysis for the convergence of NN-BLMA is carried out based on mean square error, regression analysis plots, and error histograms. Moreover, statistical values are tabulated to show the accuracy and reliability of the designed technique.

Table 1. Parameters' description and their numerical values.

Parameters	Description	Values	Unit
r	Net growth rate of plants	0.3	time^{-1}
K	Carrying capacity	1	m^{-2}
λ	Infected vector to healthy plant disease transmission rate	0.025	vector^{-1}time^{-1}
μ	Plants natural mortality rate	0.1	time^{-1}
m_1	Mortality of infected plants	0.01	time^{-1}
Π	Vector population's overall growth rate due to immigration or births	40	time^{-1}
β	Transmission rate between diseased vector and healthy plant	0.03	plant^{-1}time^{-1}
d	Vector mortality rate	0.1	time^{-1}

2. Problem Formulation

This section develops a mathematical model for persistent vector-borne viral plant disease dynamics. The model considers plant and vector populations without explicitly including the mosaic virus. $H(t)$ signifies healthy plants, while the infected plants are represented by $I(t)$, $Q(t)$ represents uninfected, and W(t) represents the infected whiteflies population.

Due to restricted plantation space and natural resources, logistic growth r and carrying capacity K are considered for healthy plants [2]. A healthy plant becomes infected when it comes into contact with an infected vector. When an infected vector and a susceptible plant are present, λ is the transmission rate, and λHW is the number of sensitive individuals moving from the susceptible compartment to the infected compartment.

An insect pest, such as a whitefly, shifts its host in response to changing biological and environmental conditions. They generally move between fields of crops [40,41]. They breed in the fields. The Holling type III survival curve describes their life course because of the high death rate they experience early on [41]. Whiteflies (adults and nymphs) can transmit illness.

Crops are typically planted and reaped at specific times of the year. Most crops are reaped a few months after they are produced. A few vectors travel from close or distant patches and reproduce in the vegetation. Vectors grow by migrating from another patch because of reproducing in the same patch or vegetative area. For the same reason, seasonal fluctuations in vector populations are ignored [35].

An open system is considered in this model. Assume Π is the rate of vector birth and migration into the system. No vertical virus transmission is allowed, and a vector cannot infect another vector. Viruses do not destroy or defend vectors. The vector retains the virus and does not recover. However, the infective insects do not get sick from the virus [31]. Let the mortality rate of plants and vectors be represented by μ and d, respectively. Infection-related plant death is expected to be higher than average plant mortality. m_1 is the infection-related mortality of infected plants. Thus, the overall plant mortality rate is $m = \mu + m_1$. Consider β to be the conversion between uninfected vectors (i.e., Q) and the infected plant (i.e., I). So, βQI signifies entering the number of uninfected vectors Q into the infected vectors W compartment.

In truth, both plant and vector infection takes time. Let $\tau_1 \in R^+$ be the healthy plant's incubation time following successful infection. At time t, the disease transmission is given by the expression $\lambda e^{-m\tau_1} H(t-\tau_1)W(t-\tau_1)$, where the positive constants described previously are λ and μ. The term $e^{-m\tau_1}$ denotes the chance of a healthy plant surviving through the incubation time $[t-\tau_1, t]$, i.e., the number of susceptible plants that came into touch with an infected vector at time $t-\tau_1$ and lived up to time t to become infected plants.

Again the latent period in a vector is $\tau_2 \in R^+$. At time t, the expression $\beta e^{-d\tau_2} Q(t-\tau_2)$ $I(t-\tau_2)$ describes the transmission of infection, where $e^{-d\tau_2}$ reflects the vector's survival probability across the latent time $[t-\tau_2, t]$. The number of uninfected vectors met an infected vector at time $t-\tau_2$ and survived until time t to become infected [35]. Based on the given assumptions, the mathematical model is

$$\begin{aligned}
\frac{dH}{dt} &= rH\left[1-\frac{H+I}{K}\right] - \lambda HW,\\
\frac{dI}{dt} &= \lambda e^{-m\tau_1} H(t-\tau_1)W(t-\tau_1) - mI,\\
\frac{dQ}{dt} &= \Pi - \beta QI - dQ,\\
\frac{dW}{dt} &= \beta e^{-d\tau_2} Q(t-\tau_2)I(t-\tau_2) - dW,
\end{aligned} \tag{1}$$

The initial biological conditions are
$H(t) > 0, I(t) > 0, Q(t) > 0, W(t) > 0; t \in [-\tau, 0], \tau = max[\tau_1, \tau_2]$,
The parameters used in the model (1) assigned some numerical values for solving the model numerically, and Table 1 shows its description and numerical values.

3. Design Methodology

This section examines artificial neural networks (ANN) using a novel approach to machine learning by focusing on the supervised neuronal learning mechanisms of these networks to utilize the study of the model for persistent vector-borne viral plant disease dynamics.

3.1. Artificial Neural Network (ANN)

An artificial neural network is a network of interconnected core components known as neurons that receives various inputs and generates only one output; each neuron represents a mapping. A neuron's output is a function of the total of its inputs produced by the activation function.

3.2. Activation Function

To introduce nonlinear properties, an activation function is used in an ANN. In a neural network, (Xi, Wi) stands for inputs, weights, and $f(Xi)$, which is the input function that is sent to the network's output. This output function can then be used as an input for any additional layers or the final output [42–44].

The number of hidden units can be optimized using a multilayer perceptron (MLP). Both the weights and biases of the connections were enhanced as well. The construction of a standard MLP with one hidden layer is as follows:

$$H_j = \sum_{i=1}^{n} W_{ij} X_i + b_j, \tag{2}$$

X_i represents the inputs, where W_{ij} and b_j represent connection weights and biased vectors, respectively. Here, a log–sigmoid function is used as an activation function in the feed-forward neural network model, which is given below.

$$f_j(x) = \frac{1}{1 + e^{-H_j}}. \tag{3}$$

The MLP, also known as the feed-forward neural network (FNN), is a type of neural network with a hidden layer between the input and output layers. This layer is called the "hidden layer." The number below the hidden layer represents the number of neurons used inside the network. Figure 1 shows an artificial neural network controller.

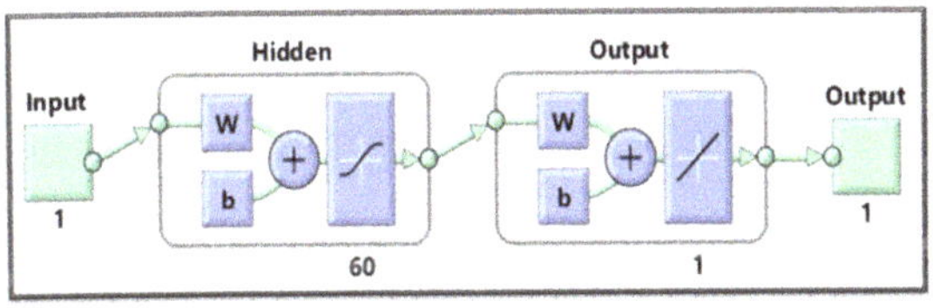

Figure 1. Architecture of an artificial neural network controller.

A backpropagated Levenberg—Marquardt method is used to train the feed-forward neural network. Local minima can be found using the LM algorithm, which is built-in in many applications.

Additionally, NN-BLMA is implemented in two phases. Figure 2 depicts the Algorithm's whole workflow, including all of its steps.

- For collecting the initial reference data set, we solve the model (1) numerically by using a state-of-the-art technique. Here we use the $RK-4$ method, which commonly gave better results, in Mathematica using the "NDSolve" package. The numerical technique generates 5001 in the range of $[0, 50]$ with a 0.01 stepsize.
- After that, the NN-BLMA is executed by using "nftool", a built-in MATLAB tool, to train, validate and test the targets (reference data set). The design technique uses 60% of the targets for training and 20% each for validation and testing. The maximum iteration is set to 1000 with a 60 number of neurons. Table 2 presents the parameters for the design scheme execution, and Algorithm 1 is the pseudo-code of the designed NN-BLMA.

Table 2. The NN-BLMA parameters settings for implementation.

Index	Learning Methodology	Training	Validation	Testing	Hidden Neuron	Max. Iteration
Description	Levenberg—Marquardt	60%	20%	20%	60	1000

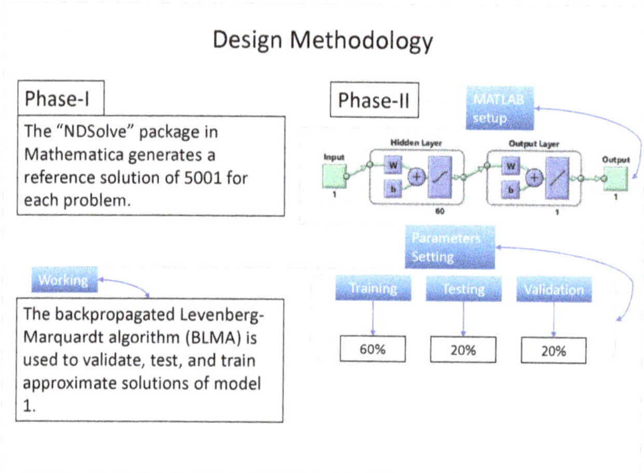

Figure 2. Working mechanism of the NN-BLMA for solving the nonlinear model of vector-borne viral plant disease dynamics.

The novel machine learning of NN-BLMA is easy to apply, handles nonlinear problems, and is also a gradient-free technique that converges faster than other machine learning technique [45–48].

Algorithm 1 Pseudocode of NN-BLMA:

Starting of NN-BLMA Construction: Construct inputs and reference data set using RK-4 method in Mathematica
Data selection: Input and target data must be selected in non-linear format, i.e., matrices.
Startup: Taking number of neurons and distributing the reference data set into training, testing and validation

- 60 Hidden neurons
- 60% data for training
- 20% data for testing
- 20% data for validation

Architecture: Each input is given a weight, and the input to the transfer function is formed by adding the weights of all of the inputs together along with the bias.
Stopping criteria: If all of the conditions listed below are met, the previous process will end automatically.

- Mu reach to its maximum value
- Number of iteration reaches to maximum
- Performance value reaches to minimum
- Validation's performance became less then maximum fail
- Gradient's performance dropped below minimum gradient

The network is generalised using training data. If the outputs are good, proceed to Saving Output; otherwise, retrain the network.
Retraining: Change the startup parameters and train the network again
Saving outputs: End the process by saving the results graphically as well as numerically
Ending of NN-BLMA

4. Numerical Experimentation and Discussion

To study the design algorithms' performance and efficiency, we discuss various cases of the nonlinear model of vector-borne viral plant disease dynamics. The cases are based on variation in two parameters (i.e., plants' natural mortality rate, μ, and vector mortality rate, d). We set the same numerical value for both parameters in the first case. In case two, there is a slight decrease in the μ parameter and a slight increase in the parameter d, while in the third case, there is an increase in the parameter μ and a decrease in the d parameter compared with the first case. Figure 3 illustrates the mathematical model and the cases detail for vector-borne viral plant disease dynamics.

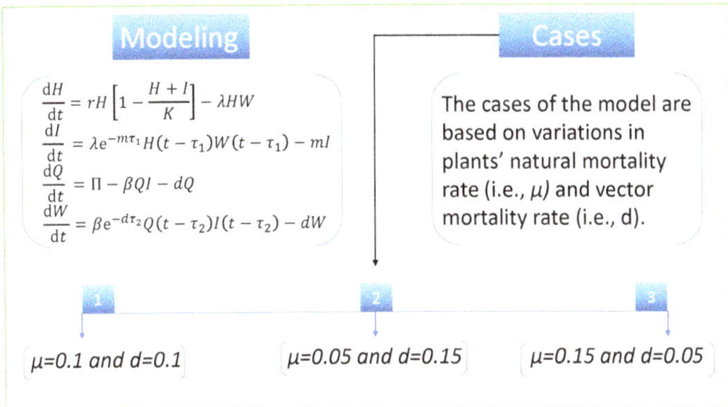

Figure 3. Vector-borne viral plant disease dynamics' model with its different cases.

The design technique generates output data sets with probabilities of 60% of the sample data for testing, 20% for training, and 20% for validation. The performance graph of the design technique shows us its mean squared error (MSE). Figures 4–6 depict the best validation performance provided by the design technique because the error is minimized after some epochs of training but may increase on the validation data set as the network begins to overfit the training data. The training is halted after six consecutive rises in the validation error, and the best performance is picked from the epoch with the lowest validation error. The case 1 performance values are in the range of 2.9721×10^{-9}, 7.1129×10^{-9}, 3.0066×10^{-8} and 2.8222×10^{-5}. Similarly, the case 2 and case 3 performance values are in the range of 1.3057×10^{-9}, 3.6923×10^{-11}, 1.17878×10^{-9}, 1.9703×10^{-4}, and 9.9788×10^{-11}, 3.1230×10^{-9}, 2.4709×10^{-8}, 2.7474×10^{-4}, respectively.

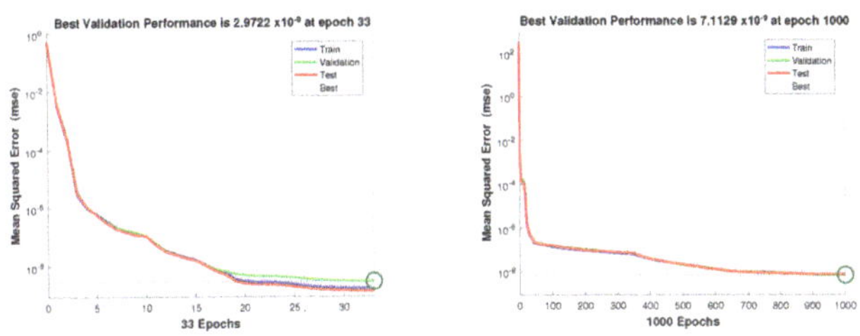

(a) (b)

Figure 4. *Cont.*

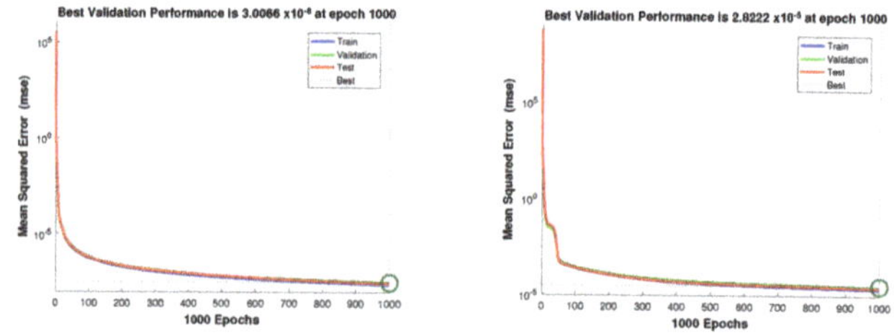

(**c**) (**d**)

Figure 4. NN-BLMA MSE for healthy and infected plants, and infected and uninfected whitefly of case 1. (**a**) $H(t)$. (**b**) $I(t)$. (**c**) $Q(t)$. (**d**) $W(t)$.

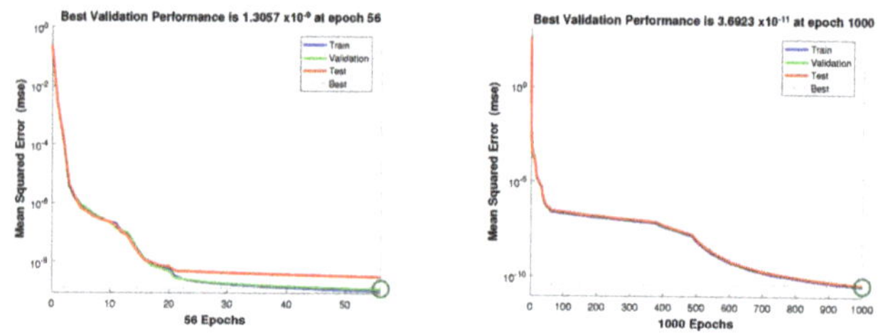

(**a**) (**b**)

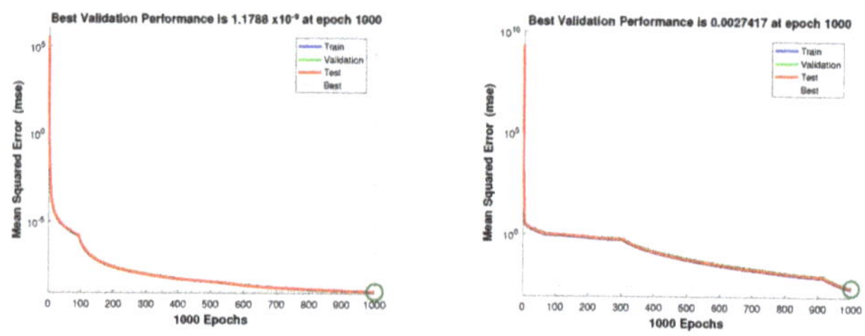

(**c**) (**d**)

Figure 5. NN-BLMA MSE for healthy and infected plants, and infected and uninfected whitefly of case 2. (**a**) $H(t)$. (**b**) $I(t)$. (**c**) $Q(t)$. (**d**) $W(t)$.

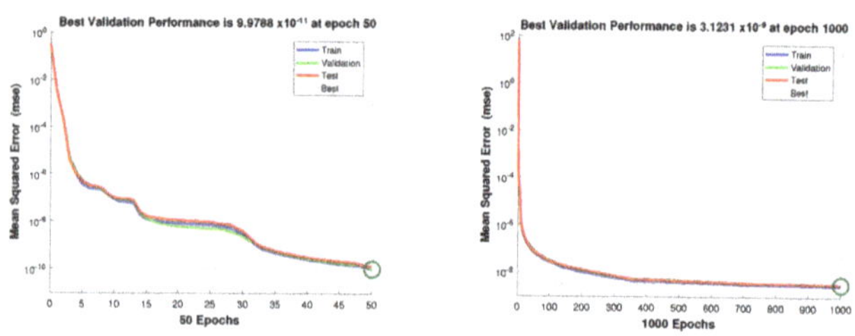

(**a**) (**b**)

Figure 6. *Cont.*

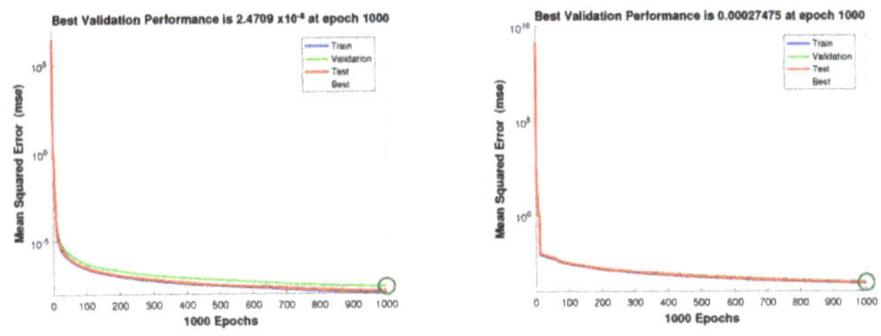

(c) (d)

Figure 6. NN-BLMA MSE for healthy and infected plants, and infected and uninfected whitefly of case 3. (**a**) $H(t)$. (**b**) $I(t)$. (**c**) $Q(t)$. (**d**) $W(t)$.

The statistical performance of all the cases in gradient, mu, and validation failures are illustrated in Figures 7–9. The gradient values for the case 1 lie in between 8.2149×10^{-8}, 2.4163×10^{-6}, 2.3721×10^{-4} and 0.2785, whereas the values for case 2 and case 3 are 9.2809×10^{-8}, 1.5761×10^{-6}, 1.4908×10^{-4}, 27.6472, and 1.1132×10^{-8}, 4.0741×10^{-6}, 3.0463×10^{-4}, 0.49652, respectively. The mu values for all the cases lie in the range 10^{-4} to 10^{-13}. The network output concerning the target for the training, validation, and test sets is shown on the regression plot. The data must fall on a 45-degree line where the network outputs and targets are equal for a perfect match. When the data fall on a 45 degree, the regression plot gives us a value of $R = 1$. This article shows the regression analysis of all the cases in Figures 10–12. From the figures, regression values are 1 for all cases, which perfectly matches the network and the targets.

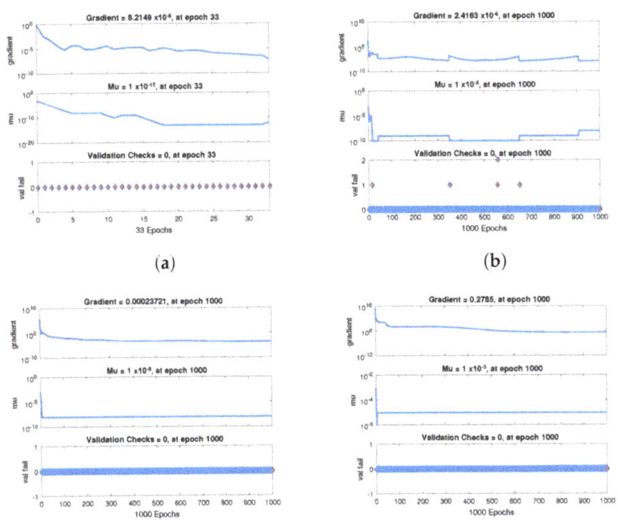

Figure 7. Value of gradient, mu and validation check of NN-BLMA for case 1. (**a**) $H(t)$. (**b**) $I(t)$. (**c**) $Q(t)$. (**d**) $W(t)$.

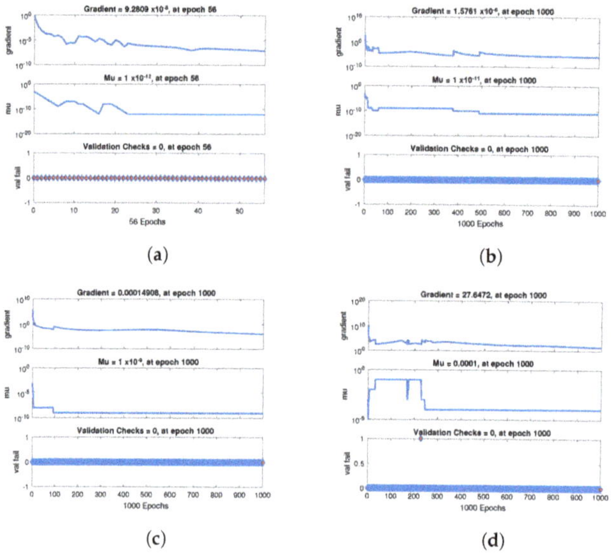

Figure 8. Value of gradient, mu and validation check of NN-BLMA for case 2. (**a**) $H(t)$. (**b**) $I(t)$. (**c**) $Q(t)$. (**d**) $W(t)$.

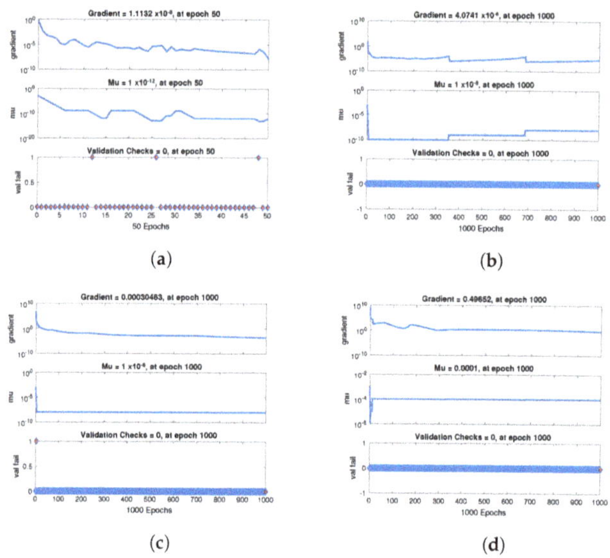

Figure 9. Value of gradient, mu and validation check of NN-BLMA for case 3. (**a**) $H(t)$. (**b**) $I(t)$. (**c**) $Q(t)$. (**d**) $W(t)$.

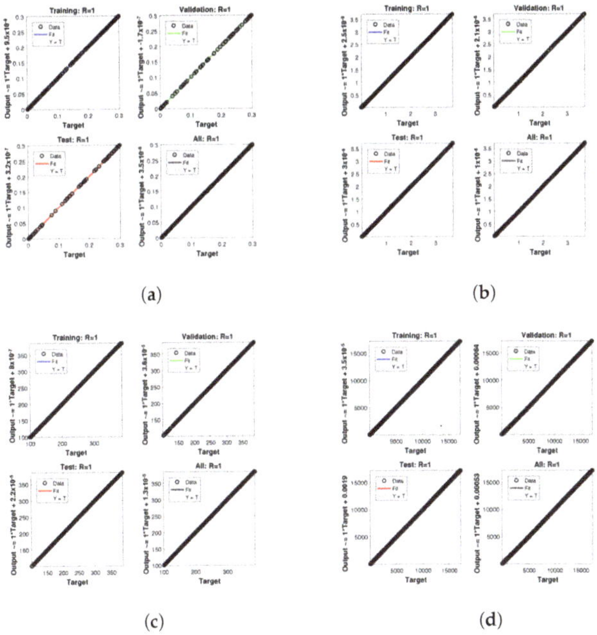

Figure 10. Analysis of regression of the design NN-BLM algorithm for case 1. (**a**) $H(t)$. (**b**) $I(t)$. (**c**) $Q(t)$. (**d**) $W(t)$.

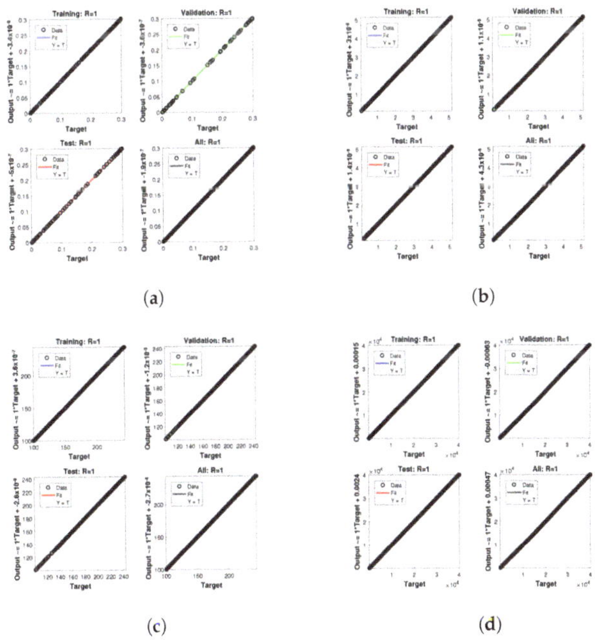

Figure 11. Analysis of regression of the design NN-BLM algorithm for case 2. (**a**) $H(t)$. (**b**) $I(t)$. (**c**) $Q(t)$. (**d**) $W(t)$.

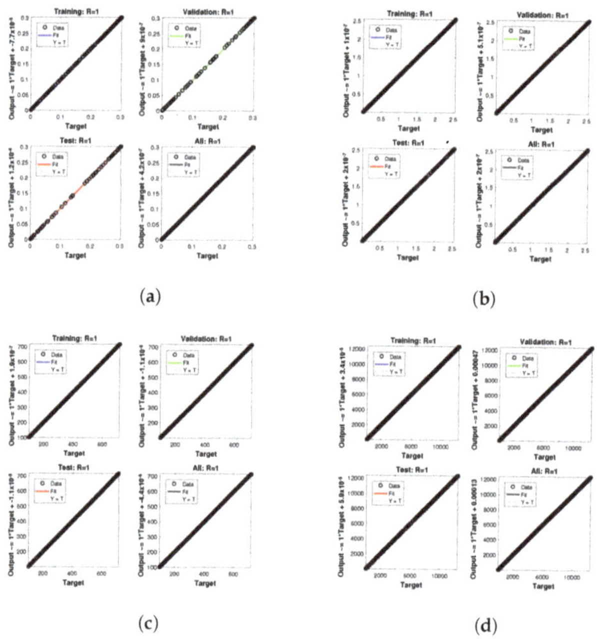

Figure 12. Analysis of regression of the design NN-BLM algorithm for case 3. (**a**) $H(t)$. (**b**) $I(t)$. (**c**) $Q(t)$. (**d**) $W(t)$.

The tables below provide the data information provided by the computing system. The tables show the best performance values in training, testing, validation, etc. Table 3 displays the best performance data for case 1, while the best performance data for case 2 and case 3 are displayed in Tables 4 and 5, respectively. These tables also show the hidden neuron count, iterations, and time spent.

Table 3. Performance values of the design NN-BLMA, and time spent by the computing system to obtain solutions for case 1.

	$H(t)$	$I(t)$	$Q(t)$	$W(t)$
Training	1.68×10^{-9}	7.50×10^{-9}	2.29×10^{-8}	2.10×10^{-5}
Validation	2.97×10^{-9}	7.11×10^{-9}	3.01×10^{-8}	2.82×10^{-5}
Testing	1.35×10^{-9}	7.61×10^{-9}	2.67×10^{-8}	2.43×10^{-5}
Gradient	8.2149×10^{-8}	2.4163×10^{-6}	2.3721×10^{-4}	0.2785
Mu	1×10^{-12}	1×10^{-8}	1×10^{-8}	1×10^{-5}
Epoches	33	1000	1000	1000
Regression	1	1	1	1
Time	30	30	30	30

Table 4. Performance values of the design NN-BLMA, and time spent by the computing system to obtain solutions for case 2.

	$H(t)$	$I(t)$	$Q(t)$	$W(t)$
Training	1.12×10^{-9}	3.41×10^{-11}	1.15×10^{-9}	1.79×10^{-4}
Validation	1.31×10^{-9}	3.69×10^{-11}	1.18×10^{-9}	1.97×10^{-4}
Testing	3.38×10^{-9}	4.01×10^{-11}	1.18×10^{-9}	2.01×10^{-4}
Gradient	9.2809×10^{-8}	1.5761×10^{-6}	1.4908×10^{-4}	27.6472
Mu	1×10^{-12}	1×10^{-11}	1×10^{-9}	1×10^{-4}
Epoches	56	1000	1000	1000
Regression	1	1	1	1
Time	30	30	30	30

Table 5. Performance values of the design NN-BLMA, and time spent by the computing system to obtain solutions for case 3.

	$H(t)$	$I(t)$	$Q(t)$	$W(t)$
Training	9.84×10^{-11}	2.84×10^{-9}	1.04×10^{-8}	2.53×10^{-4}
Validation	9.98×10^{-11}	3.12×10^{-9}	2.47×10^{-8}	2.75×10^{-4}
Testing	1.38×10^{-10}	3.16×10^{-9}	1.23×10^{-8}	2.69×10^{-4}
Gradient	1.1132×10^{-8}	4.0741×10^{-6}	3.0463×10^{-4}	0.49652
Mu	1×10^{-12}	1×10^{-8}	1×10^{-8}	1×10^{-4}
Epochs	50	1000	1000	1000
Regression	1	1	1	1
Time	30	30	30	30

The histogram of errors between targets and outputs after training a neural network is shown in Figures 13–15. Different color bars show the errors in the training, validation, and testing data. The error bars in which most of the points lie are very close to the zero error line, which means targets and the outputs are well matched and have the fewest errors, which shows the accuracy of our design technique. The error values for case 1 lie in the range 10^{-3} to 10^{-4}, 10^{-4} to 10^{-6}, 10^{-4} to 10^{-6}, and 10^{-2} to 10^{-3}. For case 2 and case 3 the error values lie in the range 10^{-4} to 10^{-5}, 10^{-5} to 10^{-7}, 10^{-4} to 10^{-5}, 10^{-3} to 10^{-4}, 10^{-2} to 10^{-3}, 10^{-4} to 10^{-6}, 10^{-4} to 10^{-6}, 10^{-3} to 10^{-5}, and 10^{-2} to 10^{-3}, respectively.

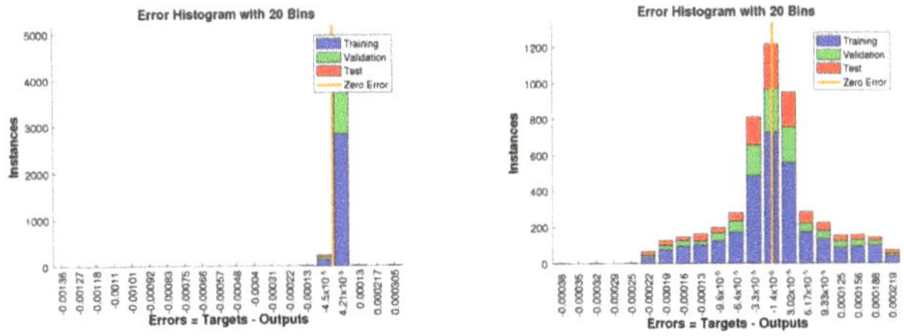

(a) (b)

Figure 13. *Cont.*

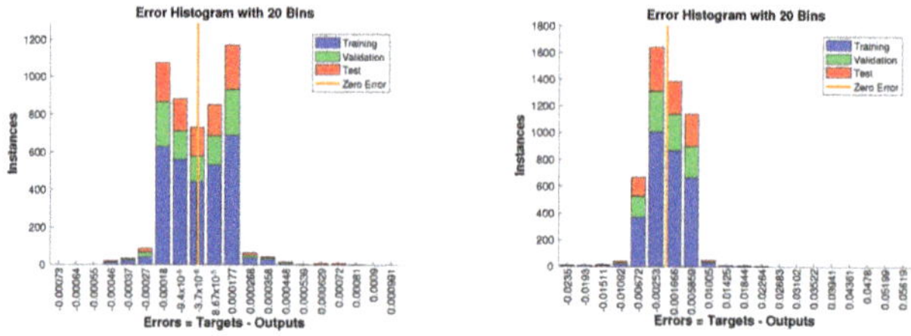

(c) (d)

Figure 13. Analysis of the error histogram in terms of the target data and the approximate solutions for case 1. (**a**) $H(t)$. (**b**) $I(t)$. (**c**) $Q(t)$. (**d**) $W(t)$.

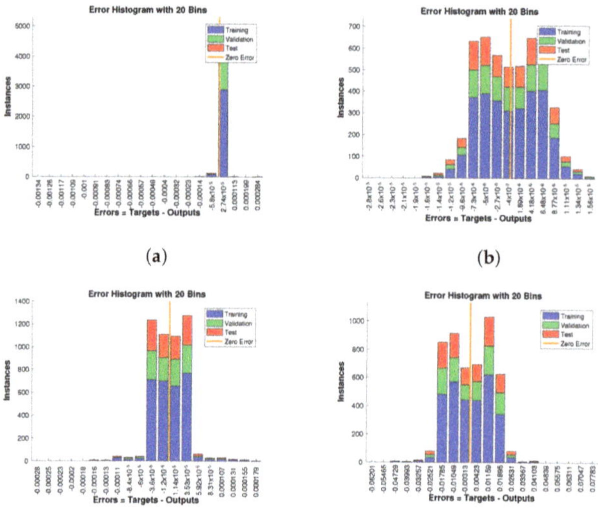

(a) (b)

(c) (d)

Figure 14. Analysis of the error histogram in terms of the target data and the approximate solutions for case 2. (**a**) $H(t)$. (**b**) $I(t)$. (**c**) $Q(t)$. (**d**) $W(t)$.

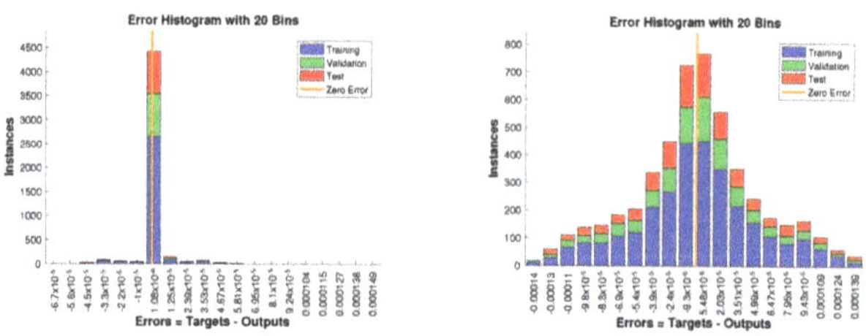

(a) (b)

Figure 15. *Cont.*

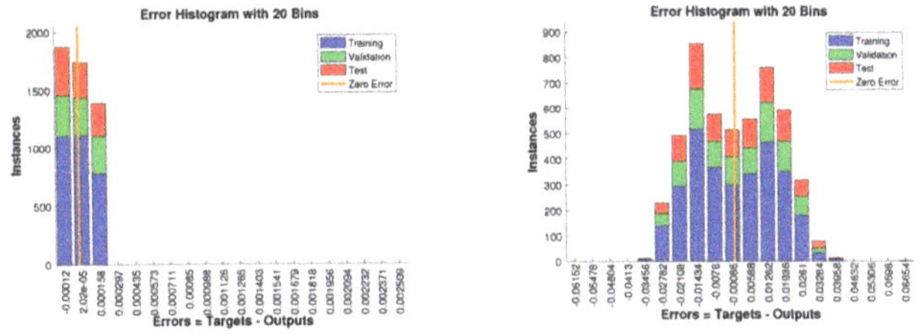

(c) (d)

Figure 15. Analysis of the error histogram in terms of the target data and the approximate solutions for case 3. (**a**) $H(t)$. (**b**) $I(t)$. (**c**) $Q(t)$. (**d**) $W(t)$.

Further, Figure 16 compares the numerical solution of the model obtained by the "ND-Solve" package in Mathematica (targets) to the solution obtained by executing NN-BLMA (outputs). The solid lines show the solution obtained by solving the model numerically by the "NDSolve" package in Mathematica, while the circles show the solution by NN-BLMA. In the figure, we see that the solutions obtained from NN-BLMA come exactly on the targets' solutions lines, which shows how accurate our design technique is. These figures also indicate the model's variation due to some parameters in the model. It is obvious from the figures that healthy plants and uninfected whiteflies rise when there is an increase in plant mortality rate and a drop in vector mortality rate. In contrast, a drop in plant mortality rate and an increase in vector mortality rate leads to a rise in infected plants and whiteflies. The comparison of statistical data given by the 'NDSlove' package in Mathematica with the outputs of NN-BLMA is illustrated in the tables below. Table 6 illustrates the comparative analysis of both the solutions for case 1, while the comparison for case 2 and case 3 are illustrated in Tables 7 and 8, respectively.

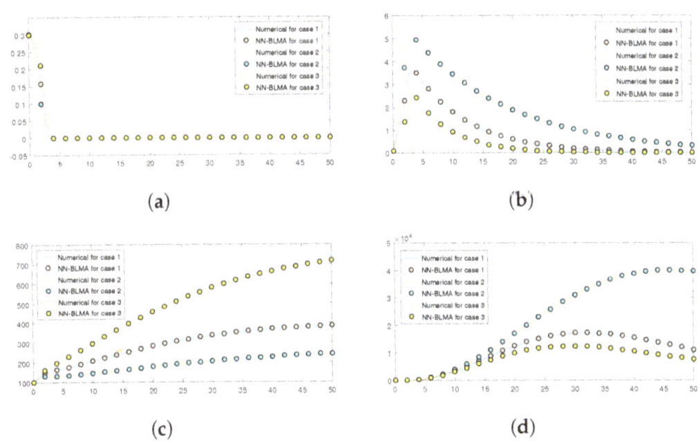

(a) (b)

(c) (d)

Figure 16. Numerical solutions' comparison of NN-BLMA with the solution obtained with other numerical methods. (**a**) Healthy plants $H(t)$. (**b**) Infected plants $I(t)$. (**c**) Uninfected whiteflies $Q(t)$. (**d**) Infected whiteflies $W(t)$.

Table 6. Comparative analysis of numerical solution with the solutions obtained from NN-BLMA for case 1.

	$H(t)$		$I(t)$		$Q(t)$		$W(t)$	
t	**Numerical**	**NN-BLMA**	**Numerical**	**NN-BLMA**	**Numerical**	**NN-BLMA**	**Numerical**	**NN-BLMA**
0	0.3	0.301402499	0.1	0.100121	100	100.0008	5	4.941716
0.5	0.295659	0.295796496	0.659197	0.659184	114.0068	114.0061	4.828814	4.811659
1	0.270171	0.270148397	1.121609	1.121648	126.3683	126.3679	5.584575	5.598602
1.5	0.225534	0.225481705	1.610816	1.610781	137.0851	137.0855	9.607768	9.620728
2	0.158531	0.158487646	2.28979	2.289811	145.892	145.892	21.79219	21.80038
2.5	0.078258	0.078217137	3.113732	3.113741	152.3632	152.3629	51.27621	51.28554
3	0.020596	0.020563851	3.647005	3.646977	156.6824	156.6827	108.024	108.0293
3.5	0.002145	0.002150992	3.673769	3.673765	159.9924	159.9922	193.49	193.4844
4	6.66×10^{-5}	6.67153×10^{-5}	3.500676	3.646977	163.2073	163.2072	303.6808	303.6772

Table 7. Comparative analysis of numerical solution with the solutions obtained from NN-BLMA for case 2.

	$H(t)$		$I(t)$		$Q(t)$		$W(t)$	
t	**Numerical**	**NN-BLMA**	**Numerical**	**NN-BLMA**	**Numerical**	**NN-BLMA**	**Numerical**	**NN-BLMA**
0	0.3	0.301386	0.1	0.100008961	100	100.0002874	5	4.91848547
0.5	0.2877	0.287758	1.021413	1.021411462	111.136459	111.1363243	4.74700621	4.704905202
1	0.243996	0.243941	1.778091	1.778087194	120.0058588	120.0057731	5.886795798	5.906826879
1.5	0.178663	0.178615	2.615206	2.615204348	126.6983972	126.6984875	11.85176957	11.86964906
2	0.099554	0.099532	3.721032	3.721035712	130.919435	130.9195312	29.55952416	29.56956409
2.5	0.033451	0.033442	4.748642	4.748643211	132.5695704	132.5695372	69.40014059	69.40036895
3	0.005173	0.005141	5.142219	5.142218201	132.6625165	132.6624534	137.4934921	137.5035161
3.5	0.000284	0.000276	5.084467	5.084461504	132.5157456	132.515754	231.7495915	231.7676383
4	4.2×10^{-6}	1.5×10^{-5}	4.939332	4.939329075	132.6092218	132.6092389	350.3411185	350.3583379

Table 8. Comparative analysis of numerical solution with the solutions obtained from NN-BLMA for case 3.

	$H(t)$		$I(t)$		$Q(t)$		$W(t)$	
t	**Numerical**	**NN-BLMA**	**Numerical**	**NN-BLMA**	**Numerical**	**NN-BLMA**	**Numerical**	**NN-BLMA**
0	0.3	0.29984487	0.1	0.10015	100	100.0001	5	4.930092
0.5	0.300552	0.30057977	0.43618653	0.43611909	116.834648	116.8346	4.92687225	4.91208
1	0.287377	0.28740484	0.71468896	0.71462628	132.625296	132.6252	5.47558404	5.499553
1.5	0.259799	0.25980546	0.99348118	0.99348179	147.331811	147.3318	8.22851282	8.216368
2	0.21123	0.21120628	1.36866609	1.3687244	160.767638	160.7677	16.453172	16.44464
2.5	0.136805	0.13676428	1.88979607	1.88976049	172.537448	172.5374	36.7607881	36.7496
3	0.056486	0.05643249	2.38574334	2.38574809	182.351472	182.3515	79.3186069	79.29844
3.5	0.010986	0.01094595	2.54620998	2.54620843	190.689067	190.6891	150.762051	150.7577
4	0.000742	0.00072402	2.42175547	2.42176342	198.537627	198.5377	248.404668	248.4245

5. Conclusions

In this paper, we analyzed a mathematical model for persistent vector-borne viral plant disease dynamics. The model includes equations for healthy and infected plants and uninfected and infected whiteflies. The selected set of parameters for numerical simulation is for the cause of the mosaic disease in cassava. To see the impact of variation in the mortality parameters on the model, we made different cases in which we vary both plant and vector mortality parameters. The reference data (targets) for NN-BLMA were generated by solving the model numerically for all the cases in Mathematica. The designed technique uses the targets to train, test, and validate the ANN and to see the impact of variation in plants' natural and vectors' mortality rates. The key points concluded from the study are given below.

- From the study, we see an increase in the mortality rate of plants, along with a decrease in the mortality rate of vectors, increases in healthy plants and uninfected whiteflies, and decreases in infected plants and infected whiteflies. In contrast, a drop in the mortality rate of plants and an increase in the mortality rate of vectors results in a decrease in healthy plants and uninfected whiteflies and an increase in the number of infected plants and infected whiteflies.
- Further, the accuracy of the design technique is illustrated through extensive graphical and tabular data, which include the best performance in terms of the mean squared error, histogram, and regression analyses.

Author Contributions: All authors contributed equally to this paper. All authors have read and agreed to the published version of the manuscript.

Funding: The study was funded by the Deanship of Scientific Research at Umm Al-Qura University, Makkah, Saudi Arabia (Grant Code: 20UQU0067DSR); and Taif University Researchers Supporting Project (TURSP- 2020/107), Taif University, Taif, Saudi Arabia.

Institutional Review Board Statement: Not applicable.

Informed Consent Statement: Not applicable.

Data Availability Statement: The data supporting this study's findings are available upon reasonable request from the corresponding author. The MATLAB code and a data file generated by the NDSolve Mathematica packager are available at author's GitHub account: https://github.com/sulaiman513/AWKUM-Optimization-Lab, accessed on 1 September 2022.

Acknowledgments: The authors would like to thank the Deanship of Scientific Research at Umm Al-Qura University for supporting this work by Grant Code: (20UQU0067DSR); and Taif University Researchers Supporting Project (TURSP- 2020/107), Taif University, Taif, Saudi Arabia.

Conflicts of Interest: The authors declare no conflict of interest.

Nomenclature

ANN	Artificial neural network
NN-BLMA	Backpropagated neural network based Levenberg—Marquardt Algorithm
MLP	Multilayer perceptron
FNN	Feed-forward neural network
BMV	Beet mosaic virus
ACMV	African cassava mosaic virus
TMV	Tobacco mosaic virus
BGMV	Bean golden mosaic virus
DDE	Delay differential equations
MSE	Mean squared error
H	Healthy plants
I	Infected plants
Q	Uninfected whiteflies
W	Infected whiteflies
r	Net growth rate of plants
k	Carrying capacity
λ	Rate of disease transmission from infected vector to healthy plant
μ	Plants natural mortality rate
m_1	Mortality of infected plants
Π	Cumulative birth or immigration rate of vector population
β	Transmission rate between diseased vector and healthy plant
d	Vector mortality rate

References

1. Plank, J. *Plant Diseases-Epidemics and Control*; Academic Press: New York, NY, USA, 1963.
2. Holt, J.; Jeger, M.; Thresh, J.; Otim-Nape, G. An epidemilogical model incorporating vector population dynamics applied to African cassava mosaic virus disease. *J. Appl. Ecol.* **1997**, *34*, 793–806. [CrossRef]
3. Yoshimura, A.; Kawasaki, K.; Takasu, F.; Togashi, K.; Futai, K.; Shigesada, N. Modeling the spread of pine wilt disease caused by nematodes with pine sawyers as vector. *Ecology* **1999**, *80*, 1691–1702. [CrossRef]
4. Bruhn, J.; Fry, W. Analysis of potato late blight epidemiology by simulation modeling. *Phytopathology* **1981**, *71*, 612–616. [CrossRef]
5. Villa, F. New computer architectures as tools for ecological thought. *Trends Ecol. Evol.* **1992**, *7*, 179–183. [CrossRef]
6. Rana, T.; Imran, M.A.; Baz, A. A Component Model with Verifiable Composition for the Construction of Emergency Management Systems. *Arab. J. Sci. Eng.* **2020**, *45*, 10683–10692. [CrossRef]
7. Hohn, T. Plant virus transmission from the insect point of view. *Proc. Natl. Acad. Sci. USA* **2007**, *104*, 17905–17906. [CrossRef]
8. Anguelov, R.; Lubuma, J.; Dumont, Y. Mathematical analysis of vector-borne diseases on plants. In Proceedings of the 2012 IEEE 4th International Symposium on Plant Growth Modeling, Simulation, Visualization and Applications, Shanghai, China, 31 October–3 November 2012; pp. 22–29.
9. Dietzgen, R.G.; Mann, K.S.; Johnson, K.N. Plant virus–insect vector interactions: Current and potential future research directions. *Viruses* **2016**, *8*, 303. [CrossRef]
10. Madden, L.V.; Hughes, G.; Van Den Bosch, F. *The Study of Plant Disease Epidemics*; American Phytopathological Society: St. Paul, MN, USA, 2007.
11. Kern, H. Problems of incubation in plant diseases. *Annu. Rev. Microbiol.* **1956**, *10*, 351–368. [CrossRef]
12. Smrž, J. The effect of the transmission of the beet mosaic virus on the variability of its incubation period. *Biol. Plant.* **1972**, *14*, 1–10. [CrossRef]
13. Legg, J. African Cassava Mosaic Disease. In *Encyclopedia of Virology*, 3rd ed.; Mahy, B.W., Van Regenmortel, M.H., Eds.; Academic Press: Oxford, UK, 2008; pp. 30–36. [CrossRef]
14. Siegel, A.; Ginoza, W.; Wildman, S.G. The early events of infection with tobacco mosaic virus nucleic acid. *Virology* **1957**, *3*, 554–559. [CrossRef]
15. Haber, S.; Maxwell, D.; Gilbertson, R. Bean golden mosaic. In *Compendium of Bean Diseases*; American Phytopathological Society: Eagan, MN, USA, 1991; pp. 42–43.
16. Rimbaud, L.; Dallot, S.; Delaunay, A.; Borron, S.; Soubeyrand, S.; Thébaud, G.; Jacquot, E. Assessing the mismatch between incubation and latent periods for vector-borne diseases: The case of sharka. *Phytopathology* **2015**, *105*, 1408–1416. [CrossRef]
17. Duffus, J.E. Whitefly transmission of plant viruses. In *Current Topics in Vector Research*; Springer: Berlin/Heidelberg, Germany, 1987; pp. 73–91.
18. Suffert, F.; Thompson, R.N. Some empirical arguments demonstrating that the latent period varies over the course of a plant disease epidemic. *BioRxiv* **2018**, *67*, 1831–1840. [CrossRef]
19. Jiao, J.; Chen, L. Global attractivity of a stage-structure variable coefficients predator-prey system with time delay and impulsive perturbations on predators. *Int. J. Biomath.* **2008**, *1*, 197–208. [CrossRef]
20. Ruan, S. Absolute stability, conditional stability and bifurcation in Kolmogorov-type predator-prey systems with discrete delays. *Q. Appl. Math.* **2001**, *59*, 159–173. [CrossRef]
21. Pal, P.J.; Saha, T.; Sen, M.; Banerjee, M. A delayed predator–prey model with strong Allee effect in prey population growth. *Nonlinear Dyn.* **2012**, *68*, 23–42. [CrossRef]
22. Cooke, K.L. Stability analysis for a vector disease model. *Rocky Mt. J. Math.* **1979**, *9*, 31–42. [CrossRef]
23. Wang, Z.; Ma, S.; Sun, B.; Wang, F.; Huang, J.; Wang, X.; Bao, Q. Effects of thermal properties and behavior of wheat starch and gluten on their interaction: A review. *Int. J. Biol. Macromol.* **2021**, *177*, 474–484. [CrossRef] [PubMed]
24. Zhang, T.; Meng, X.; Song, Y.; Li, Z. Dynamical analysis of delayed plant disease models with continuous or impulsive cultural control strategies. In *Abstract and Applied Analysis*; Hindawi: London, UK, 2012; Volume 2012, pp. 1–25.
25. Meng, X.; Li, Z. The dynamics of plant disease models with continuous and impulsive cultural control strategies. *J. Theor. Biol.* **2010**, *266*, 29–40. [CrossRef] [PubMed]
26. Munyasya, A.N.; Koskei, K.; Zhou, R.; Liu, S.T.; Indoshi, S.N.; Wang, W.; Zhang, X.C.; Cheruiyot, W.K.; Mburu, D.M.; Nyende, A.B.; et al. Integrated on-site & off-site rainwater-harvesting system boosts rainfed maize production for better adaptation to climate change. *Agric. Water Manag.* **2022**, *269*, 107672.
27. Buonomo, B.; Cerasuolo, M. Stability and bifurcation in plant–pathogens interactions. *Appl. Math. Comput.* **2014**, *232*, 858–871. [CrossRef]
28. Miao, R.; Qiu, X.; Guo, M.; Musa, A.; Jiang, D. Accuracy of space-for-time substitution for vegetation state prediction following shrub restoration. *J. Plant Ecol.* **2018**, *11*, 208–217. [CrossRef]
29. Chan, M.S.; Jeger, M.J. An analytical model of plant virus disease dynamics with roguing and replanting. *J. Appl. Ecol.* **1994**, *31*, 413–427. [CrossRef]
30. Shi, R.; Zhao, H.; Tang, S. Global dynamic analysis of a vector-borne plant disease model. *Adv. Differ. Equ.* **2014**, *2014*, 59. [CrossRef]
31. Jackson, M.; Chen-Charpentier, B.M. Modeling plant virus propagation with delays. *J. Comput. Appl. Math.* **2017**, *309*, 611–621. [CrossRef]

32. Li, Q.; Dai, Y.; Guo, X.; Zhang, X. Hopf bifurcation analysis for a model of plant virus propagation with two delays. *Adv. Differ. Equ.* **2018**, *2018*, 259. [CrossRef]
33. Banerjee, M.; Takeuchi, Y. Maturation delay for the predators can enhance stable coexistence for a class of prey–predator models. *J. Theor. Biol.* **2017**, *412*, 154–171. [CrossRef] [PubMed]
34. Venturino, E.; Roy, P.K.; Al Basir, F.; Datta, A. A model for the control of the mosaic virus disease in Jatropha curcas plantations. *Energy Ecol. Environ.* **2016**, *1*, 360–369. [CrossRef]
35. Al Basir, F.; Adhurya, S.; Banerjee, M.; Venturino, E.; Ray, S. Modelling the effect of incubation and latent periods on the dynamics of vector-borne plant viral diseases. *Bull. Math. Biol.* **2020**, *82*, 94. [CrossRef] [PubMed]
36. Hafstein, S.F. Numerical ODE solvers and integration methods in the computation of CPA Lyapunov functions. In Proceedings of the 2019 18th European Control Conference (ECC), Naples, Italy, 28–29 June 2019; pp. 1136–1141.
37. Haefner, J.W. Numerical Techniques. In *Modeling Biological Systems: Principles and Applications*; Springer: Berlin/Heidelberg, Germany, 2005; pp. 107–122.
38. Martín, M.P.; Candler, G.V. A parallel implicit method for the direct numerical simulation of wall-bounded compressible turbulence. *J. Comput. Phys.* **2006**, *215*, 153–171. [CrossRef]
39. Kumar, A.; Unny, T. Application of Runge-Kutta method for the solution of non-linear partial differential equations. *Appl. Math. Model.* **1977**, *1*, 199–204. [CrossRef]
40. Horowitz, A.R. Population dynamics of Bemisia tabaci (Gennadius): With special emphasis on cotton fields. *Agric. Ecosyst. Environ.* **1986**, *17*, 37–47. [CrossRef]
41. Henneberry, T.; Castle, S. Bemisia: Pest status, economics, biology, and population dynamics. In *Virus-Insect-Plant Interactions*; Elsevier: Amsterdam, The Netherlands, 2001; pp. 247–278.
42. Gadekallu, T.R.; Rajput, D.S.; Reddy, M.; Lakshmanna, K.; Bhattacharya, S.; Singh, S.; Jolfaei, A.; Alazab, M. A novel PCA–whale optimization-based deep neural network model for classification of tomato plant diseases using GPU. *J. Real-Time Image Process.* **2021**, *18*, 1383–1396. [CrossRef]
43. Kumar, R.; Baz, A.; Alhakami, H.; Alhakami, W.; Agrawal, A.; Khan, R.A. A hybrid fuzzy rule-based multi-criteria framework for sustainable-security assessment of web application. *Ain Shams Eng. J.* **2021**, *12*, 2227–2240. [CrossRef]
44. Baz, A.; Alhakami, H.; Alshareef, E. A framework of computational model for predicting the spread of COVID-19 pandemic in Saudi Arabia. *Int. J. Intell. Eng. Syst.* **2020**, *13*, 463–475. [CrossRef]
45. Khan, N.A.; Sulaiman, M.; Alshammari, F.S. Heat transfer analysis of an inclined longitudinal porous fin of trapezoidal, rectangular and dovetail profiles using cascade neural networks. *Struct. Multidiscip. Optim.* **2022**, *65*, 251. [CrossRef]
46. He, G.; Liu, X.; Cui, Z. Achieving global food security by focusing on nitrogen efficiency potentials and local production. *Glob. Food Secur.* **2021**, *29*, 100536. [CrossRef]
47. Alghamdi, H.; Alsubait, T.; Alhakami, H.; Baz, A. A review of optimization algorithms for university timetable scheduling. *Eng. Technol. Appl. Sci. Res.* **2020**, *10*, 6410–6417. [CrossRef]
48. Iqbal, A.; Raza, M.S.; Ibrahim, M.; Baz, A.; Alhakami, H.; Saeed, M.A. An Improved Approach for Finding Rough Set Based Dynamic Reducts. *IEEE Access* **2020**, *8*, 173008–173023. [CrossRef]

entropy

Article

Quantifying Parameter Interdependence in Stochastic Discrete Models of Biochemical Systems

Samaneh Gholami * and Silvana Ilie

Department of Mathematics, Toronto Metropolitan University, Toronto, ON M5B 2K3, Canada; silvana@torontomu.ca
* Correspondence: samaneh.gholami@torontomu.ca

Abstract: Stochastic modeling of biochemical processes at the cellular level has been the subject of intense research in recent years. The Chemical Master Equation is a broadly utilized stochastic discrete model of such processes. Numerous important biochemical systems consist of many species subject to many reactions. As a result, their mathematical models depend on many parameters. In applications, some of the model parameters may be unknown, so their values need to be estimated from the experimental data. However, the problem of parameter value inference can be quite challenging, especially in the stochastic setting. To estimate accurately the values of a subset of parameters, the system should be sensitive with respect to variations in each of these parameters and they should not be correlated. In this paper, we propose a technique for detecting collinearity among models' parameters and we apply this method for selecting subsets of parameters that can be estimated from the available data. The analysis relies on finite-difference sensitivity estimations and the singular value decomposition of the sensitivity matrix. We illustrated the advantages of the proposed method by successfully testing it on several models of biochemical systems of practical interest.

Keywords: stochastic simulation algorithm; stochastic biochemical systems; sensitivity analysis; finite-difference methods; parameter subset selection; estimability analysis

check for updates

Citation: Gholami, S.; Ilie, S. Quantifying Parameter Interdependence in Stochastic Discrete Models of Biochemical Systems. *Entropy* **2023**, *25*, 1168. https://doi.org/10.3390/e25081168

Academic Editor: Pavel Kraikivski

Received: 3 July 2023
Revised: 31 July 2023
Accepted: 3 August 2023
Published: 5 August 2023

1. Introduction

Mathematical and computational modeling have become widespread in the study of complex dynamical systems, particularly in investigating cellular processes and biochemical networks [1]. Frequently, mathematical modeling of chemical reaction systems relies on deterministic differential equations and mass action kinetics. However, biochemical systems in the cell are intrinsically noisy [2,3], and thus stochastic models must be employed to account for the random fluctuations observed experimentally, especially when some species have low molecular counts [4,5]. One of the most popular stochastic discrete models of biochemically reacting systems is the Chemical Master Equation [6,7]. This model is utilized to describe the dynamics of systems for which molecular populations of some species are low or noise is significant. It assumes that the system state is a Markov process [6]. It is generally impracticable to solve this model analytically, except for very simple systems.

Gillespie developed the Stochastic Simulation Algorithm (SSA) [8,9], a Monte Carlo technique for simulating statistically exact realizations of the stochastic process whose distribution is governed by the Chemical Master Equation. The random time change representation of the stochastic process depicting the system state was introduced in [10]. Based on this representation, Rathinam et al. [11] designed an exact Monte Carlo method for the Chemical Master Equation, the Random Time Change algorithm. Other simulation strategies for stochastic models of biochemically reacting systems were presented in the literature (for references see, e.g., [12–15]).

The biochemical networks arising in applications may be quite complex, involving many reactions and/or species, which means that their mathematical models have many parameters. Some of the values of a model's kinetic parameters may not be known [16,17] and they may need to be estimated from the available data. Also, certain parameters have a substantial influence on the system's output. Thus, it is essential to study the system's behavior when these parameters are perturbed. While stochastic discrete models of biochemical systems capture the inherent randomness observed in cellular processes, they pose challenges with regard to their parameter estimation and identification. Hence, developing efficient and accurate methods for identifying and estimating their parameters would be a key advance in studying these models.

Practical identifiability (or estimability) analysis aims to establish if the parameters can be accurately and reliably estimated from the available data [18]. In this context, identifiable parameters are those which can be determined with high confidence from the observed behavior of the system; otherwise, the parameters are unidentifiable. Using practical identifiability, one can select subsets of parameters that significantly impact the behavior of the system. If the parameters in such a subset are not interdependent, then they are identifiable. These parameters can be accurately estimated when sufficient and quality data is available, and their accurate estimation is crucial for building the model. Also, these parameters may provide insight into the key underlying mechanisms of the biochemical system. Furthermore, the identifiability analysis helps select the unidentifiable parameters, which have a negligible impact on the model behavior and can be eliminated, thus guiding model reduction. There exist numerous studies of identifiability analysis for deterministic models, such as the reaction rate equations [19–26]. Nonetheless, much less work has been dedicated to parameter estimability of stochastic models of biological processes.

One important method for practical identifiability is to utilize sensitivity analysis. Local sensitivity analysis assesses the change in the system's behavior caused by a small variation in the value of a certain parameter. Insignificant changes in the system dynamics indicate that the specific parameter is not important, and thus it is not identifiable. Also, a parameter is not identifiable if it is correlated with other parameters, such that a variation in its value can be compensated by suitable adjustments in other parameters. For stochastic models, finite-difference methods can be used to estimate the sensitivity of the expected value of the given function of the system state. In the class of finite-difference sensitivity estimators for the Chemical Master Equation, those employing exact Monte Carlo simulation methods are the Coupled Finite Difference method of Anderson [27], the Common Reaction Path scheme (based on the Random Time Change algorithm) and the Common Random Number strategy (utilizing the SSA) of Rathinam et al. [11]. These estimators utilize coupled perturbed and unperturbed trajectories to approximate sensitivities. The coupling lowers the variance of the estimator so that the method requires fewer realizations to achieve the same accuracy of the estimation. Due to this, the computational time of the algorithm is reduced, for a prescribed accuracy. Of the three strategies, the Coupled Finite Difference algorithm has the lowest variance of the estimator [28]. These schemes perform best for non-stiff models. For stiff problems, finite-difference techniques can be applied with various coupled tau-leaping strategies to increase the efficiency of the simulation [29].

In this work, we consider the problem of practical parameter identifiability for stochastic discrete biochemical networks modeled with the Chemical Master Equation. This is a critical problem, and a direct extension of the techniques developed for ordinary differential equations to stochastic discrete models is not possible. Our contribution is generalizing a method by Gábor et al. [30] to find the highest parameter identifiable sets for models of biochemical systems, from the continuous deterministic to the stochastic discrete models of well-stirred biochemical systems, which is a difficult task. The proposed method identifies the subsets of parameters that are independent and significant for the model's behavior, based on the existing data, and thus are identifiable. We utilize local sensitivity estimations to study parameter estimability. For approximating sensitivities, we apply finite-difference techniques, namely the Coupled Finite Difference [27], the Common Reaction Path, and the Common Random Number methods [11]. We make use of the normalized sensitivity matrix to develop several identifiability metrics, which adapt existing techniques for the reaction rate equations [19,20] to the more challenging Chemical Master Equation model. In addition, we apply the singular value decomposition of the non-dimensional sensitivity matrix, to determine its rank. This analysis helps gain insight into the interrelations between parameters. Furthermore, the proposed methodology can be employed to decide which parameters can be reliably estimated from the available data, for the Chemical Master Equation, and may assist experimental design for more accurate parameter approximations. It is worth noting that, in general, the expected value of the system state governed by the Chemical Master Equation may not satisfy the deterministic reaction rate equations, when some reactions are of second or higher order [14].

This paper is structured as follows. Section 2 is dedicated to the background on stochastic discrete models for well-stirred biochemical networks and their simulation methods, parametric sensitivity schemes for stochastic and deterministic models, and practical identifiability techniques, including the new algorithm for selecting subsets of identifiable parameters. The proposed algorithm is tested on various stochastic models arising in applications in Section 3. Section 4 presents a summary of our results.

2. Materials and Methods

2.1. Background

Suppose a system has N biochemical species, denoted by $S_1, S_2, \ldots, S_N$, that undergo M distinct chemical reactions. It is maintained at a constant temperature, in a constant volume. Provided that the biochemical network is well-stirred, it may be represented by a state vector,

$$X(t) = [X_1(t), X_2(t), \ldots, X_N(t)]^T,$$

where $X(t)$ has entries $X_i(t)$, the amount of S_i molecules in the system at time t. A reaction R_j produces a variation in the system state, which is given by the state change vector $\nu_j \in \mathbb{R}^N$,

$$\nu_j = [\nu_{1j}, \nu_{2j}, \ldots, \nu_{Nj}]^T ,$$

where ν_{ij} is the perturbation in the molecular amount of S_i after the reaction fires. If one reaction R_j happens during the time interval $[t, t + \Delta t]$, then the resulting state is $X(t + \Delta t) = X(t) + \nu_j$. The array having ν_j as the j-th column is called the stoichiometric matrix. Also associated with the reaction R_j, we can define the propensity function a_j, by $a_j(x)dt =$ the probability that a single reaction R_j occurs between $[t, t+dt)$, assuming that the system state at time t is x. The form of the propensity function a_j is determined by the type of reaction. For a first-order reaction, $S_m \xrightarrow{c_j} products$, the propensity is expressed as $a_j(X(t)) = c_j X_m(t)$. For a second-order reaction, $S_m + S_n \xrightarrow{c_j} products$, the propensity is $a_j(X(t)) = c_j X_m(t) X_n(t)$, if $m \neq n$ and $a_j(X(t)) = \frac{1}{2} c_j X_m(t)(X_m(t) - 1)$, if $m = n$.

2.1.1. Chemical Master Equation

To study the behavior of the well-stirred biochemical system, we need to determine $P(x, t|x_0, t_0)$, the probability of the system state being $X(t) = x$ at time t, if at t_0 it was $X(t_0) = x_0$. This probability satisfies the Chemical Master Equation [6,7]

$$\frac{d}{dt} P(x, t|x_0, t_0) = \sum_{j=1}^{M} \left[a_j(x - \nu_j) P(x - \nu_j, t|x_0, t_0) - a_j(x) P(x, t|x_0, t_0) \right] . \quad (1)$$

This is a stochastic discrete model. It is a linear system of ordinary differential equations, each equation describing the probability of the system being in a particular state x. The biochemical system state $X(t)$ is a discrete in space and continuous in time Markov process. The space of all possible states is typically quite large, and in such cases the Chemical Master Equation is of very high dimension. Therefore, it is challenging to solve it directly, except for some simple systems.

As an alternative to solving the Chemical Master Equation directly, it is possible to generate correct trajectories one by one. Gillespie [8,9] proposed a Monte Carlo strategy to compute such trajectories, which are in exact agreement with the probability distribution associated with the discrete stochastic model (1). The strategy, also referred to as the Stochastic Simulation Algorithm (SSA), has been broadly employed for solving stochastic models in Systems Biology [3,14,31]. The SSA is described below.

Gillespie's Algorithm

1. Initialize the time $t \leftarrow t_0$ and the state of the system, $X(t) \leftarrow x_0$.
2. While $t < T$
3. Calculate each propensity $a_j(X(t))$ for $j = 1, \ldots, M$ and the sum $a_0(X(t)) \leftarrow \sum_{r=1}^{M} a_r(X(t))$
4. Sample two uniform random variables over $[0,1]$, to obtain η_1, η_2.
5. Evaluate the time τ and the index j of the next occurring reaction, according to
   (a) $\tau \leftarrow -(\ln \eta_1)/a_0(x)$
   (b) $j \leftarrow$ the smallest integer fulfilling $\sum_{r=1}^{j} a_r(x) > \eta_2 a_0(x)$
6. Update the state $X(t+\tau) \leftarrow X(t) + \nu_j$ and the time $t \leftarrow t + \tau$.
7. End while.

The Random Time Change (RTC) algorithm [11], based on the Random Time Change representation [10], is another exact Monte Carlo simulation strategy for the Chemical Master Equation. We refer the reader to [11] for details on this algorithm.

2.1.2. Chemical Langevin Equation

An intermediate model between the Chemical Master Equation and the reaction rate equation is the Chemical Langevin Equation [32]. This is a system of stochastic differential equations of size equal to the number of reacting species. The Chemical Langevin Equation is a reduction in the Chemical Master Equation model assuming that the biochemical system has a macroscopically infinitesimal scale in time step such that, over δt, every reaction occurs multiple times and, at the same time, its propensity function does not vary significantly. Under these assumptions, the system state is governed by

$$dX(t,c) = \sum_{j=1}^{M} \nu_j a_j(X(t,c),c)dt + \sum_{j=1}^{M} \nu_j \sqrt{a_j(X(t,c),c)}dW_j(t) \tag{2}$$

where W_j are independent Wiener processes for $j = 1, \ldots, M$. The state $X(t)$ may be approximated by a Markov process continuous in space. Equation (2) represents the Chemical Langevin Equation.

2.1.3. Reaction Rate Equation

A coarser level of resolution in modeling biochemically reacting networks is provided by the continuous deterministic model of chemical kinetics. This model, known as the reaction rate equations, is valid under the assumption of the thermodynamic limit. In the thermodynamic limit, the molecular amounts for all species and the system volume tend towards infinity, as the concentrations of species within the system remain constant. Hence, the stochastic terms in the Chemical Langevin Equation are much smaller than the deterministic terms. As a result, the Chemical Langevin Equation model reduces to the reaction rate equations, in the thermodynamic limit. This condition is satisfied when all S_i molecular counts are very large. The reaction rate equations (RRE) are of the form

$$\frac{dX(t,c)}{dt} = \sum_{j=1}^{M} \nu_j a_j(X(t,c),c)\,. \tag{3}$$

Equation (3) is a set of ordinary differential equations, with one equation for each biochemical species. In the event that all reactions in the system are of order at most one, the reaction rate equation can be obtained from the Chemical Master Equation (1), by considering the expected value of the system state. However, in general, the evolution of the mean trajectory in the Chemical Master Equation model does not obey the continuous deterministic model. Then, the RRE does not properly depict the true behavior of the biochemical network. In fact, there are numerous cellular networks for which noise significantly influences the system dynamics [12,31,33].

2.2. *Parametric Correlations*

Sensitivity analysis plays a central role in constructing models [24]. It assesses how changes in parameters cause variations in a model's output. If a negligible adjustment in a parameter leads to significant alterations in the outcomes, we consider the model to be sensitive to that specific parameter. Precise estimations are not necessary for parameters with low sensitivity. Conversely, parameters with high associated sensitivity become key control points for the behavior of the system. In what follows, we shall focus on the sensitivity analysis of system outputs with respect to rate parameters.

2.2.1. Parametric Sensitivity for the Chemical Master Equation

Let f be a function of interest of the system state and c a model parameter. In the stochastic setting, the local sensitivity with respect to a parameter c is defined as $\frac{\partial}{\partial c}\mathbb{E}[f(X(t,c))]$ where $\mathbb{E}(\cdot)$ is the expected value. Popular methods for estimating local sensitivities with respect to the model's parameters for the Chemical Master Equation often rely on finite-difference schemes and Monte Carlo methods for generating the perturbed and unperturbed trajectories. By forward finite-difference schemes, one can estimate $\frac{\partial}{\partial c}\mathbb{E}[f(X(t,c))] \approx \{\mathbb{E}[f(X(t,c+\theta))] - \mathbb{E}[f(X(t,c))]\}/\theta$, where θ is a small perturbation of the parameter of interest, c. To efficiently approximate the sensitivity by Monte Carlo methods, the trajectories for $X(t,c+\theta)$ and $X(t,c)$ are generated using common random numbers. Among such methods are the Common Random Number (CRN), the Common Reaction Path (CRP) algorithms [11], and the Coupled Finite-Difference (CFD) algorithm [27].

2.2.2. Common Random Number

The Common Random Number presented in [11] is a finite-difference numerical method for estimating parametric sensitivities for the stochastic discrete model (1). It reuses random numbers to generate the perturbed and unperturbed paths. In doing so, it reduces the variance of the sensitivity estimator, and thus it has increased efficiency compared to a strategy based on independent random numbers. For the r-th iteration, it computes two SSA trajectories, $X_{[r]}(t,c+\theta)$ -the perturbed and $X_{[r]}(t,c)$ -the unperturbed path, each employing the same stream of uniform $(0,1)$ random numbers. Usually, the coupling of the CRN technique is less efficient than that of the CRN and CFD schemes [27]. The sensitivity of the r-th path is approximated by

$$Z_{[r]}(t,c) = \frac{f(X_{[r]}(t,c+\theta)) - f(X_{[r]}(t,c))}{\theta}, \tag{4}$$

while an estimate of the sensitivity is obtained from the sample mean $(\sum_{i=1}^{R} Z_{[r]}(t,c))/R$, R being the number of paired trajectories simulated.

2.2.3. Common Reaction Path

The Common Reaction Path technique is also a finite-difference sensitivity estimator for the Chemical Master Equation [11]. The CRP strategy applies the RTC algorithm to simulate sample paths. In this method, coupling of the processes involves some independent unit-rate Poisson processes, $\{Y_j\}_{1\le j\le M}$. The coupling of the perturbed—$X(\cdot,c+\theta)$ and unperturbed—$X(\cdot,c)$ processes is achieved using the random time change representation

$$
\begin{aligned}
X(t,c) &= x_0 + \sum_{j=1}^{M} \nu_j Y_j\Big(\int_0^t a_j(X(s,c),c)ds\Big) \\
X(t,c+\theta) &= x_0 + \sum_{j=1}^{M} \nu_j Y_j\Big(\int_0^t a_j(X(s,c+\theta),c+\theta)ds\Big)
\end{aligned}
\tag{5}
$$

The r-th iteration of the CRP algorithm generates the paired trajectories $X_{[r]}(t,c+\theta)$ and $X_{[r]}(t,c)$ with the RTC algorithm, each using the same M independent streams of unit-rate exponential random numbers. As before, the sensitivity of the r-th trajectory is estimated by (4). This coupling has been shown to be typically stronger than that of the CRN method, leading to a lower variance of the estimation [11,27].

2.2.4. Coupled Finite-Difference

Another finite-difference sensitivity estimator for the stochastic discrete model is the Coupled Finite-Difference scheme [27]. The CFD method relies on the random time change representation of the unperturbed and perturbed processes

$$
\begin{aligned}
X(t,c) &= x_0 + \sum_{j=1}^{M} \nu_j Y_j^{(1)}\Big(\int_0^t \min(a_j(X(s,c),c), a_j(X(s,c+\theta),c+\theta))ds\Big) \\
&+ \sum_{j=1}^{M} \nu_j Y_j^{(2)}\Big(\int_0^t [a_j(X(s,c),c) - \min(a_j(X(s,c),c), a_j(X(s,c+\theta),c+\theta)]ds\Big) \\
X(t,c+\theta) &= x_0 + \sum_{j=1}^{M} \nu_j Y_j^{(1)}\Big(\int_0^t [\min(a_j(X(s,c),c), a_j(X(s,c+\theta),c+\theta)]ds\Big) \\
&+ \sum_{j=1}^{M} \nu_j Y_j^{(3)}\Big(\int_0^t [a_j(X(s,c+\theta),c+\theta) - \min(a_j(X(s,c),c), a_j(X(s,c+\theta),c+\theta)]ds\Big)
\end{aligned}
\tag{6}
$$

where $\{Y_j^{(1)}\}_{1\le j\le M}$, $\{Y_j^{(2)}\}_{1\le j\le M}$. and $\{Y_j^{(3)}\}_{1\le j\le M}$ are independent unit-rate Poisson processes. Furthermore, the CFD strategy uses a version of the Next Reaction Method to compute the coupled perturbed and unperturbed trajectories, $X_{[r]}(t,c+\theta)$ and $X_{[r]}(t,c)$, and (4) to approximate the local sensitivity of the r-th path. Among the finite-difference sensitivity estimators with exact underlying simulation techniques for the CME, the CFD performs the best, followed by the CRP and the CRN [27,28]. Indeed, the CFD achieves the smallest variance of the sensitivity estimator of the three methods described above [28]. As a consequence, for the same number of trajectories simulated, we shall consider in our investigations the CFD sensitivity approximations to be the most accurate and reliable.

2.2.5. Parametric Sensitivity for the Chemical Langevin Equations

Glasserman [34] developed a technique for computing pathwise parametric sensitivities for certain problems modeled by stochastic differential equations. This method was applied to the Chemical Langevin Equation (CLE) model in [33]. For computing the sensitivity of each path, we differentiate Equation (2) with respect to parameter c and obtain

$$
\begin{aligned}
d(\frac{\partial X}{\partial c}) &= \sum_{j=1}^{M} \nu_j \Big[\frac{\partial a_j(X)}{\partial X}\frac{\partial X}{\partial c} + \frac{\partial a_j(X)}{\partial c}\Big](t)dt \\
&+ \sum_{j=1}^{M} \nu_j \Big[\frac{1}{2\sqrt{a_j(X)}}\Big(\frac{\partial a_j(X)}{\partial X}\frac{\partial X}{\partial c} + \frac{\partial a_j(X)}{\partial c}\Big)\Big](t)dW_j\,.
\end{aligned}
\tag{7}
$$

Solving the coupled system of Equations (2) and (7) for $(X, \partial X/\partial c)$ will determine the pathwise sensitivities. At time $t=0$, the local sensitivities with respect to the rate parameters are zero. The Chemical Langevin Equation is, in general, valid when all molecular amounts are sufficiently large. Effective simulation strategies for this model require adaptive time-stepping methods [35,36].

2.2.6. Parametric Sensitivity for the Reaction Rate Equations

In the deterministic scenario, the behavior of the biochemical system is governed by the reaction rate Equation (3). To find the local sensitivity for this model, the derivative with respect to the desired kinetic parameter is applied to Equation (3), yielding

$$\frac{d}{dt}\mathcal{S} = \sum_{j=1}^{M} \nu_j \left(\frac{\partial a_j(X(t,c),c)}{\partial c} + \sum_{i=1}^{N} \frac{\partial a_j(X(t,c),c)}{\partial X_i} \mathcal{S}_i \right). \tag{8}$$

Here, $\mathcal{S} = \partial X(t,c)/\partial c$ is the sensitivity with respect to parameter c. The sensitivity is computed by solving for $(X, \mathcal{S})$ the system of ordinary differential Equations (3) and (8), with the initial conditions $X(0,c) = x_0$ and $\mathcal{S}(0) = 0$. The deterministic model is applicable when all reacting molecular populations are very large. Nonetheless, when low molecular counts of some species exist or noise plays a significant role, this approach may fail in accurately capturing the characteristics of the biochemical system. Then, deterministic techniques for sensitivity-based identifiability analysis are not valid.

2.3. *Practical Identifiability Analysis*

When a model's performance is investigated, it is important to evaluate the accuracy of the parameter values. Still, poor or noisy data, interdependence of parameters, or weak dependence of the system dynamics on certain parameters may hinder the accurate estimation of parameter values. As a result, it is possible for these values to change significantly, without influencing the model's output. Consequently, the concept of identifiability is essential for the analysis of a mathematical model [19,24].

Identifiability can be classified into two main categories: structural identifiability and practical identifiability. For a structurally identifiable model, there exists a unique parameterization for any specified output of the model (see, e.g., [21,26]). On the other hand, practical identifiability involves detecting non-identifiable parameters by fitting the model to data that closely resemble the available observations (see, e.g., [18,19,22,25] for analyses of deterministic models). For this type of identifiability, it is helpful to study the parametric sensitivity of the model. In this work, we use sensitivity-based identifiability for the Chemical Master Equation. We determine identifiability and collinearity indexes by generalizing methods for deterministic models [19] to the more challenging case of stochastic discrete biochemical systems.

2.3.1. Sensitivity-Based Identifiability Analysis

Several identifiability strategies for deterministic models exist in the literature. One such approach by Brun et al. [19] is based on local sensitivity analysis of deterministic models. Sensitivity analysis quantifies the impact of parameter variations on the system's dynamics.

Below, we review some techniques for identifiability analysis of deterministic models relying on local parametric sensitivity. These techniques can be applied to the reaction rate Equation (3). Denote by

$$S_{ik}(X,t,c) = \frac{\partial X_i(t,c)}{\partial c_k} \tag{9}$$

the local sensitivity of the molecular amount $X_i(t,c)$ at time t, with respect to the kinetic parameter c_k. For time t, the parametric sensitivity matrix is $S(X,t,c) = \dfrac{\partial}{\partial c}X(t,c) =$

$\{S_{ik}(X,t,c)\}_{1\leq i\leq N, 1\leq k\leq M}$. In addition, the non-dimensional sensitivity coefficient corresponding to the i-th species and the parameter c_k at time t is

$$s_{ik}(t) = \frac{c_k}{X_i(t,c)} \frac{\partial X_i(t,c)}{\partial c_k} \,. \tag{10}$$

Here, $c = [c_1, \ldots, c_M]$ is the vector of kinetic parameters associated to reactions $\{R_j\}_{1\leq j\leq M}$. Furthermore, let $t_1 < t_2 < \cdots < t_L$ be a sequence of time-points spanning the integration interval $[0, T]$. Ideally, some of these time-points should be inside the interval corresponding to the biochemical network's transient behavior, when applicable. Also, consider the concatenated non-dimensional sensitivity matrix, for all the time-points in the grid, and apply the normalization (10) for each entry,

$$s(X,c) = \begin{pmatrix} s_{11}(t_1) & \cdots & s_{1M}(t_1) \\ \vdots & \ddots & \vdots \\ s_{N1}(t_L) & \cdots & s_{NM}(t_L) \end{pmatrix} . \tag{11}$$

To rank the parameters of the model, we utilize the non-dimensional sensitivity matrix of size $(NL) \times M$ from (11). The k-th column in this matrix measures the sensitivities with respect to c_k, the rate parameter of reaction R_k. Let us calculate the norm of each column in the sensitivity matrix (11) to obtain a parameter ranking. The norm of each column $s_k(X,c) = [s_{1k}(t_1), \ldots, s_{Nk}(t_1), \ldots, s_{1k}(t_L), \ldots, s_{Nk}(t_L)]^T$ serves as a measure of the significance of parameter c_k on the dynamics of the system. A higher norm indicates that altering that parameter value has a substantial impact on the system state. Parameters can be arranged in order of their significance. The following sensitivity measure is employed for evaluating the significance of the parameters, based on the sensitivity matrix (adapted after [19])

$$\delta_k^{msqr} = \sqrt{\frac{1}{n} \sum_{i=1}^{n} s_{ik}^2} \,. \tag{12}$$

The larger the measure δ_k^{msqr}, the more significant the parameter c_k is (for $1 \leq k \leq M$).

2.3.2. Parameter Collinearity

Extensive research has been conducted to examine the collinearity in various problems. Brun et al. [19] introduces a strategy for identifying parameter relationships based on collinearity analysis, in the deterministic framework, and presents a novel approach to explore the connections between parameters. Note that the columns of a matrix B are nearly linearly dependent (or near collinear) if a non-zero vector $z = [z_1, \ldots, z_M]^T$ exists such that $Bz \approx 0$, where B has M columns. If the $Bz = 0$ holds and $z \neq 0$, the columns of B are linearly dependent (or collinear).

Now, take the normalized sensitivity matrix $\tilde{S}$, having as the m-th column the vector

$$\tilde{s}_m(X,c) = \frac{s_m(X,c)}{||s_m(X,c)||_2} \,,$$

for $1 \leq m \leq M$. It is useful to first normalize these vectors, to prevent biases due to differences in the absolute value of local sensitivities for various parameters. A large norm of $||s_m||_2$ indicates that a small variation in parameter c_m can significantly impact the system's behavior; thus, this parameter is important. For this parameter to be identifiable, it should not be correlated with other parameters.

Let us consider any subsets K of k parameters $(k \leq M)$ from the set of parameters $\{c_1, c_2, \ldots, c_M\}$ and the corresponding sub-matrix $\tilde{S}_K(X,c)$ of the normalized sensitivity matrix, with columns the k sensitivity vectors. A measure of collinearity of the subset K of parameters, with corresponding matrix $\tilde{S}_K$, is given by

$$CI_K = \frac{1}{\min\limits_{\|z\|_2=1} \|\tilde{S}_K z\|_2} = \frac{1}{\sqrt{\lambda_k}} \tag{13}$$

where λ_k is the minimum eigenvalue of the matrix $\tilde{S}_K^T \tilde{S}_K$ and $\|\cdot\|_2$ is the norm-2 of a vector. The measure (13) is known as the collinearity index of the subset K [19,30]. The closest the columns of the matrix $\tilde{S}_K$ are to a linearly dependent set of vectors, the smallest $\min\limits_{\|z\|_2=1} \|\tilde{S}_K z\|_2$ is. Thus, a large collinearity index CI_K indicates a high level of collinearity of the parameters in the set. This implies that changes in the model dynamics due to small perturbations in one of the parameters of the almost collinear set may be prevented by suitable variations in the other parameters of the set. As a consequence, if a set of parameters is collinear, it is not identifiable. According to [19], a subset of parameters is considered identifiable if the associated collinearity index satisfies $CI_K < 20$. With this observation, it is possible to uncover the subsets of model parameters that can be identified as well as those that cannot be identified. The collinearity index may be computed for all the subsets K of the parameter space, to determine the parameter subsets that are not collinear. When a group of parameters has a high collinearity index, any set containing it as a subset will also have a high collinearity index.

Another technique to assess the identifiability of the model parameters is to use the singular value decomposition (SVD) of a matrix. In general, the SVD [37,38] of an $n \times M$ matrix s is

$$s = \mathbf{U}\boldsymbol{\Sigma}\mathbf{V}^T\,, \tag{14}$$

where the $\mathbf{U}$ is an $n \times n$ unitary matrix, $\mathbf{V}$ is an $M \times M$ unitary matrix and $\boldsymbol{\Sigma}$ is an $n \times M$ non-negative diagonal matrix with the diagonal entries

$$\sigma_1 \geq \sigma_2 \geq \ldots \sigma_r > \sigma_{r+1} = \cdots = \sigma_M = 0\,.$$

The values $\{\sigma_m^2\}_{1 \leq m \leq M}$ are the eigenvalues of the matrix $s^T s$. The index r measures the rank of the matrix s and it is the largest number of linearly independent columns of this matrix. Numerically, the singular values $\sigma_{r+1}, \cdots, \sigma_M$, which are below a specified small tolerance are considered practically zero. In this work, we use the singular value decomposition of the matrix s to determine its rank. This rank is a reliable measure of the number of rate parameters that are not collinear. Furthermore, zero or very close to zero singular values show that the group of all the reaction rate parameters of the model are collinear. Therefore, there are some model parameters that cannot be estimated from the available data.

Brun et al. [20] also introduced a determinant measure

$$\rho_k = \det(S_K^T S_K)^{1/2k} \tag{15}$$

to find the appropriate number of parameters to estimate.

The metrics considered above can be utilized to determine the identifiability of parameter sets as follows. The sensitivity measure δ_k^{msqr} is used to evaluate the importance of each parameter c_k. On the other hand, the collinearity index measures whether the set K of parameters are independent, whenever $CI_K < 20$. In the case that both conditions are satisfied, (a) the parameters in the subset K are not collinear and (b) each parameter in the group is important, the parameters in K are identifiable. Finally, the determinant ρ_K can be employed to compare the identifiability of various groups of parameters.

2.3.3. Method for Selecting Subsets of Identifiable Parameters

The practical identifiability methods presented above were developed for continuous deterministic models [19,20], and are thus applicable for the reaction rate equation model. However, this model may fail to faithfully represent the behavior of biochemical systems, which involve low molecular counts of some species. Consequently, new methodologies are required for the parameter identifiability of stochastic discrete models of biochemical systems. In this work, we develop novel strategies for determining sets of identifiable parameters for the Chemical Master Equation. We generalize the work of Gábor et al. [30] on identifying subsets of identifiable parameters in deterministic models, to address the much more challenging case of stochastic discrete models of well-stirred biochemical systems. This generalization is essential as stochasticity plays a significant role in accurately modeling real-world biological systems, and our approach allows for an in-depth study of more complex biochemical networks encountered in applications.

The measures presented above were designed for deterministic models. We aim to adapt these measures to systems modeled by the Chemical Master Equation. For this model, the sensitivity coefficients are computed as

$$S_{ik}(\mathbb{E}[X],t) = \frac{\partial}{\partial c_k}\mathbb{E}[X_i(t,c)].$$

Then, we shall compute the sensitivity matrix for the CME according to

$$S(t) = \frac{\partial \mathbb{E}[X(t,c)]}{\partial c} = \begin{pmatrix} \frac{\partial}{\partial c_1}\mathbb{E}(X_1(t,c)) & \cdots & \frac{\partial}{\partial c_M}\mathbb{E}(X_1(t,c)) \\ \vdots & \ddots & \vdots \\ \frac{\partial}{\partial c_1}\mathbb{E}(X_N(t,c)) & \cdots & \frac{\partial}{\partial c_M}\mathbb{E}(X_N(t,c)) \end{pmatrix}. \tag{16}$$

Take a sequence of time-points $0 = t_1 < t_2 < \ldots < t_L = T$, relevant to the biochemical system under consideration. The fully normalized (non-dimensional) sensitivity coefficient of the i-th species with respect to the c_k parameter at time t_ℓ is

$$s_{ik}(t_\ell) = \frac{c_k}{\mathbb{E}[X_i(t_\ell,c)]}\frac{\partial}{\partial c_k}\mathbb{E}[X_i(t_\ell,c)] \quad \text{for } 1 \le i \le N,\ 1 \le k \le M. \tag{17}$$

The concatenated non-dimensional sensitivity matrix over these discrete time-points with entries (17) is

$$s(\mathbb{E}[X],c) = \begin{pmatrix} s_{11}(t_1) & \cdots & s_{1M}(t_1) \\ \vdots & \ddots & \vdots \\ s_{N1}(t_L) & \cdots & s_{NM}(t_L) \end{pmatrix}. \tag{18}$$

Normalizing the ℓ-th column of matrix (18), namely $s_\ell(\mathbb{E}[X],c)$, gives

$$\tilde{s}_\ell(\mathbb{E}[X],c) = \frac{s_\ell(\mathbb{E}[X],c)}{\|s_\ell(\mathbb{E}[X],c)\|_2}. \tag{19}$$

Finally, the normalized sensitivity matrix $\tilde{S}$ has $\tilde{s}_\ell(\mathbb{E}[X],c)$ as it is ℓ-th column. For the Chemical Master Equation, the sensitivity measure δ_k^{msqr} and the collinearity index CI_K are computed using (12) and (13), respectively, for the sensitivity matrix of the expected value $\mathbb{E}[X]$ rather than the system state X, as was the case for the reaction rate equation.

Moreover, we will employ the finite-difference methods described above to estimate parametric sensitivities. Recall that a finite-difference estimate of the sensitivity with respect to parameter c_k, over R coupled perturbed and unperturbed paths, is

$$\frac{\partial}{\partial c_k}\mathbb{E}[X(t,c)] \approx Z_R = \frac{1}{R}\sum_{r=1}^{R}\frac{X_{[r]}(t,c_k+\theta) - X_{[r]}(t,c_k)}{\theta}.$$

While we compute the coupled trajectories using the CFD, CRP, or CRN strategies, our method can be applied to other finite-difference sensitivity estimators [29].

The measure (12) can be calculated to rank parameters from most to least influential. Small values of δ^{msqr} correspond to parameters with a small influence on the model. We select those parameters that show the value of δ^{msqr} larger than 0.2 [39]. With an initial ranked list, we compute the collinearity indices for this list. This method can be applied to models of moderate size.

Algorithm 1 calculates the normalized sensitivity matrix, as follows. A grid with L time-points ranging from 0 to T is selected. We choose equally distributed time steps, such that data is collected from all important regions of the interval of integration. This depends on the particular model. We note that an adaptive time-stepping procedure can be included instead. Then, the sensitivity matrices $S(t_l)$ from Equation (16) are approximated with a specific finite-difference sensitivity estimator. Afterwards, we compute the concatenated non-dimensional sensitivity matrix s. We normalize each column of s individually to ensure consistency and comparability. The normalization implies dividing each column s_k by its vectorial norm-2. Column normalization yields a matrix denoted by $\tilde{S}$. This matrix has as its k-th column $\{\tilde{s}_k\} = s_k/||s_k||_2$. Also, for each parameter c_k we compute the sensitivity measure δ_k^{msqr} from Equation (12), using the entries of the k-th columns of the sensitivity matrices $S(t_\ell)$.

Algorithm 1 Computing the Normalized Sensitivity Matrix

Initialize: Time grid: $0 = t_1 < t_2 < \ldots < t_L = T$.
Input: Estimates of sensitivity matrices $S(t_\ell)$ from (16).
Compute the concatenated non-dimensional sensitivity matrix s from (18) with entries (17)
for $k = 1$ to M **do**
  normalize $\tilde{s}_k = \dfrac{s_k}{||s_k||_2}$ where s_k is the k-th column of s and $||\cdot||_2$ is norm-2
end for
Compute normalized matrix $\tilde{S} = \{\tilde{s}_k\}_{1\le k\le M}$
for $k = 1$ to M **do**
  Compute sensitivity measure δ_k^{msqr} according to (12) for parameter c_k
end for

In Algorithm 2, we introduce a method for the selection of identifiable parameter subsets based on sensitivity measures and collinearity indices. This procedure extends and refines a methodology by Gábor et al. [30] from the deterministic to the more difficult case of stochastic biochemical networks. The goal of Algorithm 2 is to iteratively assess the practical identifiability of subsets of model parameters. A threshold value is set for the collinearity indices, which measure the level of collinearity between parameter groups. The threshold value determines the acceptable level of collinearity. With a normalized sensitivity matrix obtained from Algorithm 1 as input, the following steps are considered. The parameters are ranked according to their sensitivity measure, those with a sensitivity measure below a critical value (chosen here as 0.2) are considered unimportant and may be discarded. If the ranked list of parameters is of moderate size, combinations of parameters are generated. For each combination, the algorithm computes the corresponding collinearity index. This involves calculating the collinearity indices for pairs, triples, etc. These indices quantify the degree of collinearity between the parameters of a certain group. When the computed collinearity index for a parameter subset is below the threshold value, that subset of parameters is deemed identifiable. By applying this algorithm, a subset of parameters with low collinearity and high identifiability can be selected. This allows for the reduction in model complexity and for the accurate and reliable estimation of the most important parameters, from the input data.

Algorithm 2 Selecting a Subset of Identifiable Parameters

Input: Normalized sensitivity matrix;
Input: Set threshold value of collinearity index: $CI_{cr} = 20$
Require: Rank parameters c_j based on $\delta_j^{msqr} > 0.2$
if Ranked list is of moderate size **then**
  1: Number of all combinations: $\mathcal{C} = Length(combnk)$
  2: Compute collinearity indices for all combinations of the ranked list of parameters:
  for k = 1 to $\mathcal{C}$ **do**
    For every combination of the ranked list of parameters, calculate the collinearity indices:
    $CI_2 = collinearityindex(pairs), CI_3 = collinearity(triples)$, etc.
    $L_2 = pair\ combination,\ L_3 = triple\ combination$, etc.
  end for
end if
if $CI_k \leq CI_{cr}$ **then**
  The corresponding combination recorded as an identifiable set
end if

3. Results

In this section, we apply our method to select subsets of practically identifiable parameters in the Chemical Master Equation on three realistic models. We observe that the collinearity indices play a significant role in finding the subsets of estimable parameters, using local stochastic sensitivities. The parametric sensitivities of the stochastic discrete model of well-stirred biochemical systems are approximated by finite-difference schemes, namely the Common Random Number, Common Reaction Path, and Coupled Finite Difference techniques. By applying perturbation in each of these finite-difference techniques, we can assess the sensitivity of the model outputs to changes in the model's parameters. The choice of perturbation size for finite-difference approximations is essential for obtaining accurate and reliable results while minimizing computational effort. The specific perturbation sizes, representing 5%, 1%, 2% of the parameter value, are often chosen based on a trade-off between accuracy and numerical stability. In addition, we find the parameters with high sensitivities. Those with low sensitivity have a reduced impact on the model outputs and cannot be estimated accurately. In the stochastic context, we consider the SVD of the normalized sensitivity matrix to determine its rank. This rank gives the number of model parameters that are not collinear.

For validation of the methods introduced above, we compare the results obtained with the Chemical Master Equation, with those derived with the Chemical Langevin Equation and those for the reaction rate equations, on two models of biochemically reacting systems. Still, we emphasize the importance of considering stochastic discrete models of biochemical networks to accurately describe the dynamics of these systems, particularly when some molecular populations are small or noise is driving the system behavior. The parametric sensitivities estimated for the reaction rate equations or the Chemical Langevin Equations may not yield accurate estimability results, in general. For each model, we generated 10,000 coupled trajectories to approximate the parametric sensitivities of the Chemical Master Equation by finite-difference schemes. The CFD strategy is considered to be more accurate and reliable than the CRN and the CRP methods [28]. The case studies tested are an infectious disease network [40], the Michaelis–Menten system and a genetic toggle-switch model [11].

3.1. *Infectious Disease Model*

An infectious disease model [40] considers two species: S_1—the infected particles and S_2—the particles which can be infected. These species, which may depict molecules, cells, or humans, participate in five reactions. The first two reactions represent the death of species S_1 and S_2, respectively, while the third and fourth reactions describe the birth or production of particles of the S_1 and S_2 type. The two species interact through the fifth reaction, in which an infected particle S_1 infects a particle S_2. The initial conditions are $S_1(0) = 20$ and $S_2(0) = 40$. The system is studied on the time-interval $[0, 10]$. For our simulations, 10,000 trajectories were generated to estimate the solution of the Chemical Master Equation.

Table 1 provides information on the reaction channels of the biochemical system and the values of their rate parameters. It includes the reaction channels denoted by R_1, R_2, R_3, R_4, and R_5. Each reaction is described by its reactants and products. The last column lists the parameter values corresponding to the rates at which the reactions occur. These parameter values are specified for the stochastic model considering molecular numbers, rather than for the deterministic reaction rate equations expressed in terms of concentrations. A sample trajectory of the number of the infected S_1 particles and of the susceptible S_2 particles as functions of time, computed using Gillepie's algorithm, is given in Figure 1.

Table 1. Infectious disease model: the list of reactions and the corresponding rate parameter values.

	Reaction Channel	Rate Parameter Value
R_1:	$S_1 \longrightarrow \emptyset$	$c_1 = 2.0$
R_2:	$S_2 \longrightarrow \emptyset$	$c_2 = 0.1$
R_3:	$\emptyset \longrightarrow S_1$	$c_3 = 25$
R_4:	$\emptyset \longrightarrow S_2$	$c_4 = 75$
R_5:	$S_1 + S_2 \longrightarrow S_1 + S_1$	$c_5 = 0.05$

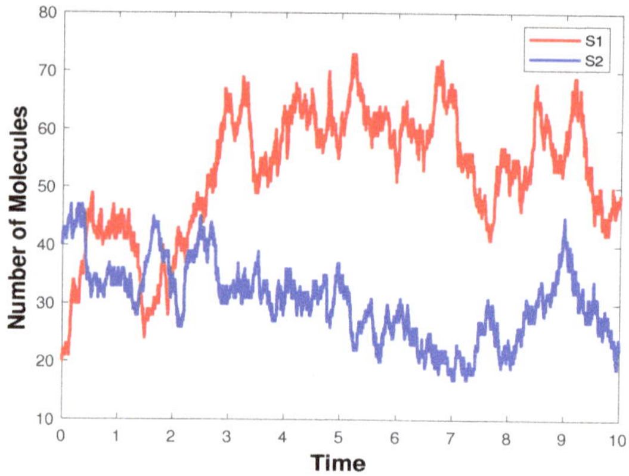

Figure 1. Infectious disease model: the evolution in time of the number of molecules of the species S_1—infected individuals and S_2—individuals which can be infected, generated with Gillespie's algorithm, on the interval $[0, 10]$.

The finite-difference sensitivity estimations are calculated with 10,000 trajectories using the CFD, the CRN, and the CRP strategies, with a perturbation of 5% of the parameter value. The path-wise sensitivities for the Chemical Langevin Equation are computed over 10,000 trajectories, with the Euler-Maruyama scheme applied to the Equations (2) and (7), and are utilized to estimate the sensitivities of the expected value of the state vector. Also, the parametric sensitivities are approximated for the reaction rate equations. These estimations are used to calculate the collinearity indices for all parameter combinations, for the Chemical Master Equation, the Chemical Langevin Equation, and the RRE models. The results are presented in Tables 2–6. The sensitivity measures are reported in Table 2, showing that c_2 is the least significant among all the parameters.

Table 2. Infectious disease model: comparison of δ^{msqr} for a 5% perturbation.

Parameter	δ^{msqr} of CFD Sensitivity	δ^{msqr} of CRP Sensitivity	δ^{msqr} of CRN Sensitivity	δ^{msqr} of RRE Sensitivity	δ^{msqr} Path-Wise Sensitivity
c_1	0.97	0.96	0.94	0.97	0.98
c_2	0.02	0.02	0.1	0.02	0.02
c_3	0.26	0.29	0.26	0.26	0.26
c_4	0.55	0.66	0.54	0.55	0.55
c_5	0.68	0.69	0.67	0.71	0.71

Tables 3–6 reveal that the collinearity indices for the reaction rate equation and the Chemical Langevin Equation models exhibit greater consistency with the collinearity indices for the Chemical Master Equation, computed using with the CFD sensitivity estimator, compared to the CRN and the CRP estimators. Notably, the pair subset $\{c_1, c_3\}$ has the highest collinearity index; however, it is relatively low for the CRP and the CRN schemes in comparison with the other estimations. This is due to the lower accuracy of the CRP and the CRN schemes when compared to the CFD technique. For pair sets, the subset $\{c_1, c_3\}$, for the triple sets, the subset $\{c_3, c_4, c_5\}$ and among the quadruple ones, the subset $\{c_2, c_3, c_4, c_5\}$ have high value of collinearity indices in relation to the other subsets.

Table 3. Infectious disease model: collinearity indices for pair subsets. The CME sensitivities are estimated over 10,000 trajectories with the CFD, CRN, and CRP algorithms and a 5% perturbation.

Parameter Subset	Collinearity Index of CFD Sensitivity	Collinearity Index of CRP Sensitivity	Collinearity Index of CRN Sensitivity	Collinearity Index of RRE Sensitivity	Collinearity Index of Path-Wise Sensitivity
c_4 c_5	1.18	1.13	1.18	1.2	1.19
c_3 c_5	1.93	1.17	1.94	1.92	1.95
c_3 c_4	1.339	1.25	1.32	1.32	1.31
c_2 c_5	1.103	1.15	1.13	1.18	1.17
c_2 c_4	4.69	2.37	1.16	9.77	9.96
c_2 c_3	1.43	1.27	1.02	1.34	1.33
c_1 c_5	1.86	1.89	1.9	1.85	1.86
c_1 c_4	1.35	1.28	1.33	1.34	1.34
c_1 c_3	10.816	3.04	7.2	11.34	11.22
c_1 c_2	1.466	1.31	1.00	1.36	1.35

There is no subset with high collinearity indices (>20) in pair subsets (Table 3) but there is a parameter subset of size 3 with collinearity index greater than 20 (Table 4). In fact, the parameter subset $\{c_3, c_4, c_5\}$ is not identifiable with the Coupled Finite Difference sensitivity estimator, the Chemical Langevin Equation, or the deterministic sensitivities. However, the Common Random Number and the Common Reaction Path sensitivities show different results. In Table 5, two parameter subsets of size 4 show a collinearity index greater than 20 with the deterministic, stochastic continuous, and CFD sensitivity estimations. All subsets containing the parameters $\{c_3, c_4, c_5\}$ are collinear, which is in agreement with the results in Table 4. This indicates that these parameter subsets are poorly identifiable. Consequently, the sensitivity-based estimability analysis performed on the RRE, the CLE, and the CME models are in agreement, thus validating the proposed method for the more general discrete stochastic model. The Common Random Number and the Common Reaction Path techniques could not provide an accurate assessment of the identifiability of various subsets, with only 10,000 realizations, being thus less reliable.

Table 4. Infectious disease model: collinearity indices for triple subsets. The CME sensitivities are estimated over 10,000 trajectories with the CFD, CRN, and CRP algorithms and a 5% perturbation.

Parameter Subset	Collinearity Index of CFD Sensitivity	Collinearity Index of CRP Sensitivity	Collinearity Index of CRN Sensitivity	Collinearity Index of RRE Sensitivity	Collinearity Index of Path-Wise Sensitivity
$c_3\ c_4\ c_5$	21.19	2.63	9.6	21.3	21.77
$c_2\ c_4\ c_5$	5.0444	2.38	1.2	9.97	10.15
$c_2\ c_3\ c_5$	7.7768	2.91	2.01	10.48	10.51
$c_2\ c_3\ c_4$	4.88	2.38	1.43	9.83	10.01
$c_1\ c_4\ c_5$	9.92	3.65	9.4	10.83	10.98
$c_1\ c_3\ c_5$	11.07	3.12	7.2	11.68	11.73
$c_1\ c_3\ c_4$	10.87	3.05	7.2	11.46	11.45
$c_1\ c_2\ c_5$	7.44	4.8	2	7.87	7.95
$c_1\ c_2\ c_4$	4.95	2.38	1.43	9.82	10.01
$c_1\ c_2\ c_3$	11.02	3.06	7.3	11.45	11.44

Table 5. Infectious disease model: collinearity indices for quadruple subsets. The CME sensitivities are estimated over 10,000 trajectories with the CFD, CRN, and CRP algorithms and a 5% perturbation.

Parameter Subset	Collinearity Index of CFD Sensitivity	Collinearity Index of CRP Sensitivity	Collinearity Index of CRN Sensitivity	Collinearity Index of RRE Sensitivity	Collinearity Index of Path-Wise Sensitivity
$c_1\ c_2\ c_3\ c_4$	11.509	3.06	7.3	11.53	11.49
$c_1\ c_2\ c_3\ c_5$	11.092	4.88	7.3	13.65	13.53
$c_1\ c_2\ c_4\ c_5$	10.2347	4.94	9.4	13.82	14.20
$c_1\ c_3\ c_4\ c_5$	22.6313	3.87	10.54	22.19	22.49
$c_2\ c_3\ c_4\ c_5$	21.4369	2.91	9.6	25.71	27.77

Table 6. Infectious disease model: collinearity indices for the set of all kinetic parameters. The CME sensitivities are estimated over 10,000 trajectories with the CFD, CRN, and CRP algorithms and a 5% perturbation. The singular values for the CFD, the CLE, and the RRE sensitivity estimations show that the number of parameters that are not collinear is four.

Parameter Subset	Collinearity Index of CFD Sensitivity	Collinearity Index of CRP Sensitivity	Collinearity Index of CRN Sensitivity	Collinearity Index of RRE Sensitivity	Collinearity Index of Path-Wise Sensitivity
$c_1\ c_2\ c_3\ c_4\ c_5$	22.65	5.01	10.54	26.17	28.09
singular values	16.31, 9.48, 1.06, 0.21, 0.06	16.27, 10.35, 2.98, 1.79, 0.14	15.86, 9.28, 1.31, 1.1, 0.52	36.73, 21.86, 2.21, 0.48, 0.09	37.03, 21.76, 2.19, 0.48, 0.09

3.2. Michaelis–Menten Model

The second model we analyze is the Michaelis–Menten biochemical system, which involves four species—a substrate S_1, an enzyme S_2, a complex S_3 and a product S_4—and three reactions. We denote by Y_i the number of molecules of the species S_i. With this notation, the initial conditions for the number of molecules are $Y_1(0) = \lfloor 5 \times 10^{-7} n_A vol \rfloor$, $Y_2(0) = \lfloor 2 \times 10^{-7} n_A vol \rfloor$ and $Y_3(0) = Y_4(0) = 0$, where $n_A = 6.023 \times 10^{23}$ is Avogadro's number and $vol = 10^{-15}$ denotes the volume of the system. The reactions and the values of the rate parameters are included in Table 7. This model is integrated on the interval $[0, 50]$. Figure 2 depicts a realization of the system state, simulated with Gillespie's algorithm.

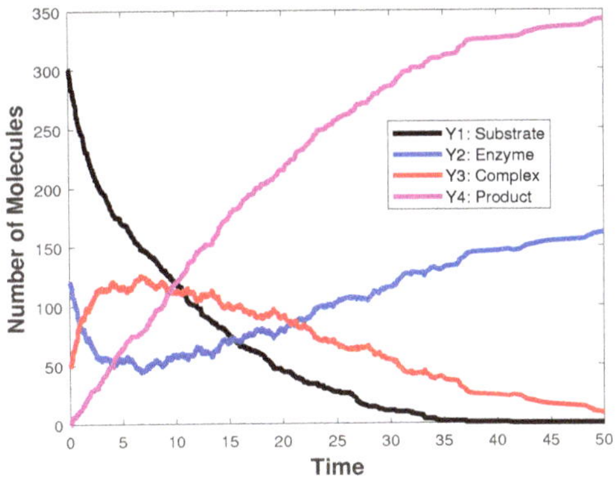

Figure 2. Michaelis–Menten model: the evolution in time of the number of molecules of a substrate, an enzyme, a complex and a product, generated with Gillespie's algorithm, on the interval $[0, 50]$.

Table 7. Michaelis–Menten model: the list of reactions and the corresponding rate parameter values.

	Reaction Channel	Rate Parameter Value
R_1:	$S_1 + S_2 \longrightarrow S_3$	$c_1 = 10^6/(n_A vol)$
R_2:	$S_3 \longrightarrow S_1 + S_2$	$c_2 = 10^{-4}$
R_3:	$S_3 \longrightarrow S_4 + S_2$	$c_3 = 10^{-1}$

We start by approximating the parametric sensitivities for the Chemical Master Equation. The finite-difference sensitivity estimations obtained with the CFD, the CRP, and the CRN algorithms use a perturbation which represents 1% and 5%, respectively, of the value of the parameter of interest. The sensitivity measures provided in Table 8 indicate that c_2 may not be estimated as accurately as the other parameters. The collinearity indices obtained for the perturbation value 1% with each sensitivity estimator for pairs of parameters are reported in Table 9, while the indices for the set of all parameters are recorded in Table 10. For each subset, the results for the stochastic Michaelis–Menten model demonstrate low collinearity indices, below 20. The choice of the finite-difference sensitivity estimator does not significantly affect the parameter identifiability. The stochastic discrete modeling approach to identifiability analysis yields parameter subsets that are not collinear for the Michaelis–Menten system. Additionally, the Tables include the RRE identifiability metrics to validate the CME estimability results. The collinearity indices for the perturbation value of 5% can be found in the Appendix A, and they are consistent with the results obtained using a perturbation of 1%.

Table 8. Michaelis–Menten model: comparison of δ^{msqr} for a 1% perturbation.

Parameter	δ^{msqr} of CFD Sensitivity	δ^{msqr} of CRP Sensitivity	δ^{msqr} of CRN Sensitivity	δ^{msqr} of RRE Sensitivity
c_1	1.11	1.1	1.07	1.07
c_2	0.002	0.01	0.003	0.002
c_3	1.31	1.30	1.29	1.29

Table 9. Michaelis–Menten model: collinearity indices for pair subsets. The CME sensitivities are estimated over 10,000 trajectories with the CFD, CRN, and CRP algorithms and a 1% perturbation.

Parameter Subset	Collinearity Index of CFD Sensitivity	Collinearity Index of CRP Sensitivity	Collinearity Index of CRN Sensitivity	Collinearity Index of RRE Sensitivity
$c_1\ c_2$	2.9	1.35	1.47	4.85
$c_1\ c_3$	2.21	2.17	2.17	2.17
$c_2\ c_3$	1.56	1.21	1.2	1.87

Table 10. Michaelis–Menten model: collinearity indices for the triple subset. The CME sensitivities are estimated over 10,000 trajectories with the CFD, CRN, and CRP algorithms and a 1% perturbation.

Parameter Subset	Collinearity Index of CFD Sensitivity	Collinearity Index of CRP Sensitivity	Collinearity Index of CRN Sensitivity	Collinearity Index of RRE Sensitivity
$c_1\ c_2\ c_3$	3.92	2.25	2.43	5.3

3.3. Genetic Toggle Switch Model

The last biochemical system investigated is the genetic toggle switch [11,28]. Multi-stable stochastic switches arise in modeling key biological processes. The model considers two gene pairs, whose interaction creates a bistable switch, as each gene negatively regulates the synthesis of the other gene. Due to the presence of noise, the system can transition between the states represented by an abundance of one species and an almost total absence of the other. In this genetic switch system, the two species U and V take part in four reactions. Table 11 specifies the reaction channels and their propensities. We examine the system using the following parameter values [11]

$$\alpha_1 = 50,\ \beta = 2.5,\ \alpha_2 = 16,\ \gamma = 1\,, \tag{20}$$

and the initial conditions $X_V(0) = X_U(0) = 0$. Figure 3 displays a sample path for the molecular numbers of the two species, simulated with Gillespie's algorithm (left) along with the standard deviation of the CFD, CRP, and CRN sensitivity estimators as functions of time (right).

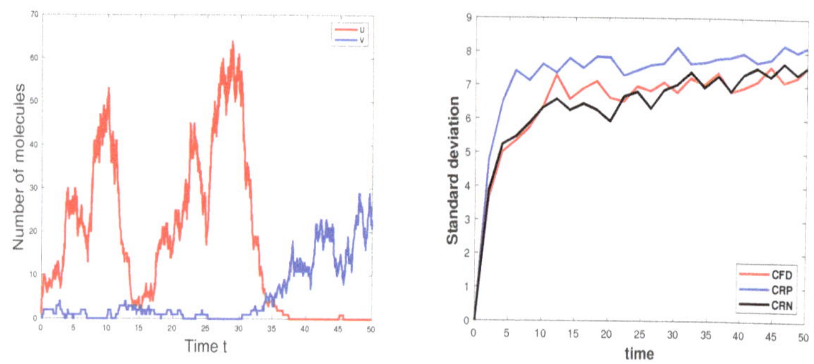

Figure 3. Genetic toggle switch model: (**Left**): the evolution in time of the number of molecules of the species U and V, generated with Gillespie's algorithm, on the interval $[0, 50]$. (**Right**): standard deviations of the three estimators, CFD, CRP, and CRN.

Table 11. Genetic toggle switch model: the list of reactions and their propensity functions.

	Reaction Channel	Propensity Function
R_1:	$\emptyset \longrightarrow U$	$a_1 = \dfrac{\alpha_1}{1+X_V^{\beta}}$
R_2:	$U \longrightarrow \emptyset$	$a_2 = X_U$
R_3:	$\emptyset \longrightarrow V$	$a_3 = \dfrac{\alpha_2}{1+X_U^{\gamma}}$
R_4:	$V \longrightarrow \emptyset$	$a_4 = X_V$

The reaction rate equation model cannot capture the stochastic transitions between the states, and thus the deterministic tools for analyzing this system are not applicable. We perform an estimability analysis of the Chemical Master Equation model for the genetic toggle switch, on the interval $[0, 50]$. To assess how variations in the parameter values affect the dynamics of the system, we approximate the local sensitivities with respect to the parameters whose values are given by (20). We simulate 10,000 coupled sample paths with the CFD, and the CRP methods. The finite-difference sensitivity estimators are applied with a perturbation $\theta = 10^{-4}$ for each parameter value. The sensitivity measures are provided in Table 12 and those calculated using the CFD method show that all parameters have $\delta^{msqr} > 0.2$, being thus important enough, while the RRE sensitivity measures indicate that the parameters β and γ are insignificant.

Table 12. Genetic toggle switch model: comparison of δ^{msqr}.

Parameter	δ^{msqr} of CFD Sensitivity	δ^{msqr} of RRE Sensitivity
$\alpha 1$	2.22	0.89
β	0.6762	0
$\alpha 2$	4.21	0.31
γ	4.3	0

Employing the local sensitivity approximations, we compute the collinearity indices for all the subsets of the parameter set $\{\alpha_1, \alpha_2, \beta, \gamma\}$. Tables 13–15 record the collinearity indices for the pair, triple and quadruple subsets, respectively. No subset of parameters exhibits collinearity based on the CFD, the CRP, and the CRN sensitivity estimations. We conclude that all four parameters are identifiable for the stochastic discrete model. These results are confirmed by the singular values computed with the CFD sensitivity estimator, which are [32.21; 29; 12.18; 4]. Different values of the parameters for this model may yield different results for estimability in the stochastic genetic toggle-switch system.

Table 13. Genetic toggle switch model: collinearity indices for pair subsets.

Parameter Subset	Collinearity Index of CFD Sensitivity	Collinearity Index of CRP Sensitivity	Collinearity Index of CRN Sensitivity	Collinearity Index of RRE Sensitivity
$\alpha_1\ \alpha_2$	1	1.01	1.72	2.22
$\gamma\ \alpha_2$	1.32	1.08	1.12	*
$\gamma\ \alpha_1$	1.27	1.17	1.07	*
$\beta\ \alpha_2$	1.01	1.1	1.56	*
$\beta\ \alpha_1$	1.00	1.35	2.13	*
$\beta\ \gamma$	1.19	1.25	1.1	*

The CME sensitivities with respect to parameters are estimated over 10,000 with the CFD and CRP methods and perturbation $\theta = 10^{-4}$. *: Collinearity index does not exist.

Table 14. Genetic toggle switch model: collinearity indices for triple subsets.

Parameter Subset	Collinearity Index of CFD Sensitivity	Collinearity Index of CRP Sensitivity	Collinearity Index of CRN Sensitivity	Collinearity Index of RRE Sensitivity
$\beta\ \gamma\ \alpha_1$	1.38	1.37	2.46	*
$\beta\ \gamma\ \alpha_2$	1.42	1.25	1.80	*
$\beta\ \alpha_1\ \alpha_2$	1.01	1.38	2.18	*
$\gamma\ \alpha_1\ \alpha_2$	1.52	1.19	1.73	*

The CME sensitivities with respect to parameters are estimated over 10,000 with the CFD and CRP methods and perturbation $\theta = 10^{-4}$. *: Collinearity index does not exist.

Table 15. Genetic toggle switch model: collinearity indices for the quadruple subset.

Parameter Subset	Collinearity Index of CFD Sensitivity	Collinearity Index of CRP Sensitivity	Collinearity Index of CRN Sensitivity	Collinearity Index of RRE Sensitivity
$\beta\ \gamma\ \alpha_1\ \alpha_2$	1.64	1.39	2.45	*

The CME sensitivities with respect to parameters are estimated over 10,000 with the CFD and CRP methods and perturbation $\theta = 10^{-4}$. *: Collinearity index does not exist.

4. Discussion

Stochastic models of well-stirred biochemical processes provide a valuable framework for capturing inherent variability at the cellular level when some molecular species have low amounts. Chemical Master Equation is a frequently adopted stochastic discrete model for such processes. By contrast, deterministic approaches are often not suitable for modeling cellular systems as they fail to capture the intrinsic randomness observed experimentally. Many models of realistic biochemical processes depend on a fairly large number of parameters. The values of some of these parameters may be unknown and have to be estimated. Parameter estimation is a critical step in modeling biochemical systems. However, determining appropriate parameter values for stochastic discrete models of biochemical networks poses many challenges. It is essential to determine the key parameters which are identifiable from the experimental data, as well as those that cannot be reliably estimated. For a subset of parameters to be practically identifiable, each parameter of the subset should have a significant contribution to the system dynamics as well as the parameters of the subset should not be correlated.

In this work, we propose a method for detecting collinearity in subsets of parameters for the stochastic discrete model of the Chemical Master Equation, with the goal of finding the parameter sets that exert the greatest influence on the biochemical system state. In addition, we introduce a technique for determining the highest parameter identifiable sets for stochastic biochemical systems, by extending methods from deterministic models to stochastic models. Our analysis is based on estimating the local sensitivities of the system state with respect to the model's parameters. This is achieved by utilizing finite-difference approximations of the parameter sensitivities, specifically the Coupled Finite Difference, the Common Reaction Path, and the Common Random Number schemes. Furthermore, we examine the role of the singular value decomposition of the sensitivity matrix in identifying parameters that are not collinear in stochastic models of biochemical systems. On one hand, we showed that our practical identifiability method is accurate, by comparing the results obtained in the deterministic and stochastic scenarios, on two biochemical systems of practical importance, for which the deterministic model accurately describes the evolution of the expected value of the stochastic system state. Excellent agreement among the various approaches was obtained for these biochemical networks. On the other hand, we wish to emphasize that, in general, a stochastic strategy for selecting identifiable parameter sets should be considered, as it relies on more accurate and reliable estimations of the parametric sensitivities for the widely applicable model of the Chemical Master Equation, compared to the deterministic reaction rate equations. The advantages of our approach over the deterministic one were illustrated by the tests performed on a third model, a genetic toggle switch system exhibiting an interesting multistable behavior. For this model, our stochastic identifiability strategies display excellent performance, while the deterministic techniques show their limitations, by not being able to assess the estimability of the model parameters.

We expect the method to perform best on stochastic biochemical models with a moderate number of reaction rate parameters. Specifying identifiable parameter subsets with the tools provided above may be used to refine models, improve predictions, and study the underlying biological processes under consideration.

Author Contributions: Conceptualization, S.G. and S.I.; methodology, S.G. and S.I.; software, S.G.; validation, S.I.; investigation, S.G. and S.I.; writing—original draft preparation, S.G. and S.I.; writing—review and editing, S.I.; visualization, S.G. and S.I.; supervision, S.I.; funding acquisition, S.I. All authors have read and agreed to the published version of the manuscript.

Funding: This research was supported by a grant from the National Sciences and Engineering Research Council of Canada (NSERC)—Grant No. RGPIN-2020-05469, and Toronto Metropolitan University.

Institutional Review Board Statement: Not applicable.

Informed Consent Statement: Not applicable.

Data Availability Statement: No new data were created.

Acknowledgments: The authors wish to thank the anonymous referees for their suggestions which helped improve the presentation and Monjur Morshed for discussions.

Conflicts of Interest: The authors declare no conflict of interest.

Appendix A

Table A1. Michaelis–Menten model: comparison of δ^{msqr} for a 5% perturbation.

Parameter	δ^{msqr} of CFD Sensitivity	δ^{msqr} of CRP Sensitivity	δ^{msqr} of CRN Sensitivity	δ^{msqr} of RRE Sensitivity
c_1	1.03	1.04	1.04	1.07
c_2	0.002	0.005	0.003	0.002
c_3	1.22	1.23	1.23	1.29

Table A2. Michaelis-Menten model: collinearity indices for pair subsets. The CME sensitivities are estimated over 10,000 trajectories with the CFD, CRN and CRP algorithms and a 5% perturbation.

Parameter Subset	Collinearity Index of CFD Sensitivity	Collinearity Index of CRP Sensitivity	Collinearity Index of CRN Sensitivity	Collinearity Index of RRE Sensitivity
c_1 c_2	3.43	2.18	2.59	4.85
c_1 c_3	2.21	2.13	2.13	2.17
c_2 c_3	1.67	1.48	1.49	1.87

Table A3. Michaelis–Menten model: collinearity indices for the triple subset. The CME sensitivities are estimated over 10,000 trajectories with the CFD, CRN and CRP algorithms and a 5% perturbation.

Parameter Subset	Collinearity Index of CFD Sensitivity	Collinearity Index of CRP Sensitivity	Collinearity Index of CRN Sensitivity	Collinearity Index of RRE Sensitivity
c_1 c_2 c_3	4.08	2.78	3.4	5.3

References

1. Kitano, H. Computational systems biology. *Nature* **2002**, *420*, 206–210. [CrossRef] [PubMed]
2. Maheshri, N.; O'Shea, E.K. Living with noisy genes: How cells function reliably with inherent variability in gene expression. *Annu. Rev. Biophys. Biomol. Struct.* **2007**, *36*, 413–434. [CrossRef]
3. Raj, A.; van Oudenaarden, A. Nature, nurture, or chance: Stochastic gene expression and its consequences. *Cell* **2008**, *135*, 216–226. [CrossRef] [PubMed]
4. Ozbudak, E.M.; Thattai, M.; Kurtser, I.; Grossman, A.D.; van Oudenaarden, A. Regulation of noise in the expression of a single gene. *Nat. Genet.* **2002**, *31*, 69–73. [CrossRef]
5. Raser, J.M.; O'Shea, E.K. Noise in gene expression: Origins, consequences, and control. *Science* **2005**, *309*, 2010–2013. [CrossRef]
6. Gillespie, D.T. A rigorous derivation of the chemical master equation. *Stat. Mech. Its Appl.* **1992**, *188*, 404–425. [CrossRef]
7. McQuarrie, D.A. Stochastic approach to chemical kinetics. *J. Appl. Probab.* **1967**, *4*, 413. [CrossRef]
8. Gillespie, D.T. A general method for numerically simulating the stochastic time evolution of coupled chemical reactions. *J. Comput. Phys.* **1976**, *22*, 403–434. [CrossRef]
9. Gillespie, D.T. Exact stochastic simulation of coupled chemical reactions. *J. Phys. Chem.* **1977**, *81*, 2340–2361. [CrossRef]
10. Ethier, S.N.; Kurtz, T.G. *Markov Processes: Characterization and Convergence*; Wiley: New York, NY, USA, 1986.
11. Rathinam, M.; Sheppard, P.W.; Khammash, M. Efficient computation of parameter sensitivities of discrete stochastic chemical reaction networks. *J. Chem. Phys.* **2010**, *132*, 034103–034116. [CrossRef]
12. El Samad, H.; Khammash, M.; Petzold, L.; Gillespie, D.T. Stochastic modelling of gene regulatory networks. *Int. J. Robust Nonlinear Control* **2005**, *15*, 691–711. [CrossRef]
13. Strehl, R.; Ilie, S. Hybrid stochastic simulation of reaction-diffusion systems with slow and fast dynamics. *J. Chem. Phys.* **2015**, *143*, 234108. [CrossRef] [PubMed]
14. Wilkinson, D.J. *Stochastic Modelling for Systems Biology*; Taylor & Francis: Boca Raton, FL, USA, 2019.
15. Thanh, V.H.; Zunino, R.; Priami, C. On the rejection-based algorithm for simulation and analysis of large-scale reaction networks. *J. Chem. Phys.* **2015**, *142*, 244106. [CrossRef]
16. Barrows, D.; Ilie, S. Parameter estimation for the reaction-diffusion master equation. *AIP Adv.* **2023**, *13*, 065318. [CrossRef]
17. Petre, I.; Mizera, A.; Hyder, C.L.; Meinander, A.; Mikhailov, A.; Morimoto, R.I.; Sistonen, L.; Eriksson, J.E.; Back, R.J. A simple mass-action model for the eukaryotic heat shock response and its mathematical validation. *Nat. Comput.* **2011**, *10*, 595–612. [CrossRef]
18. Vajda, S.; Rabitz, H.; Walter, E.; Lecourtier, Y. Qualitative and quantitative identifiability analysis of nonlinear chemical kinetic models. *Chem. Eng. Commun.* **1989**, *83*, 191–219. [CrossRef]
19. Brun, R.; Reichert, P.; Künsch, H.R. Practical identifiability analysis of large environmental simulation models. *Water Resour. Res.* **2001**, *37*, 1015–1030. [CrossRef]
20. Brun, R.; Kühni, M.; Siegrist, H.; Gujer, W.; Reichert, P. Practical identifiability of ASM2d parameters—Systematic selection and tuning of parameter subsets. *Water Res.* **2002**, *36*, 4113–4127. [CrossRef] [PubMed]
21. Chis, O.T.; Banga, J.R.; Balsa-Canto, E. Structural identifiability of systems biology models: A critical comparison of methods. *PLoS ONE* **2011**, *6*, e27755. [CrossRef] [PubMed]
22. Holmberg, A. On the practical identifiability of microbial growth models incorporating Michaelis-Menten type nonlinearities. *Math. Biosci.* **1982**, *62*, 23–43. [CrossRef]
23. Jacquez, J.; Greif, P. Numerical parameter identifiability and estimability: Integrating identifiability, estimability, and optimal sampling design. *Math. Biosci.* **1985**, *77*, 201–227. [CrossRef]
24. Komorowski, M.; Costa, M.J.; Rand, D.A.; Stumpf, M.P.H. Sensitivity, robustness, and identifiability in stochastic chemical kinetics models. *Proc. Natl. Acad. Sci. USA* **2011**, *108*, 8645–8650. [CrossRef] [PubMed]
25. Rodriguez-Fernandez, M.; Egea, J.A.; Banga, J.R. Novel metaheuristic for parameter estimation in nonlinear dynamic biological systems. *BMC Bioinform.* **2006**, *7*, 483. [CrossRef]
26. Villaverde, A.F. Observability and structural identifiability of nonlinear biological systems. *Complexity* **2019**, *2019*, 1–12. [CrossRef]
27. Anderson, D.F. An efficient finite-difference method for parameter sensitivities of continuous time Markov chains. *SIAM J. Num. Anal.* **2012**, *50*, 2237–2258. [CrossRef]
28. Srivastava, R.; Anderson, D.F.; Rawlings, J.B. Comparison of finite difference based methods to obtain sensitivities of stochastic chemical kinetic models. *J. Chem. Phys.* **2013**, *138*, 074110. [CrossRef] [PubMed]
29. Morshed, M. Efficient Finite-Difference Methods for Sensitivity Analysis of Stiff Stochastic Discrete Models of Biochemical Systems. Ph.D. Thesis, University of Waterloo, Waterloo, ON, Canada, 2017.
30. Gábor, A.; Villaverde, A.F.; Banga, J.R. Parameter identifiability analysis and visualization in large-scale kinetic models of biosystems. *BMC Syst. Biol.* **2017**, *11*, 54. [CrossRef] [PubMed]
31. Turner, T.E.; Schnell, S.; Burrage, K. Stochastic approaches for modelling in vivo reactions. *Comput. Biol. Chem.* **2004**, *28*, 165–178. [CrossRef]
32. Gillespie, D.T. The chemical Langevin equations. *J. Phys. Chem.* **2000**, *113*, 297–306. [CrossRef]
33. Ilie, S.; Gholami, S. Simplifying stochastic mathematical models of biochemical systems. *Appl. Math.* **2013**, *4*, 248–256. [CrossRef]
34. Glasserman, P. *Monte Carlo Methods in Financial Engineering*; Springer: Berlin/Heidelberg, Germany, 2004.

35. Ilie, S. Variable time-stepping in the pathwise numerical solution of the chemical Langevin equation. *J. Phys. Chem.* **2012**, *137*, 234110. [CrossRef] [PubMed]
36. Sotiropoulos, V.; Kaznessis, Y.N. An adaptive time step scheme for a system of stochastic differential equations with multiple multiplicative noise: Chemical Langevin equation, a proof of concept. *J. Chem. Phys.* **2008**, *128*, 014103. [CrossRef] [PubMed]
37. Corless, R.M.; Fillion, N. *An Introduction to Numerical Methods from the Point of View of Backward Error Analysis*; Springer: New York, NY, USA, 2013.
38. Golub, G.; Van Loan, C. *Matrix Computations*, 3rd ed.; The Johns Hopkins University Press: London, UK, 1996.
39. Weijers, S.R.; Vanrolleghem, P.A. A procedure for selecting best identifiable parameters in calibrating activated sludge model no. 1 to full-scale plant data. *Water Sci. Technol.* **1997**, *36*, 69–79. [CrossRef]
40. Jahnke, T. On reduced models for chemical master equation. *Multiscale Model. Simul.* **2011**, *9*, 1646–1676. [CrossRef]

entropy

Article

VeVaPy, a Python Platform for Efficient Verification and Validation of Systems Biology Models with Demonstrations Using Hypothalamic-Pituitary-Adrenal Axis Models

Christopher Parker [1,*], Erik Nelson [2] and Tongli Zhang [1,*]

[1] Department of Pharmacology & Systems Physiology, College of Medicine, University of Cincinnati, Cincinnati, OH 45221, USA

[2] Department of Psychiatry & Behavioral Neuroscience, College of Medicine, University of Cincinnati, Cincinnati, OH 45221, USA

* Correspondence: parkecp@mail.uc.edu (C.P.); tongli.zhang@uc.edu (T.Z.)

Abstract: In order for mathematical models to make credible contributions, it is essential for them to be verified and validated. Currently, verification and validation (V&V) of these models does not meet the expectations of the system biology and systems pharmacology communities. Partially as a result of this shortfall, systemic V&V of existing models currently requires a lot of time and effort. In order to facilitate systemic V&V of chosen hypothalamic-pituitary-adrenal (HPA) axis models, we have developed a computational framework named VeVaPy—taking care to follow the recommended best practices regarding the development of mathematical models. VeVaPy includes four functional modules coded in Python, and the source code is publicly available. We demonstrate that VeVaPy can help us efficiently verify and validate the five HPA axis models we have chosen. Supplied with new and independent data, VeVaPy outputs objective V&V benchmarks for each model. We believe that VeVaPy will help future researchers with basic modeling and programming experience to efficiently verify and validate mathematical models from the fields of systems biology and systems pharmacology.

Keywords: HPA axis; Major Depressive Disorder; stress test; Python; Verification & Validation; differential equations model

check for updates

Citation: Parker, C.; Nelson, E.; Zhang, T. VeVaPy, a Python Platform for Efficient Verification and Validation of Systems Biology Models with Demonstrations Using Hypothalamic-Pituitary-Adrenal Axis Models. *Entropy* **2022**, *24*, 1747. https://doi.org/10.3390/e24121747

Academic Editor: Pavel Kraikivski

Received: 20 October 2022
Accepted: 24 November 2022
Published: 29 November 2022

Publisher's Note: MDPI stays neutral with regard to jurisdictional claims in published maps and institutional affiliations.

1. Introduction

The life cycle of a computational model involves development, verification, validation, and application. Before a model can be confidently applied to help solve a problem, it must be carefully examined and evaluated. The process of evaluating a model includes two steps: verification and validation (V&V). According to Thacker et al. [1],

> *"Verification is the process of determining that a model implementation accurately represents the developer's conceptual description of the model and its solution. Validation is the process of determining the degree to which a model is an accurate representation of the real world from the perspective of the intended uses of the model."*

In certain fields of mathematical modeling, such as nuclear engineering, model V&V has been performed following well-defined procedures for decades [2,3]. In the 1980s, for instance, the International Atomic Energy Association defined standardized benchmarks for validation of models of reactor cores [2]. In a paper from 1993, Nakagawa applies the benchmarks to prove the validity of their model of a reactor core [3]. The practice of creating standardized benchmarks for V&V has persisted in nuclear engineering—as seen in a paper by Höhne et al. from 2018 [4]. The standardization of V&V procedures is not limited to nuclear engineering, and can be found in other fields of engineering, as well. For instance, the American Society of Mechanical Engineers published a set of standards for V&V in computational solid mechanics and the American Institute of Aeronautics and Astronautics also published a set of standards for computational fluid dynamics [5].

However, V&V practices in systems biology and systems pharmacology are still being improved to meet significant challenges, in part due to the individual variability and resultant complexity inherent to physiological systems [6,7]. For example, Hicks et al. [8] present best practices for V&V of neuromusculoskeletal modeling and the basic concepts presented are applicable for most models in systems biology and systems pharmacology. For instance, "creation of gold standard datasets" and ensuring that efficient tools for V&V are available are excellent goals for the physiological modeling research community, in general. In this work, we have followed these recommendations and tailored some of the specifics to meet the needs of HPA axis modeling.

In systems biology and systems pharmacology, the ideal model would describe a physiological system adequately in any situation—including exposing the system to a variety of inputs, such as stress or pharmaceuticals. In practice, it is difficult to develop a systems biology or systems pharmacology model that is generalizable to situations even slightly different from the original research. A practical challenge for many researchers using mathematical models is to quickly and efficiently determine which model from the literature is best suited to their current work—or which model could be most effectively modified to fit their needs. Unfortunately, the lack of useful tools for V&V in the field often means that developing a new model from the ground up is more efficient than constructing and testing models from the literature in search of a viable candidate.

1.1. Custom Tools to Facilitate Model V&V

In order to help future researchers carry out V&V efficiently, we have developed a Python code library, VeVaPy, with several useful modules for this purpose. The level of difficulty of model V&V represents a significant gap in the field of HPA axis research, one that we aim to fill with our tools and data. Currently, it requires a high level of programming expertise to take a model from the literature and reconstruct it for V&V—the tools available for this purpose (at least for non-stoichiometric models) are not designed for ease of use by biologists. We concede that our V&V code still requires a degree of programming know-how at this point, but we believe that it is a significant improvement over the status quo.

The four modules of the code library are called dataImport (includes several HPA axis data sets for use in model validation, with ACTH & cortisol concentration data at rest and under acute stress), DEsolver (more streamlined differential equation solver, works with ODE or DDE systems), optimize (easily facilitates parameter optimization), and visualize (generates graphs based on user specifications). We use VeVaPy to demonstrate several case studies of HPA axis model V&V—similar to the case studies presented by Hicks et al. [8].

1.2. Validation against Novel Data Collected in MDD Patients

For validation of the HPA models we present as case studies, we compare them against new and independent data collected from Major Depressive Disorder (MDD) patients undergoing stress tests. MDD is a mental disorder with severe implications for quality of life. Symptoms include weight loss/gain, hypersomnia or insomnia, slowing of speech and action, impaired concentration, depressed mood, decreased interest in work/hobbies, low self-esteem, increased feelings of guilt, and suicidal thoughts [9]. For depression to be considered MDD, the symptoms must last a minimum of two weeks and cause significant difficulties functioning at work and interacting socially [9]. There are three main subtypes of MDD with significant differences in symptoms: melancholic depression is characterized by weight loss and insomnia, atypical depression is characterized by weight gain and hypersomnia, and uncategorized depression does not fit neatly into either of those two subtypes [10]. The differences between subtypes likely extend beyond symptoms, with some authors hypothesizing that different physiological features are associated with each subtype of MDD [10,11].

MDD is linked to multiple types of physiological disruptions, for example, neuroimaging features or sleep EEG disturbances [11]. However, we are primarily interested in the link between MDD and dysregulation of the hypothalamic-pituitary-adrenal (HPA) axis. There is a lack of consensus regarding whether MDD subjects generally exhibit HPA axis dysfunction—melancholic MDD subjects, however, are more likely associated with increased HPA axis activity and hypercortisolemia [10].

The HPA axis is a neuroendocrine system involved in the body's stress response. On exposure to a stressor, the paraventricular nucleus (PVN) of the hypothalamus releases corticotropin-releasing hormone (CRH). CRH is released not into the systemic circulation, but into the hypophyseal portal system connecting the hypothalamus directly to the anterior pituitary [12]. The anterior pituitary releases adrenocorticotropic hormone (ACTH) into the systemic circulation in response to increased CRH concentration. The main target of circulating ACTH is stimulation of glucocorticoid production/secretion in the zona fasciculata of the adrenal cortex [12]. The glucocorticoid synthesized is cortisol in humans and corticosterone in rodents. During this process, very little CRH enters the systemic circulation (making collection of CRH concentration data exceedingly difficult), while levels of ACTH and cortisol are readily detectable in blood.

Cortisol acts on various tissues throughout the body by way of glucocorticoid receptors (GRs)—which are nearly ubiquitous—and mineralocorticoid receptors (MRs). Both receptor types translocate to the cell nucleus when bound to cortisol and exert their effects through stimulation or repression of gene transcription. The stress response generated by cortisol includes immune system suppression, increased gluconeogenesis, and increased metabolism of fat, protein, and carbohydrates. Another important function occurs in the hypothalamus and pituitary as cortisol-GR binding decreases the synthesis of CRH and ACTH, respectively. In this way, cortisol exerts negative feedback on its own production. However, GR binding in the hippocampus serves to stimulate CRH production, so the system has both positive and negative feedback mechanisms to consider [13].

Concentrations of cortisol and ACTH normally exhibit both circadian and ultradian oscillations. Circadian refers to the oscillations with a period of roughly 24-h. These oscillations are largely controlled by the circadian clock in the suprachiasmatic nucleus (SCN) of the hypothalamus. However, many peripheral tissues contain lesser circadian clocks, including the adrenal glands [13]. The circadian oscillation of cortisol and ACTH peaks around 8 AM, decreases until after midnight, and then increases again until the morning peak. Ultradian oscillations have a period of 60–90 min and represent 12 to 18 episodes of cortisol/ACTH secretion throughout a day, with little to no secretion between them [14]. Both forms of oscillation likely exist to facilitate more rapid and stronger stress reactions at certain times of day. It has been shown that responses to noise stress are reduced during non-secretory periods and increased during secretory periods in rats [13].

If cortisol levels are sustained at high or low levels for too long, the health consequences are typically serious. Hypercortisolism is a chronic elevation of cortisol concentration, and it is implicated in the development of depression, cardiovascular disease, and Type 2 diabetes mellitus [13]. Hypocortisolism is a chronic decrease in cortisol concentration that is associated with impaired memory formation and post-traumatic stress disorder (PTSD) [13]. Several authors suggest that hypocortisolism is likely caused by increased negative feedback of cortisol on the HPA axis while hypercortisolism is likely caused by decreased negative feedback of cortisol on the HPA axis [15,16]. According to Holsboer [16], the negative feedback caused by cortisol binding GRs is impaired in MDD, likely due to decreased sensitivity and density of GRs.

Treatment options for MDD patients include evidence-based psychotherapies such as cognitive behavioral therapy (CBT) and/or antidepressant medications such as selective serotonin reuptake inhibitors (SSRIs). According to Holsboer [16], in patients with depression, elevated CRH levels lead to hypercortisolism. SSRI therapy is associated with normalization of CRH and cortisol concentrations in these patients [16] possibly by inducing an upregulation of MRs, which are known to inhibit HPA axis activity [17].

If the hypothesis of Holsboer [16] holds and MDD patients (or at least melancholic MDD patients) have decreased sensitivity and density of GRs, we should be able to detect differences in the behavior of the HPA axis under stress when compared with healthy controls. We would expect to see the concentration of CRH, ACTH, and cortisol rapidly increase on exposure to a stressor—the same process we would see in a healthy subject. However, with cortisol at a high concentration, we hypothesize that we would see a slower return to basal concentrations in MDD patients because diminished GR activity would result in decreased inhibition of CRH and ACTH secretion. The best way to test this hypothesis is to take blood samples and measure ACTH and cortisol at short intervals while MDD patients and healthy controls experience a significant amount of stress. Our chosen method of producing a stress response in a laboratory setting is the Trier Social Stress Test (TSST)—see Section 2.3 for a description of TSST procedures.

1.3. Model Validation against Experimental Data

A model that can accurately simulate the HPA axis during a TSST will allow us to make predictions about differences between MDD patients and healthy controls, so our validation procedure for the case studies presented herein is based on their ability to match TSST data. The state of HPA axis modeling in the literature is described in the following section, and the procedure by which we chose models to use as case studies is described in Section 2.1. After model selection, we perform our verification procedure (described in Section 2.2) and our validation procedure.

For validation, we begin by running a parameter optimization algorithm on each model. This yields the optimal parameters for matching stress test data (optimal parameters are generated for data sets from seven patients and the mean concentrations of all 58 patients). To compare model data matching objectively and quantitatively for the TSST data, we compute a cost function for each model on each data set. The result of the cost function is used by the parameter optimization algorithm to determine the suitability of parameter sets, comparing many sets to each other until it finds the optimal parameters for each model on each data set. We also use the cost function to assess the validity of models, as it indicates how well the model has simulated the experimental situation. Furthermore, the cost function values allow us to compare models to each other, and this allows us to objectively determine which model fits our data sets most closely. For a discussion of how the cost function is computed, see Section 2.4.

1.4. Mathematical Models of the HPA Axis in the Literature

There are many mathematical models of the HPA axis in healthy, MDD, and PTSD subjects in the literature. These models are predominantly ordinary differential equation (ODE)-based, although there are also some delay differential equation (DDE)-based models. They primarily vary in the way circadian oscillations are generated and (if they are considered at all) how ultradian oscillations arise.

Figure 1 shows a timeline of HPA axis modeling, starting with the first ODE model of an oscillating biological system by Goodwin in 1965 [18]. This model included a negative feedback loop to produce oscillating solutions rather than the steady-state solutions seen in prior models. While not specifically modeling the HPA axis, this was a direct precursor to the modern form of HPA axis models. The first improvement was made by Veldhuis et al. in 1989 with an HPA axis model attempting to model the ultradian rhythm of cortisol [19]. However, this model was a convolution model rather than an ODE model. In 1994, Gonzalez-Heydrich et al. published an ODE model including equations for CRH, ACTH, and cortisol [20]. This was the first "modern" model of the HPA axis, and the basic structure of models has remained largely the same since. The model by Liu et al. in 1999 was an ODE model with five equations, including CRH, ACTH, free cortisol, cortisol binding globulin (CBG)-bound cortisol, and albumin-bound cortisol—CBG and albumin are the two main proteins that bind and inactivate cortisol in the blood [21]. It was able to produce ultradian oscillations in CRH, ACTH, and cortisol concentrations, but failed to produce circadian oscillations.

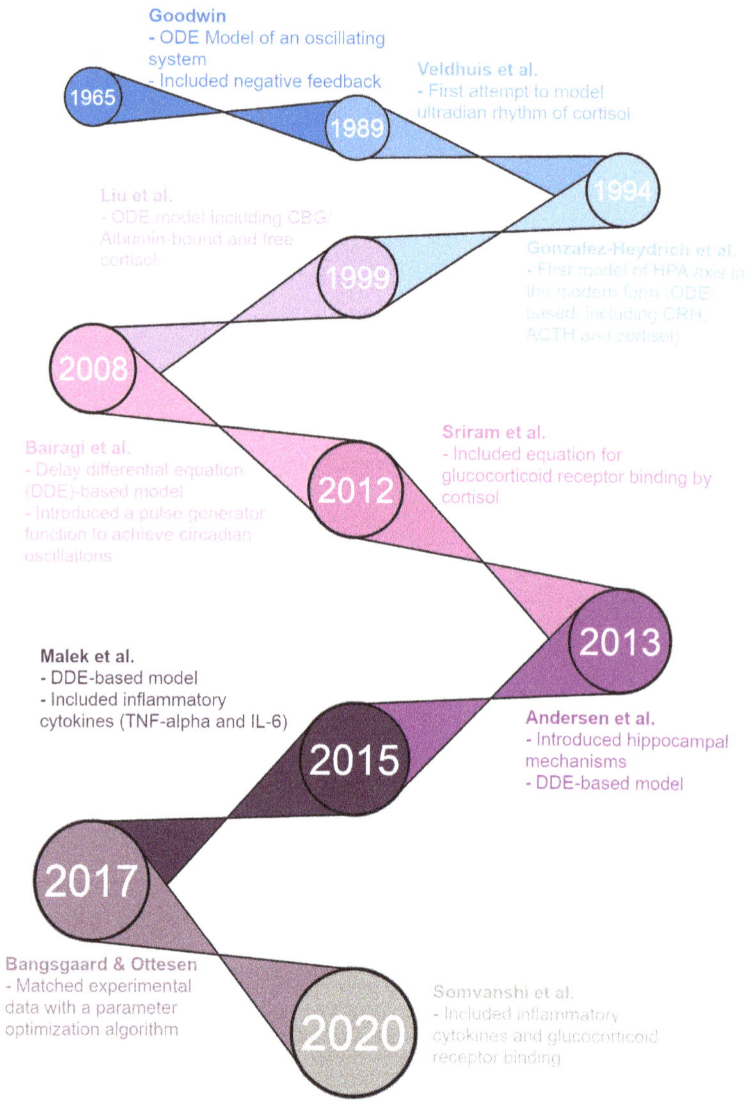

Figure 1. Timeline of hypothalamic-pituitary-adrenal (HPA) axis modeling. Models included, in chronological order: Goodwin [18], Veldhuis et al. [19], Gonzalez-Heydrich et al. [20], Liu et al. [21], Bairagi et al. [22], Sriram et al. [23], Andersen et al. [24], Malek et al. [25], Bangsgaard & Ottesen [26], and Somvanshi et al. [27].

The remaining models on the timeline are the five models we reproduce for this paper, plus the model by Andersen et al. [27] that we have included in the supplementary materials (due to a lack of valid figures for verification in the model paper). First, in 2008, Bairagi et al. produced a DDE-based model that included delays between the production of ACTH/cortisol and their action [22]. The model was able to produce both ultradian and circadian oscillations but required a pulse generator function representing the suprachiasmatic nucleus (SCN) of the hypothalamus for the circadian oscillations. In 2012, Sriram et al. published a model with four ODEs, including CRH, ACTH, cortisol, and GR availability/binding [23]. The inclusion of GR interactions allowed this model to produce both circadian and ultradian oscillations without external input from the SCN. The model by Andersen et al. in 2013 was a DDE-based model, and the authors attempted to produce oscillations by introducing hippocampal GR/MR interactions [24]. However, this model was unable to produce any oscillations for physiologically reasonable parameter values. In 2015, Malek et al. published an ODE model of the HPA axis and its interactions with inflammatory cytokines [25]. The model also ran as an HPA axis-only model by setting the initial concentrations of the cytokines to zero and was able to produce both types of oscillations desired, circadian and ultradian (through an external pulse generator function). Bangsgaard & Ottesen published an ODE model in 2017 that innovated by matching experimental data from individual patients with a parameter optimization procedure [26]. This allowed the authors to determine differences in parameters between healthy control subjects and depressed subjects. Finally, in 2020, Somvanshi et al. published another ODE model of the HPA axis and its interactions with inflammatory cytokines [27]. Their model differed from that of Malek et al. because it included equations for many other species, including GRs.

2. Methods

We used several tools to ensure that VeVaPy is easily accessible, well documented, and user-friendly, for the convenience of future researchers attempting to perform mathematical modeling of the HPA axis. The code for the demonstration models was written in Jupyter notebooks (https://www.jupyter.org, accessed 23 November 2022), which allows for Markdown text in between code segments. This allowed us to include a Table of Contents in each notebook for easy navigation, as well as well-formatted and easily readable instructions for use of the models. These Jupyter notebooks are all publicly available on Github (https://www.github.com/cparker-uc/VeVaPy, accessed 23 November 2022) and can be run on any computer with internet access and a web browser through Binder (see our Github repository for instructions; Binder homepage: https://www.mybinder.org, accessed 23 November 2022). We present further information about these tools in the Discussion.

2.1. Model Selection

We selected models for demonstration of VeVaPy by searching PubMed for "HPA Axis Mathematical Model", on 26 August 2021. This search yielded 1023 results. We selected all papers which included language in the abstract suggesting that a mathematical model was used to study some feature of the HPA axis, a total of 41 papers. Then, an initial screening analysis was performed on all search results, which eliminated all papers not related to differential equations modeling of the HPA axis—35 papers remained at this point. We then performed a more thorough analysis of the remaining papers, and selected all which met the following set of criteria:

- Included all necessary equations in dimensional form, 16 models selected
- Included all parameter values used for at least one figure
- Not substantially similar to an earlier model
- The model is based on the human HPA axis
- We have excluded our own models

Five models that satisfy all of these criteria are deemed to have high potential for successful V&V and further tested in our work.

2.2. Model Verification

We modified the general VeVaPy model template for each of the five selected models, inputting the systems of differential equations, parameter values and bounds, and initial conditions. To verify that the models were performing as the authors intended, we ran simulations to recreate figures from the model papers. This process requires the most modification of the template, because each paper includes very different figures that must be replicated. While we have developed the visualize module of VeVaPy to create graphs of variable concentrations over time, we are still working on expanding it to accommodate different formats (for instance, multiple variables on the same graph, as shown in the third figure of Malek et al. [28], see Supplementary Figures S5 and S6). The results of this process are discussed in Section 3.4.

2.3. Data Collection

The data VeVaPy used includes new patient data locally collected and data that are electronically collected from previous publications. For electronic collection, we used Lab-Notes software (http://mpf.biol.vt.edu/lab_website/Labnotes.php, accessed 23 November 2022), and the filenames indicate the data sources as follows: Bangsgaard-Ottesen-2017 [26], Bremner-2007 [28], Carroll-2007 [29], Golier-2007 [30], Yehuda-1996 [15]. These data represent basal concentrations of cortisol or both cortisol & ACTH measured at short intervals over 24 h. The other data included in VeVaPy come from patients undergoing TSST as described below.

The data were collected following clinical research procedures approved by the IRBs of University of Cincinnati and Cincinnati Children's Hospital. Briefly, subjects were initially screened with the Structured Clinical Interview for DSM-IV (SCID) and the Inventory of Depressive Symptoms clinician-rated version (IDS-C) by a trained clinician. A total of 88 subjects between the ages of 18 and 65 were selected for the study, with 22 being healthy controls and the other 66 fulfilling the following criteria: DSM-IV criteria for a major depressive episode, either meeting the modified criteria listed in Supplementary Tables S1 and S2 for melancholic or atypical depression, or not falling in any depressive subtype; a score on the IDS-C of 20 or greater. Several exclusion criteria were also defined, as listed in Supplementary Table S3. All subjects were given an opportunity to read the informed consent document and the protocol was verbally explained at the screening visit. This procedure was approved by IRB review.

Subjects returned at 5:00 PM on the first day of testing and stayed at the General Clinical Research Center (GCRC) of Cincinnati Children's Hospital until all testing was completed at 6:00 PM on the third day. Blood samples were collected at 10-min intervals to determine basal levels of cortisol and ACTH from 8:00 PM to 9:00 PM on day 1 and 8:00 AM to 9:00 AM on day 2. Subjects also had saliva samples taken every 20 min during these time intervals to serve as a measure of free cortisol.

A Trier Social Stress Test (TSST) was performed on the second day starting at 5:00 PM. The test involved subjects making an oral presentation to a panel of judges (whom the subjects were told were scientists specializing in behavior analysis), ostensibly to convince the judges that they are the most qualified candidate for a job opportunity related to their interests. Following the oral presentation, there was a question/answer session with the judges and then the subjects were given a series of mental arithmetic tasks to perform for the next five minutes. The total time for the presentation, question/answer, and mental arithmetic tasks was 20 min. Blood samples to determine cortisol and ACTH levels were drawn 30 min and 15 min before the TSST began, at the beginning of the TSST, 10 min and 20 min into the TSST, and then every 15 min for 90 min after the conclusion of the TSST. Heart rate was also measured during the TSST, and saliva samples were collected at the start of the TSST, the end of the TSST, and every 30 min for 90 min after the conclusion of the TSST. Note that, following the TSST, all subjects were informed that the panel of judges were not actually scientists and had been instructed to not react or offer positive feedback during the presentation.

The subjects also underwent a combined dexamethasone-CRH (DEX/CRH) test. This began with 1.5 mg of dexamethasone administered at 11:00 PM on day two. Saliva and blood samples were taken at 8:00 AM on day three to determine cortisol, ACTH, and dexamethasone levels (dexamethasone levels to control for differences in dexamethasone metabolism). At 2:30 PM and 3:00 PM on day three, the subjects once again had blood drawn to determine basal levels of cortisol and ACTH before the CRH test. Then, at 3:00 PM on day three, the subjects were administered 100 mcg ovine CRH (oCRH). Blood samples were taken every 15 min for the first hour and every 30 min for the second and third hours to determine cortisol and ACTH levels. Saliva samples were collected every 30 min for three hours following dosing with oCRH. This concluded the procedure, and the subjects were dismissed.

To facilitate matching the cortisol and ACTH concentration data, we excluded subjects with any data points missing. There were a total of seven MDD subjects and one control subject lacking at least one data point, so overall we had 58 subjects to use for modeling purposes. All subjects used in our modeling for this paper underwent the TSST, and the subjects included 43 diagnosed with MDD and 15 healthy control subjects. We have not included any analysis of the DEX/CRH data in this paper, however, this data will be useful for future analyses using a model modified to allow dosing with dexamethasone and oCRH.

2.4. Model Validation

Each model was run with a parameter optimization algorithm against a subset of the patients from the TSST data set and the mean cortisol and ACTH concentrations between all patients. We did not perform this process against all 58 patients due to the extreme time and computational power requirements of such an undertaking. We used the scipy.optimize.differential_evolution package for parameter optimization. The cost function used for parameter optimization involved creating splines between simulated points for ACTH and cortisol and computing the mean sum of squared errors between the splines and the data to be matched. The equation is as follows:

$$cost = \frac{\sum_i (d_{ACTH,i} - s_{ACTH}(t_i))^2 + (d_{CORT,i} - s_{CORT}(t_i))^2}{2}$$

where $d_{ACTH,i}$ & $d_{CORT,i}$ are the data points at time t_i, and $s_{ACTH}(t_i)$ & $s_{CORT}(t_i)$ are the spline functions for the simulated ACTH and cortisol, respectively, at time t_i. The splines' points were normalized to the mean concentration of the respective data set to be matched, and the data sets were normalized to their mean, as well. This normalization procedure allowed us to compare cost function values between models, even when the models operated on different time/concentration scales.

The reason for creating splines between the simulated points when computing the cost function is to handle a limitation in the ODE solver methods available in Python (and MATLAB, also, because the same differential equation solver method is commonly used in both languages). The problem is that the step size of the solver is not fixed, so we cannot guarantee that we will have a solution at the exact time point in the data being matched. Although we may get very close, the time steps are very often off by a small amount. The best solution we have found is to compute splines between each point in our solution array, and then select the points on the splines to exactly match the time points of data.

Each model was run against the concentrations averaged over all patients because it is the best example of how we expect the concentrations to behave (starts low before the stress test, peaks during and shortly after the test, returns quickly to baseline before measurement period ends). To illustrate the differences observed in individual patients, the models were also run against several individual patients (patients 1, 10, 20, 30, 40 & 50). Of particular interest, we demonstrated the results against patient 1 as an example of the data sets that fit into neither our understanding of how ACTH and cortisol concentrations should interact with each other nor how they should behave after exposure to a stressor (ACTH is decreasing over nearly the entire measurement period, yet cortisol spikes 30 min after the TSST ends). We also illustrate the model simulation results against patient 40 as a good example of how the concentrations should behave (with limited individual variation, making it distinct from the mean data set). The processes were facilitated by the ability of VeVaPy to efficiently plug in different data and model, as described below.

We tested using alternative cost functions and optimization algorithms and chose the mean sum of squared errors for the cost and differential evolution for the algorithm because they outperformed the other options. The alternative cost functions we tested involved using the maximum of the maximum distances between the ACTH & cortisol simulation spline curves and the real-world data, or the mean of the maximum distances between ACTH & cortisol simulation spline curves and the real-world data. These cost functions performed slightly worse overall when compared to the mean sum of squared errors. For alternative optimization algorithms, we tried using scipy.optimize.shgo and scipy.optimize.basinhopping. These algorithms performed worse than differential evolution, in general. However, all of these methods can be implemented easily in VeVaPy, by passing different arguments to the optimize module.

When choosing which parameters to optimize for each model, we considered the authors' intentions and tended to optimize only those parameters which were reported to vary between patient populations, at first. However, to demonstrate the maximum effectiveness of parameter optimization, we also ran simulations where we optimized every parameter (simply to determine the optimal data matching from each model). However, we did not modify the equations for any of the models used for demonstration. The only changes made from the original model publications is the parameter values we have optimized.

3. Results

First, we present a simplified description of how VeVaPy functions, leaving all technical information to Section 3.1. The tool contains a template that can be edited to add information about the model to be simulated. The information required for VeVaPy to function with a novel model includes: the system of differential equations constituting the model, the parameter values (and reasonable bounds on those parameters for optimization), and initial conditions for each variable (although these can also be optimized given reasonable bounds). The tool then uses the enclosed modules to simulate the system and optionally optimize parameters—the modules will be described in detail in the next section. The outputs from VeVaPy include the optimized parameters, the quantitative description of data matching suitability and graphs for visualization of the simulations. Figure 2 shows the simplified input/output diagram of VeVaPy.

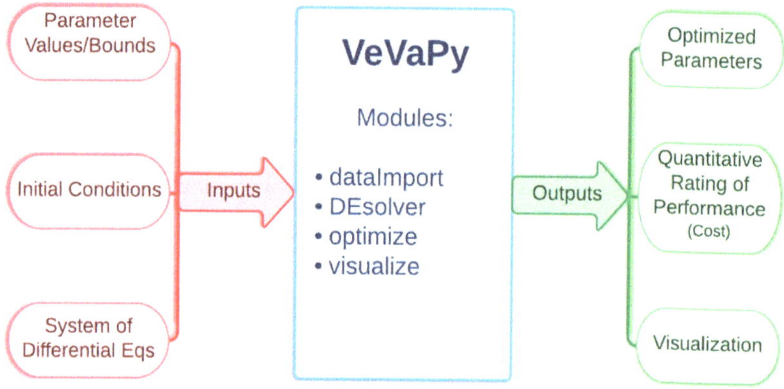

Figure 2. Simplified diagram of VeVaPy, showing inputs and outputs of the tool.

3.1. Code Diagram and Module Descriptions of VeVaPy

See Figure 3 for a diagram of the Jupyter notebook template used for each of the five demonstration models. The following section describes each step in the template in turn (following the numbering seen in the left-hand block of the diagram).

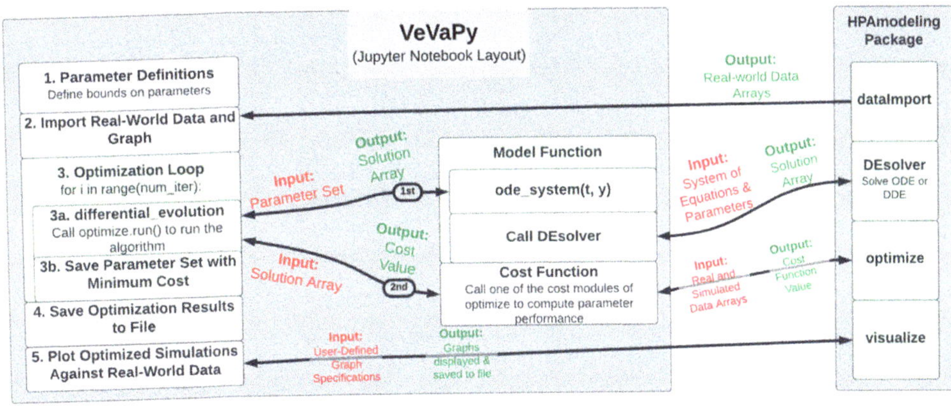

Figure 3. Code diagram of VeVaPy.

The VeVaPy template begins with Section 1. Parameter Definitions, where parameters are defined, along with bounds on each parameter for use in optimization. In this section, initial conditions (or bounds on initial conditions for optimization), time scale, and integration time length are also defined. Section 2. Import Real-World Data and Graph calls the dataImport module using the time scale defined by the user. The module imports all of the data we have gathered on the HPA axis (both basal data and TSST data) into arrays for analysis. This makes validation more streamlined and more powerful, as one can access a wide range of data sets without needing to scour the literature. We have also included code in the template to plot each data set from dataImport to allow users to easily see differences between data sets.

In Section 3. Optimization Loop, the parameter optimization is performed by the optimize module. The user adds the system of equations for their model into the model function (in the ode_system subfunction). Then, the user calls the run() method of the optimize module. This sets up and runs the parameter optimization algorithm, which repeatedly calls the model function, passing a set of parameters each time.

The model function then calls the DEsolver module using the system of equations defined by the user and the parameter set from the optimization algorithm. DEsolver allows for solving ODE and DDE systems in a user-friendly fashion. Currently, it is not straightforward to solve ODE systems in Python when using any solver other than the default (lsoda from the FORTRAN library odepack), and we are unaware of any straightforward methods for solving DDE systems. VeVaPy makes both of these possible for HPA axis models, requiring only a function defining the equations and a single call to the module.

DEsolver then returns the solution of the system, which is passed by the model function to the optimization algorithm, in turn. The algorithm then calls the cost function, passing this solution array. The cost function then calls a cost() method of the optimize module, depending on which cost function is desired (for instance, sum of standard errors is SSE_cost). This method takes arrays containing simulated data and the real-world data to use for validation and returns a single value for cost—representing the suitability of the parameter set tested. Based on this cost value, the optimization algorithm determines the most accurate parameter set that it can find, and then this set is saved and the loop repeats.

After the loop has run as many times as desired, Section 4. Save Optimization Results to File runs, and all optimized parameter sets and solution arrays are saved to an Excel file. Finally, in Section 5. Plot Optimized Simulations Against Real-World Data, the VeVaPy module visualize is used for creating graphs of each variable comparing the simulated values with the data set used for validation.

Note that when performing verification, the optimization loop is replaced with a single call to the model function, which then uses DEsolver to solve the system using the parameter values provided. The template contains code for this purpose, with modification necessary only to ensure that the graphs generated contain the same information as those the model is being verified against.

Not only are these modules useful for reproducing HPA axis models from the literature, but they can also be used for creation of new models of the HPA axis or potentially generalized to model other systems. Given the extensive experience and knowledge typically required to create a differential equation-based model of a physiological system, or even to reproduce one from the literature, we have attempted to make VeVaPy as user-friendly as possible to allow a broad audience to use it. As explained at the beginning of Section 2, all five of the models have been written in Jupyter notebooks with thorough documentation explaining the purpose of each code segment. Furthermore, the use of our custom library has been demonstrated in these models, and the code for each module in the library also includes thorough documentation. As a result, the reproduction of an HPA axis model starting from our template Jupyter notebook will be much more easily accomplished than starting from scratch.

3.2. Description of Collected Data

The data used for our validation demonstration are described below. As seen in Figure 4, the mean concentrations of ACTH and cortisol from the MDD patients before, during and after administration of a TSST follow the expected trend. Levels are steady and comparable to basal concentrations of MDD patients during the 30 min leading up to the test. During the 20 min that the subjects were participating in the TSST, levels sharply increase and then decrease back to baseline over the 90 min following the end of the test.

However, there is a large amount of variation between subjects. Of the six subjects chosen for matching, the general trend of an increase in ACTH and cortisol concentration followed by a decrease back to baseline is observable in patients 10, 20, 30, 40, and 50—although the degree to which concentrations increase and decrease varies widely. As an illustration of this point, the data for patient 1 and patient 40 are shown in Figure 5A,B, respectively. Strangely, patient 1 exhibits the largest peak in cortisol concentration at 18:20, 30 min after the conclusion of the TSST. Additionally, the ACTH concentration data for patient 1 is decreasing over nearly the entire time frame, which does not coincide with the increasing cortisol concentration. Therefore, it is to be expected that mathematical models will struggle to match the data from this subject. We expect that the mean concentration data, along with patients 10, 40, and 50 will be most successfully matched as they most closely follow the expected trend.

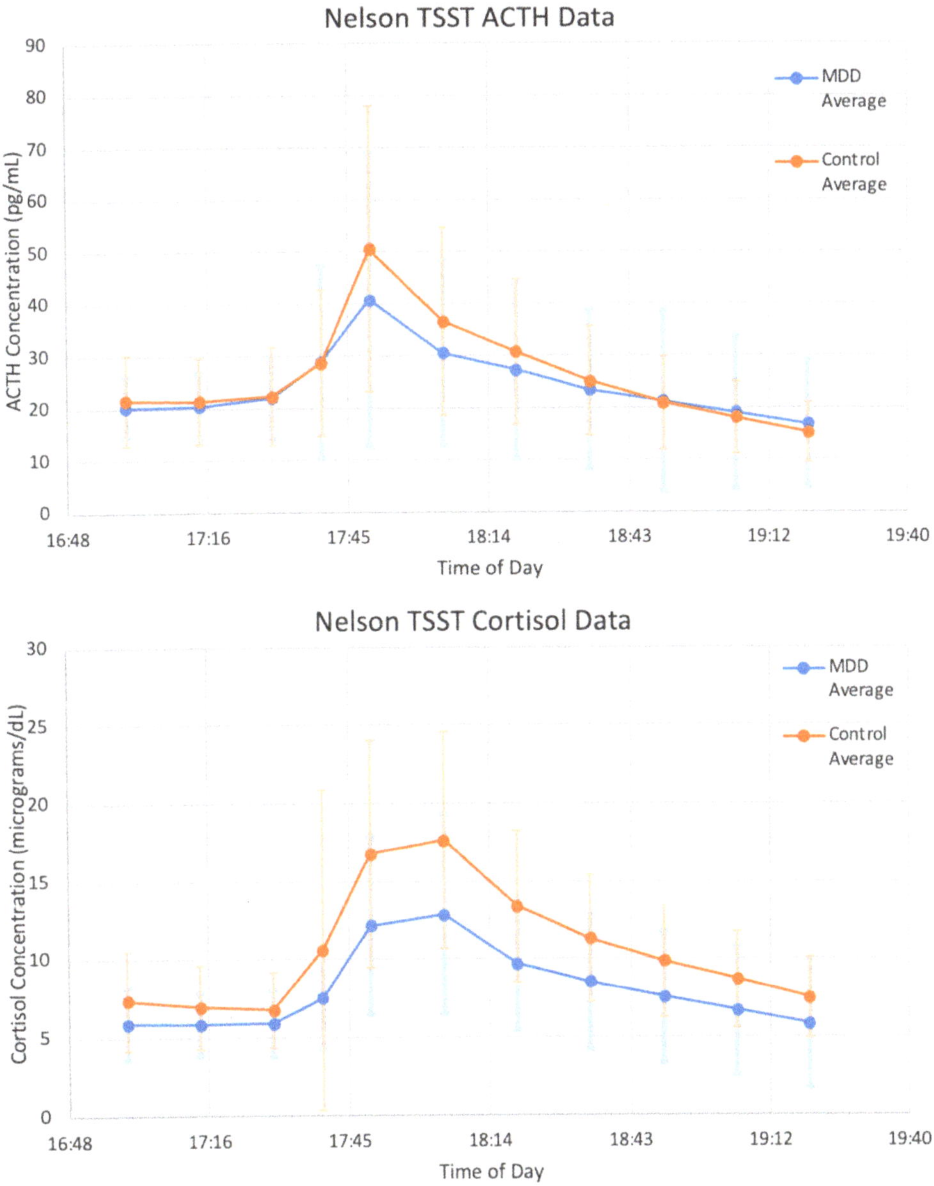

Figure 4. Mean of MDD (blue) and Control (orange) Patients' Trier social stress test (TSST) data. Adrenocorticotropic hormone (ACTH) concentration (**top**) and cortisol concentration (**bottom**) are graphed without any model simulations. Vertical bars indicate standard deviation at each data point.

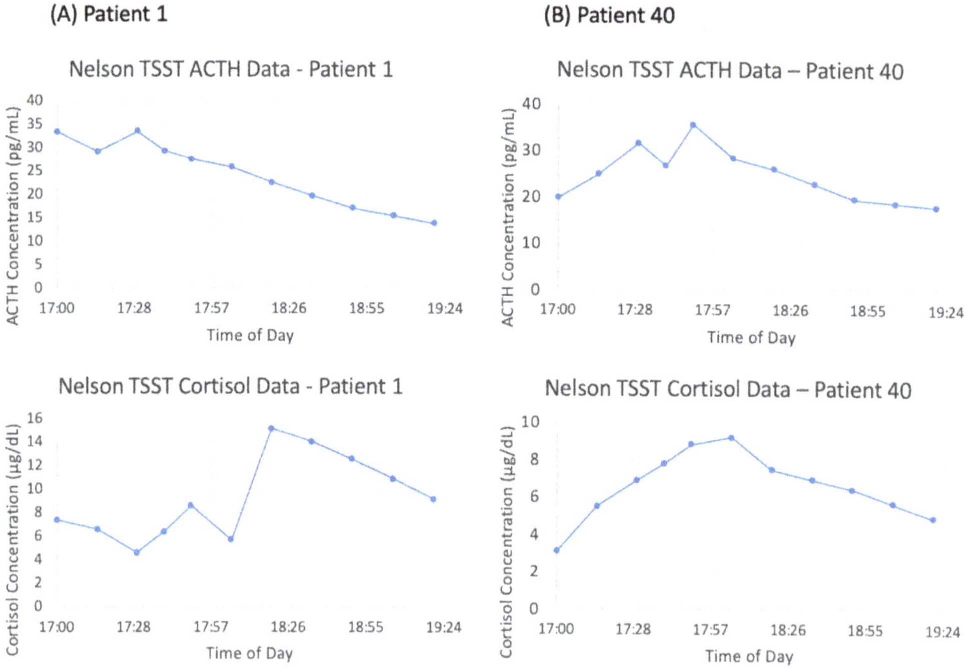

Figure 5. Trier social stress test (TSST) data from (**A**) patient 1 and (**B**) patient 40. Adrenocorticotropic hormone (ACTH) concentrations (**top**) and cortisol concentrations (**bottom**) are graphed without any model simulations.

3.3. *Summary of Selected HPA Axis Models*

The five models selected following our search of the literature include: Bairagi et al. [22], Bangsgaard & Ottesen [26], Malek et al. [25], Somvanshi et al. [27], and Sriram et al. [23]. See Table 1 for a summary of the characteristics of each model—including the number of equations, number of parameters and number of feedback loops (all of which give some indication of the amount of detail included in the system). Two of the papers were primarily interested in whether the HPA axis system itself exhibited ultradian oscillations or whether clock inputs from the brain were necessary, two of the papers used their models to study the interactions between the HPA axis and inflammatory cytokines, and the final paper was interested in determining whether PTSD patients exhibited stronger negative feedback from cortisol on the hypothalamus and pituitary than control patients.

Two of the models [23,27] replace the concentration of cortisol with the concentration of bound GRs when computing negative feedback—which is logical because cortisol must bind GRs to exert its negative feedback. In the model by Sriram et al. [23], this allows for the introduction of a positive feedback loop in the receptor binding equation, which generates bistability and therefore Hopf bifurcations in the model (which is an indication that the model can successfully generate ultradian oscillations without needing an external pulse generator function) [31].

Aside from the differences in handling negative feedback interactions, the other major difference in the models is the presence or absence of a function modeling external circadian drive from the SCN. Four of the models include a function for the SCN drive in the equation for CRH [22,25–27] while the other model does not include any circadian drive input from outside the HPA axis [23]. The model by Somvanshi et al. [27] also includes a function in the equation for ACTH to describe the adrenal circadian clock drive.

Table 1. Overview of Selected Models.

Year	Authors	Target	Number of Equations	Number of Parameters	Number of Feedback Loops	Unique Features
2008	Bairagi, Chatterjee, Chattopadhyay	Circadian & Ultradian Rhythms	3	12	1 Negative	Ultradian rhythm w/o Circadian
2012	Sriram, Rodriguez-Fernandez, Doyle	Cortisol Levels in PTSD	4	20	1 Negative, 1 Positive	Glucocorticoid Receptor Binding
2015	Malek, Ebadzadeh, Safabakhsh, Razavi, Zaringhalam	HPA Axis Relationship to Inflammatory Cytokines	5	32	0	Equations for TNF-alpha, IL-6 and endotoxin
2017	Bangsgaard, Ottesen	Comparing Differences in Optimal Parameters between Individuals	3	17	1 Negative	Used Parameter Optimization on Model to Compare Individuals
2020	Somvanshi, Mellon, Yehuda, Flory, Bierer, Makotkine, Marmar, Jett, Doyle	Relation of HPA Axis to Inflammation in Subjects with PTSD	17	92	1 Negative, 1 Positive	More Detailed Glucocorticoid Receptor and Inflammatory Cytokine Dynamics

3.4. Verification of Selected Models with VeVaPy

To verify that each model performs as the authors intended, we reproduced a figure from each original paper. All of the model papers contain at least one figure in which cortisol concentration over time is shown—these are the figures we have reproduced. Figure reproductions and the original figures are included in Supplementary Figures S1–S7.

In order to generate the figures, the VeVaPy template is edited to include the system of model equations, the parameter values and initial conditions defined in the model publication, and the time over which to integrate. The VeVaPy module DEsolver then solves the system and returns a solution array to visualize, which generates figures that are comparable to the publications. The module visualize allows users to define the variables to plot and the ranges over which to plot them. For each variable, a graph is produced showing the concentration values over the requested time range. The graphs can contain both simulation results and real-world data, but only one variable can currently be shown per graph. As mentioned in Section 2.2, this cannot accommodate all figures for verification currently (such as those in Malek et al. [28]). However, many verifications can be performed easily.

3.5. Validation of Selected Models with VeVaPy

Our demonstration of a validation procedure using novel TSST data and models from the literature illustrates how VeVaPy makes this process more streamlined. We started validation with VeVaPy by using parameter values found in the original publications of the selected models (see the model files on GitHub for details).

With the authors' published parameters and the data from patient 40, we see in Figure 6 that the original parameter values provided in Sriram et al. and data from this patient do not agree. When a new experimental procedure is used to collect data, it is likely that there will be no models specifically designed to simulate the experiment. This partially explains why the V&V process is more challenging in systems physiology. Since the data for model construction and data for model validation are often collected in different contexts, more than one set of parameter values are needed for proper estimation of the model's capacity to explain new data.

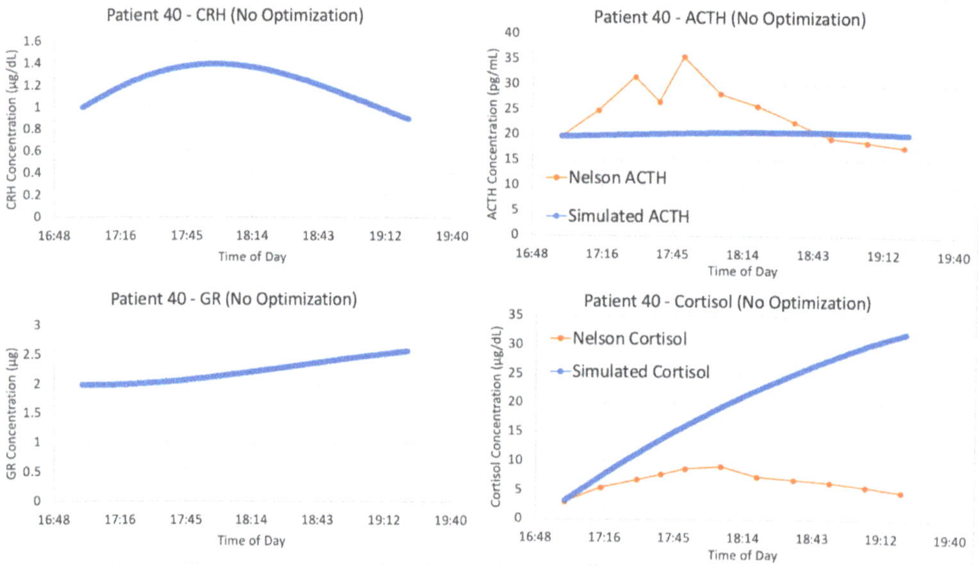

Figure 6. Sriram et al. [23] model without parameter optimization vs. Trier social stress test (TSST) data from patient 40. Graphs include model simulations of corticotropin-releasing hormone (CRH) concentration (**upper left**, blue), adrenocorticotropic hormone (ACTH) concentration (**upper right**, blue), cortisol concentration (**lower right**, blue) and bound glucocorticoid receptor (GR) concentration (**lower left**, blue) against ACTH concentration (**upper right**, orange) and cortisol concentration (**lower right**, orange) from patient 40. The blue lines represent the average of 5 iterations of the parameter optimization algorithm.

To maximize this capacity, VeVaPy includes a package for parameter optimization. This ensures that when the published parameters are inaccurate for the current experimental conditions, we can determine whether a change in parameters can yield a more accurate simulation. This is easily facilitated by VeVaPy, with several parameter optimization algorithms and cost function options easily available.

The optimized parameters for Sriram et al. improved the matching between the model and data from patient 40, as seen in Figure 7. The model fits exceptionally well when matching the data from patient 40, with an average cost function value over five iterations of the parameter optimization algorithm of 0.05827298.

We see similar results when running the model by Bangsgaard et al. in VeVaPy against the data from patient 40. The initial results using the authors' published parameters are shown in Figure 8, and as expected the simulation does a very poor job matching the experimental data.

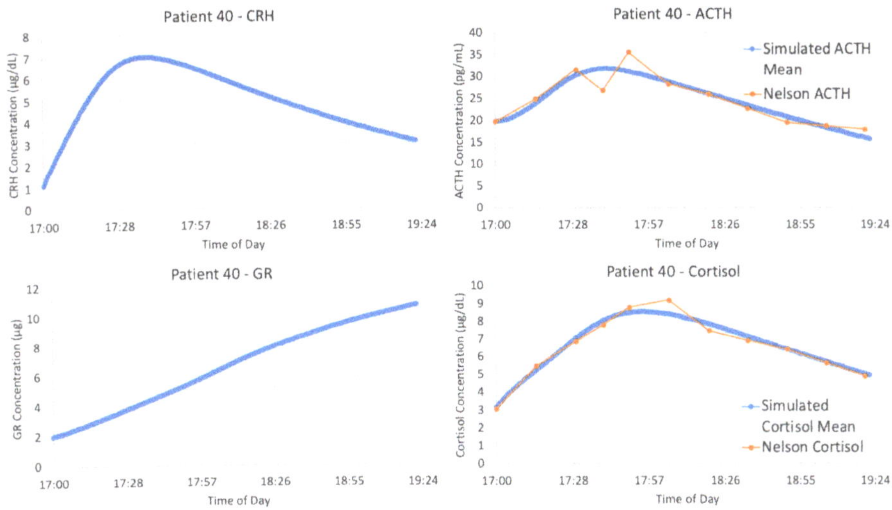

Figure 7. Sriram et al. [23] model vs. Trier social stress test (TSST) data from patient 40. Graphs include model simulations of corticotropin-releasing hormone (CRH) concentration (**upper left**, blue), adrenocorticotropic hormone (ACTH) concentration (**upper right**, blue), cortisol concentration (**lower right**, blue) and bound glucocorticoid receptor (GR) concentration (**lower left**, blue) against ACTH concentration (**upper right**, orange) and cortisol concentration (**lower right**, orange) from patient 40. The blue lines represent the average of 5 iterations of the parameter optimization algorithm.

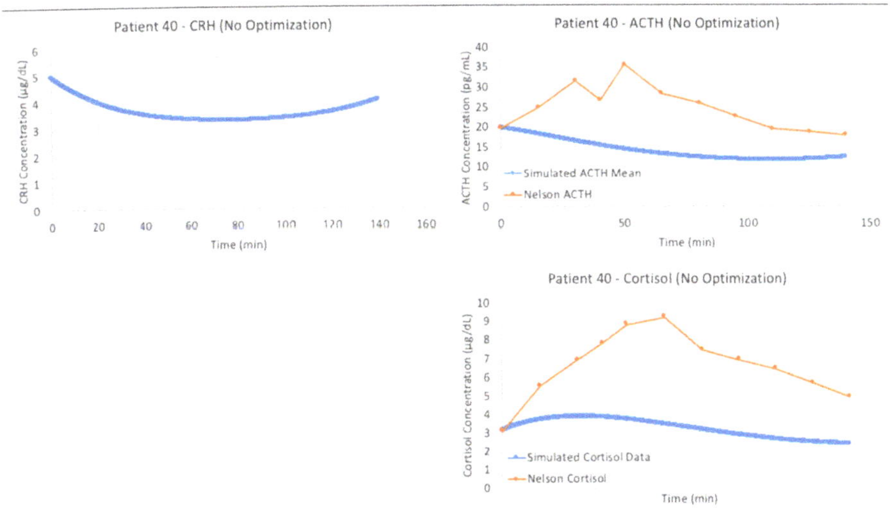

Figure 8. Bangsgaard et al. [26] model without parameter optimization vs. Trier social stress test (TSST) data from patient 40. Graphs include model simulations of corticotropin-releasing hormone (CRH) concentration (**left**, blue), adrenocorticotropic hormone (ACTH) concentration (**upper right**, blue) and cortisol concentration (**lower right**, blue) against ACTH concentration (**upper right**, orange) and cortisol concentration (**lower right**, orange) from the patient 40. The blue lines represent the average of 5 iterations of the parameter optimization algorithm.

After running the Bangsgaard model with VeVaPy's parameter optimization function, we see significantly improved fit when matching the data from patient 40. As shown in Figure 9, the model performs nearly as well as the Sriram model in this instance.

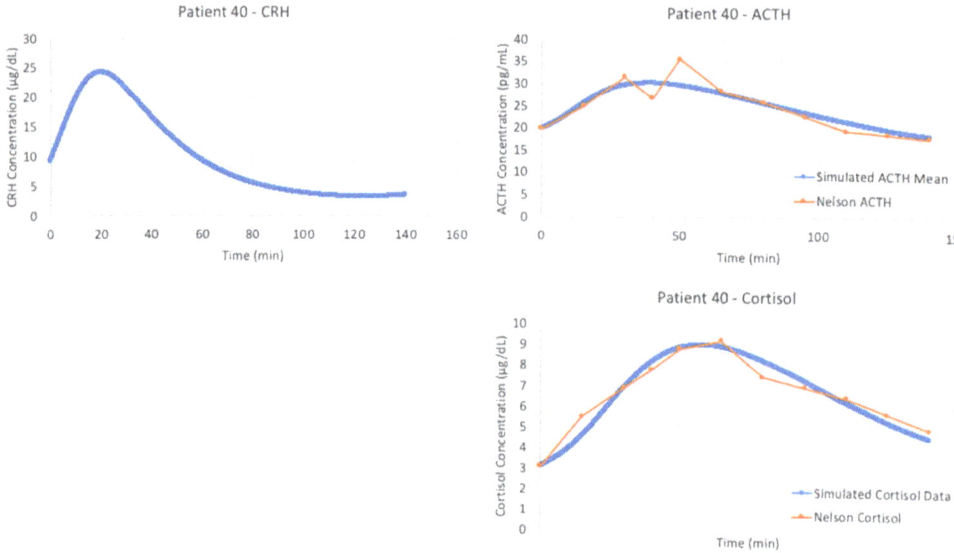

Figure 9. Bangsgaard et al. [26] model vs. Trier social stress test (TSST) data from patient 40. Graphs include model simulations of corticotropin-releasing hormone (CRH) concentration (**left**, blue), adrenocorticotropic hormone (ACTH) concentration (**upper right**, blue) and cortisol concentration (**lower right**, blue) against ACTH concentration (**upper right**, orange) and cortisol concentration (**lower right**, orange) from the patient 40. The blue lines represent the average of 5 iterations of the parameter optimization algorithm.

For the remaining three models, the results of parameter optimization are not nearly as positive, but optimization still yielded slight improvements in fit. Figure 10 shows the models with and without parameter optimization against patient 40 (as with the models by Sriram et al. [23] and Bangsgaard & Ottesen [26] above). This process clearly demonstrates that even with parameter optimization, models are often not suitable for problems outside of their initial intended use.

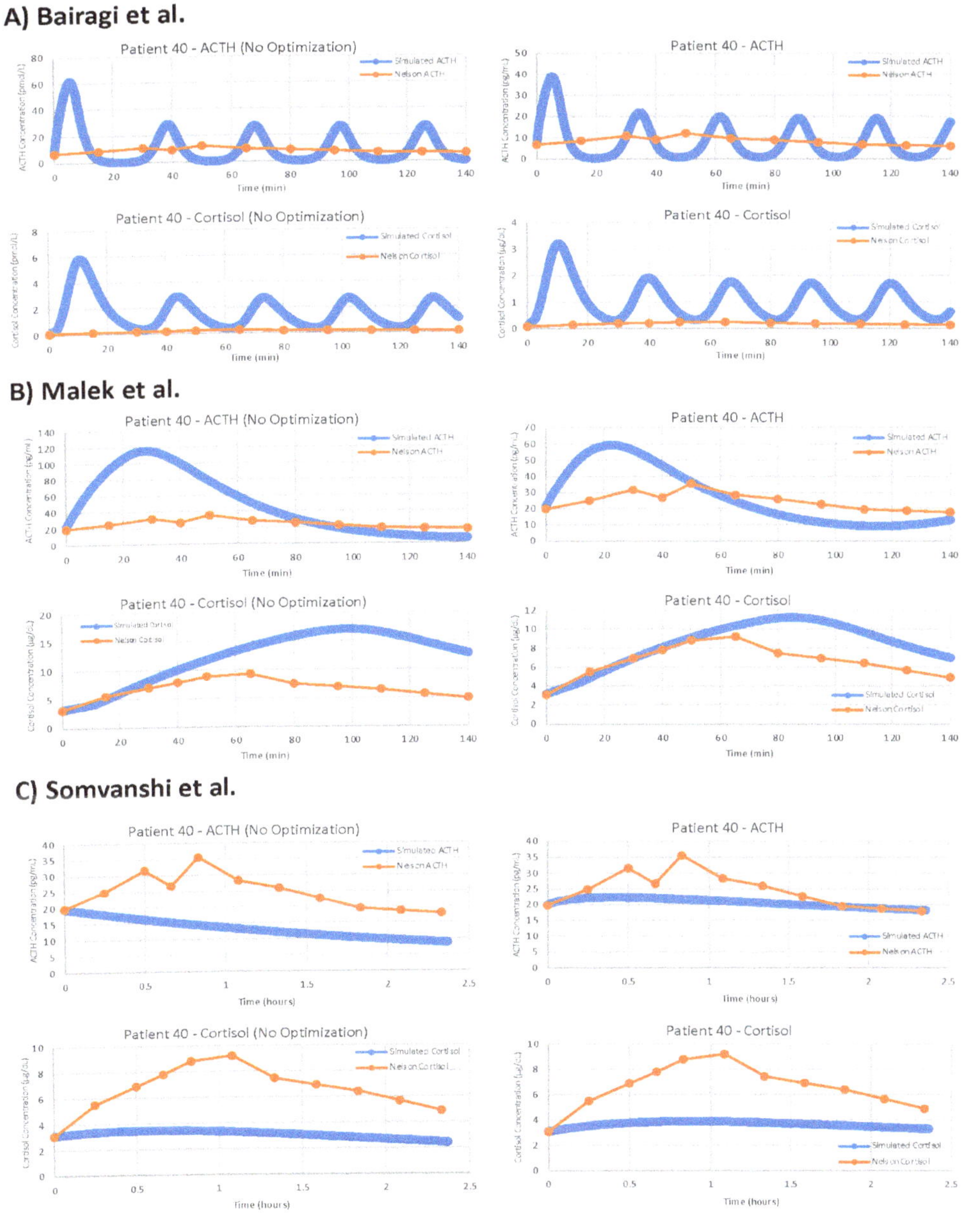

Figure 10. Further examples of models with and without parameter optimization. Simulated concentrations are in blue, patient 40 data is in orange. The left column of graphs shows the models running simulations with the parameters from the publication, while the right column of graphs shows the models running simulations with optimized parameters. Demonstrated models are (**A**) Bairagi et al. [22], (**B**) Malek et al. [25], (**C**) Somvanshi et al. [27].

For each model, we have computed the average cost function value for five iterations of the parameter optimization algorithm on each of the seven data sets tested. We then took the average of these seven average cost function values to obtain a single value representing the overall suitability of each model when matching our TSST data. The overall average cost function values given by VeVaPy for each model are shown in Table 2, alongside the best cost function value on a single patient, and the cost function values of the models without parameter optimization. It should be noted that the model by Bairagi et al. [22] required a large amount of computational power, and as such, we were only able to run one iteration of the parameter optimization algorithm for each data set. Each iteration of the model ran for approximately 36 h, which was more than 10 times longer than any of the other models we tested.

Table 2. Model Ranking Based on Cost Function Value.

Model	Overall Cost Function Value ± Standard Deviation (After Parameter Optimization)	Best Cost Function Value for a Single Patient (Patient ID)	Overall Cost Function Value (Authors' Parameters, No Optimization)	Best Cost Function Value for Single Patient (Patient ID) (Authors' Parameters, No Optimization)
Sriram et al. (2012) [23]	0.33 ± 0.19	0.058 (Patient 40)	26.03 ± 16.39	5.11 (Patient 50)
Bangsgaard & Ottesen (2017) [26]	0.58 ± 0.46	0.12 (Patient 40)	3.63 ± 1.03	2.14 (Patient 10)
Somvanshi et al. (2020) [27]	2.59 ± 0.94	1.10 (Patient 20)	6.61 ± 0.90	5.43 (Patient 20)
Malek et al. (2015) [25]	6.78 ± 7.78	1.17 (Patient 30)	11.95 ± 0.40	11.31 (Patient 20)
Bairagi et al. (2008) [22]	64.86 ± 38.85	35.84 (Patient 1)	656.26 ± 280.47	343.34 (Patient 1)

Based on the overall cost function value of each model with optimized parameters versus the authors' published parameters, we can clearly see that our procedure is yielding significant improvements in data matching. Further, between the models tested, we see widely varying levels of suitability after parameter optimization. The models by Sriram et al. [23] and Bangsgaard et al. [26] far outperform the others. The normalization performed when computing costs allows for comparison between models without needing to convert time/concentration scales beforehand—we have reported results from each model without converting all models to the same scales to demonstrate this.

3.6. VeVaPy Facilitates Efficient Validation against Individual Patients

With VeVaPy, we efficiently compared data from seven individual patient data sets against five models with five iterations per patient, and one model with a single iteration per patient, for a total of 182 runs of the parameter optimization. The main time-consuming step is the repeated integration of these models necessary for parameter optimization—especially for very complex models or those with systems of DDEs. However, VeVaPy makes it straightforward to "plug and play" individual models and data. The results of model validation against the MDD patient mean data set are shown for all five models in Figure 11.

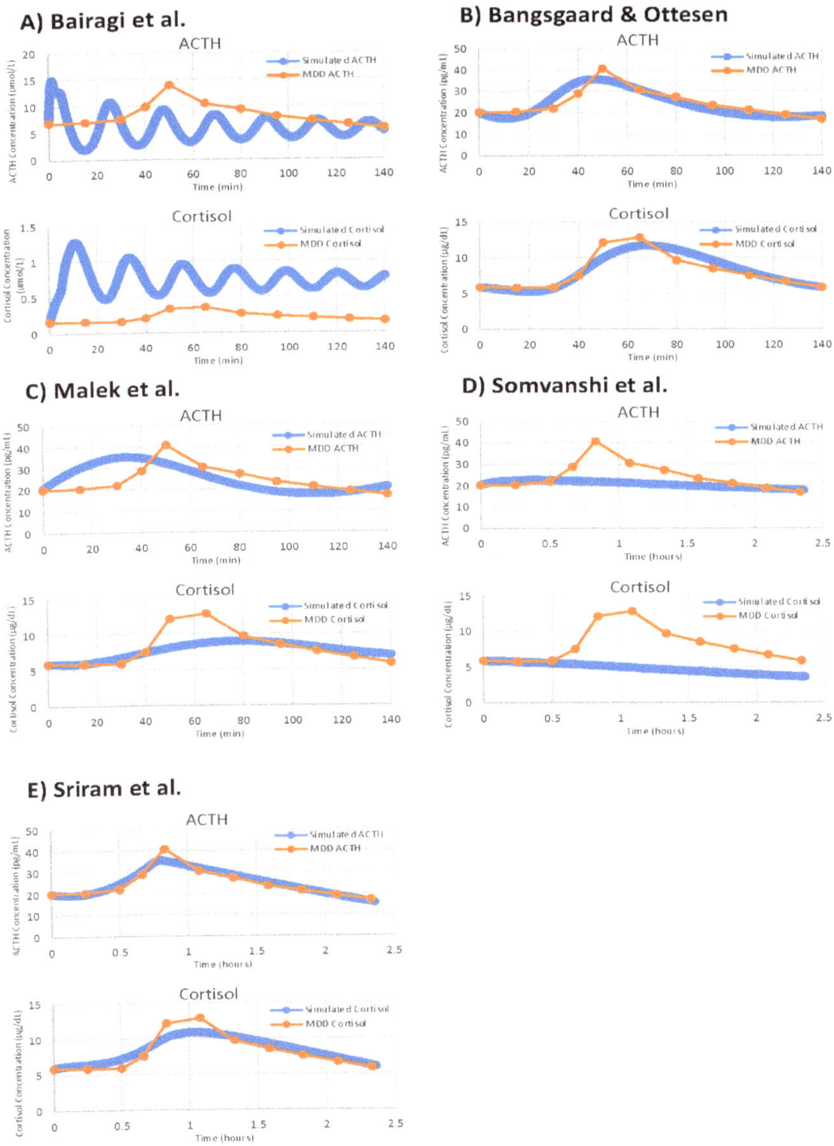

Figure 11. Model validation figures for all five demonstration models against the mean of all MDD patients in the Nelson TSST data. Models depicted are: (**A**) Bairagi et al. [22], (**B**) Bangsgaard & Ottesen [26], (**C**) Malek et al. [25], (**D**) Somvanshi et al. [27], (**E**) Sriram et al. [23].

VeVaPy is designed to facilitate this process by requiring specification of the validation data set in a single location. The tool then runs the algorithm and outputs the results to an Excel file for further analysis. There is also the option in VeVaPy to loop over multiple patients in a data set, running the parameter optimization on each individual indicated by the user. This makes it very efficient to run many different validation tests, without needing to make changes to the code.

3.7. Assessing Model Generalizability

After optimizing parameters against a data set, VeVaPy can easily test the resulting parameter set against other data sets to determine whether a model can be generalized to other situations. The optimized parameters are loaded from the file where they were saved when generated and the procedure for performing a simulation without optimization is followed. In order to demonstrate this process, we have used several of the optimal parameter sets from the model by Sriram et al. [23] to run simulations against all individual patients from the TSST data set Figure 12 highlights some of the more interesting results from this process.

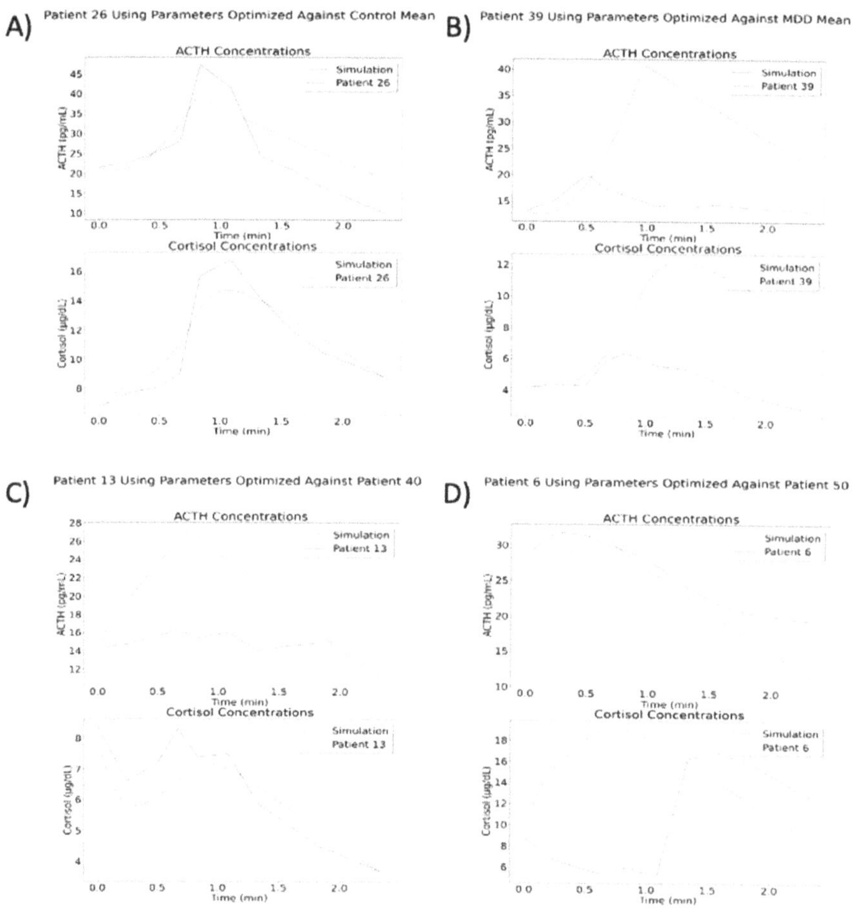

Figure 12. Results of Using Optimized Parameters in Generalized Cases. (**A**) The optimized parameter sets have some cases where they perform reasonably well (especially against patients from the same group). (**B**) Some of the parameter sets match certain patients very poorly, such as the parameters optimized against the mean of all MDD patients against patient 39 (MDD/neither subtype). (**C**) Many of the simulations matched either ACTH or cortisol but did not match the other. Parameters optimized against patient 40 (MDD/neither subtype) match the general cortisol concentration trend from patient 13 (MDD/atypical), but the simulated ACTH concentration is extremely high. (**D**) Similar to C, but with simulated cortisol concentration not matching while simulated ACTH concentration follows the correct general trend. Simulation run with parameters optimized against patient 50 (MDD/atypical), shown with data from patient 6 (MDD/atypical).

The results of these simulations vary widely, likely due to the variation observed between individual patients. Some parameter sets did yield better cost function values when matching individual patients within the same group. For instance, using parameters from optimization against all control patients gave an average cost function value of 6.007 against control patients and a value of 9.064 against MDD patients. Likewise, using parameters from optimization against patient 40 (MDD/neither subtype) gave an average cost function value of 2.899 against control patients and a value of 2.380 against MDD patients. However, using parameters from optimization against all MDD patients gave an average cost function value of 4.195 against control patients and a value of 4.421 against MDD patients. This demonstrates that while some optimized parameters are slightly generalizable to other patients or group mean concentrations, the cost function values are much higher than we would like in all of these situations. A deeper analysis of this issue falls outside of the scope of this paper, but we intend to examine this behavior more fully in a subsequent paper.

3.8. *VeVaPy Runtimes*

We have recorded the time required to run a variety of simulations with each demonstration model. These data are summarized in Table 3 below. Running simulations without parameter optimization requires only milliseconds for all five models. However, optimizing parameters for any of the models requires a significant investment in time and computing power. The models vary widely in this regard, though, with a gap of 96.5 min between the fastest model and slowest model.

Table 3. Simulation Runtimes with VeVaPy by Model.

Model	Without Optimization (Milliseconds)	With Optimization (Minutes)
Bairagi et al. [22]	483	101.625
Bangsgaard & Ottesen [26]	51.1	27.170
Malek et al. [25]	57.5	26.794
Somvanshi et al. [27]	5.1	5.128
Sriram et al. [23]	1.76	7.159

The simulations performed for the average runtimes without parameter optimization used several optimal parameter sets that were generated during the course of model validation. We ran 100 simulations with each model, and the average runtime of these 100 simulations is reported. For the runtimes with parameter optimization, we report the average time for a single iteration of the optimization algorithm for each model against two data sets from the TSST data: mean of all control patients and mean of all MDD patients. Each of these optimization runs consisted of five iterations of the differential evolution optimization algorithm using a population size of 10. Population size determines how many parameter sets are "evolved" and checked for improvement at each step of the optimization, so lower population sizes will yield faster runtimes with less accuracy at finding the minimum cost value.

One major factor increasing the runtime for the models by Malek et al. [25] and Bairagi et al. [22] is the presence of delayed variables. This requires extra computation at each step of integration in order to look up the value of the variable in a previous time step, which becomes a significant time cost when it is being performed thousands of times per iteration of the optimization algorithm. This is an area where there is likely room for improvement in the VeVaPy framework—and one which we will be working on in the future.

4. Discussion

In this work, we have shown how we have created a Python based package, VeVaPy, that can be used to efficiently verify and validate HPA axis models. We have thoroughly documented the code behind VeVaPy and published it freely on GitHub, in line with the recommended best-practices for model publication. We hope that others will find VeVaPy useful, and it can help future researchers spend less time and effort performing V&V when developing their own models or checking model papers in the literature.

In order to test and demonstrate VeVaPy, we verified five HPA axis models from the literature and validated them against novel TSST data from MDD patients. The models were ranked based on their average cost function value when running the differential evolution parameter optimization algorithm on each model against several TSST data sets. All five models are included in the VeVaPy repository and ready for use—though the validation results indicate that the models by Sriram et al. [23] and Bangsgaard & Ottesen [26] would be the strongest candidates for repurposing to explain the TSST data.

Consistent with many others in the scientific community, we have found that verification of published models was challenging [32–35]. As we will elaborate on below, we have encountered two main difficulties during the course of this research: data sets not provided alongside models, and non-standardized model development and publishing practices.

The first difficulty arises due to a lack of easily accessible data published in machine-readable formats. While many papers in the HPA axis modeling literature use cortisol concentration data to validate their models, they seldom include a supplementary file with a spreadsheet of the data used. Often, the papers cited by these modeling publications as the source of the data used do not include spreadsheets of the data either. It is often very difficult or even impossible to reach the authors of papers with useful data, especially when the papers are not from the last few years.

To prevent future researchers from experiencing this challenge, we have begun the process of curating data sets from the literature and packaging them with VeVaPy. Currently, as described in Section 2.3, we have four data sets containing cortisol data sampled at short intervals from patients at rest over 24 h (two of the data sets also contain ACTH data over the same time period). These data are useful for validation of models intended to describe the HPA axis at rest. While the number of patients per data set is rather small (n = 29–47), and the sets only contain mean concentration data, they provide a basic level of confidence that a model is valid among various patient populations. Further, for validation of models intended to simulate the HPA axis under stress, the TSST data contained in VeVaPy is unique in the literature, as far as we are aware.

The second difficulty we faced—non-standardized model development and publishing practices—warrants a more in-depth discussion. As mentioned briefly in Section 2.1, there were many more models published in the literature than the five we have reproduced here. Unfortunately, however, the majority of differential equation based HPA axis models published in the literature are non-reproducible for a variety of reasons. The problem is primarily due to a lack of information, rather than dishonesty or poor model design/performance. Very few models are published with the full code used by the authors, and others do not include necessary basic information such as the initial conditions used for each simulation, or a full list of parameter values used.

It is encouraging that both some grant agencies and some journals have acknowledged this challenge. In a statement by the director of the NIH published in 2014 [36], the root causes of the crisis (including over-emphasis on high-impact journal publications by hiring and tenure committees or the withholding of information about experimental procedures to retain a competitive edge) were discussed along with the steps the NIH was considering to address the crisis (including changes to the way grants are awarded to allow for more reproduction of published work to take place). Meanwhile, other researchers have suggested that journals must enact and enforce reproducibility standards to solve the crisis [37,38]. Some journals have taken steps in the right direction—notably, Science implemented a reproducibility policy in February 2011. According to Stodden et al. [39], the policy of Science has been successful to a degree but has not been enforced strictly enough. The rate of data availability improved from 52% to 75%, but the rate of code availability only improved from 43% to 54% [39]. However, most suggestions for making computational research more reproducible focus on individual researchers.

A survey of 1576 scientists showed favorable attitudes towards all suggested practices for improving reproducibility included in the survey (practices such as better mentorship, more robust experimental design, and journal checklists) [40]. A concern expressed by some researchers surveyed was the amount of added time and effort to ensure that an experiment is reproducible. However, as stated by Waltemath & Wolkenhauer [41], "irreproducibility hinders researchers and the scientific community by wasting time and money." Following best practices for reproducibility in computational research from the start of model development can decrease the overall amount of work required for reproducibility and in the long term it will save the community significant time and effort.

To ensure that VeVaPy is user friendly and easily extensible, we have followed the suggested best practices in the literature to the best of our ability—and we will review and summarize them here. We hope that by proposing (and following ourselves) these best practices, future computational biological scientists will not struggle with some of the challenges we have. The Physiologically Based Kinetic (PBK) Model Reporting Template presented by the Organisation for Economic Co-operation and Development (OECD) [32] is a good compilation of general best practices for model reproducibility. The template is presented in Table 4 below, and Table 5 contains our curated list of best practices suggested in the literature by various authors [32–35,37,38,41–52].

Table 4. OECD Model Reporting Template [32].

PBK Model Reporting Template Sections	
A.	Name of model
B.	Model author and contact details
C.	Summary of model characterization, development, validation and regulatory applicability
D.	Model characterization 1. Scope and purpose of the model 2. Model conceptualization (model structure, mathematical formulation) 3. Model parameterization (parameter estimation and analysis) 4. Computer implementation (solving the equations) 5. Model performance 6. Model documentation
E.	Identification of uncertainties (report for each item in D.)
F.	Model implementation details (software used, availability of code)
G.	Peer engagement (report extent of review by peers during development)
H.	Parameter tables (report all relevant inputs to the model for any simulations described)
I.	References and background information

Note that the version of the OECD PBK Model Reporting Template published in [32] includes more thorough guidance about what to include in each section. Completing this template guarantees that a model will be published in accordance with our general best practices (see Table 5), assuming the authors also provide the full code of their model as the template suggests. See the Supplementary Material for complete information on the five models covered in this paper.

Table 5. Proposed Best Practices for Model Publication.

Proposed Best Practices	**Comments/Justification**
All models should be published with all code used by the authors	This is essential to ensure models can be exactly reproduced without undue struggle
Model code must include proper documentation	Without documentation regarding how to run a model, such as thorough comments throughout the code or a readme file, it is often difficult to dissect complex code and determine how it is meant to work
Exact scope of the model should be made clear	It is important that the audience knows when it is appropriate to use the model, lest they form false assumptions based on use of the model in a context it was not designed to simulate
All input data (parameters, initial conditions, etc.) must be provided and justified	Too often models are published without a clear list of parameter and initial condition values, making them non-reproducible. Other times, the sources for parameter and initial condition values are not provided, leaving their validity in question.

We can categorize best practices suggestions based on which aspect of an experiment they address. The categories include experimental design, performing experiments and collecting data, analysis of data, and reporting data/results. The paper by Munafò et al. [45] presents general suggestions which are applicable to many areas of science, and which address all of the aforementioned categories. These suggestions include protecting against cognitive bias during experimental design and data collection (e.g., using blinding), including independent researchers with no personal stake in all steps of an experiment, study pre-registration, improving statistical analysis training, improving the quality of reporting, and promoting transparency and open science.

Due to the nature of computational research, the suggestions regarding experimental design and data collection are often not relevant. As such, the literature about reproducibility in computational science mostly focuses on the last category listed above: reporting data/results. The suggestions in this category vary in their specificity from general statements (see the suggestions listed in Table 5) to specific software recommendations (e.g., use Git for version control).

We have also made a list of suggested best practices and the software we recommend for implementing them (see Table 6). These suggestions come from several literature sources and our own experience with modeling software [33,41,42,46–50,52]. The following best practices and the suggested software for implementation will be discussed: fully document the process of model development including all simulation inputs and algorithms, share model code and the associated documentation in public repositories like Github, ensure that model code can be run on as many computers as possible, and make model code easy to understand.

Table 6. Best practices and suggested software for their implementation.

Best Practice Suggestion	Software for Implementation
Fully document the process of model development including all simulation inputs and algorithms	• Jupyter Notebooks (https://www.jupyter.org, accessed 23 November 2022) • R Notebooks (https://rmarkdown.rstudio.com, accessed 23 November 2022)
Share model code and the associated documentation publicly	• GitHub (https://www.github.com, accessed 23 November 2022) • BioModels (https://www.ebi.ac.uk/biomodels, accessed 23 November 2022)
Ensure that model code can be run on as many computers as possible	• VMWare Virtual Machine (https://www.vmware.com, accessed 23 November 2022) • Docker (https://www.docker.com, accessed 23 November 2022) • Binder (https://www.mybinder.org, accessed 23 November 2022)
Make model code easy to understand	• Systems Biology Markup Language (SBML, https://synonym.caltech.edu, accessed 23 November 2022) • CellML (https://www.cellml.org, accessed 23 November 2022)

In our experience and that of Kim et al. [42], documentation of the model development process including all simulation inputs and algorithms is most easily accomplished using computational notebooks and version-control systems. The code included with this paper is written in Python using Jupyter notebooks, as explained in Section 2.2. The version-control system we utilized is called Git, which allows for users to save all versions of a program from its creation. This allows for easily stepping back through versions to see when a change was made or to determine when an error was introduced. Through using these tools in tandem, we have fully documented the development process of VeVaPy.

Using Git for version-control makes it simple to deposit model code in a public, version-controlled repository—GitHub is a website which allows for any Git repository to be uploaded to the Internet and (optionally) made public. This is the option that we have chosen to use, and VeVaPy can all be found at https://www.github.com/cparker-uc/VeVaPy (accessed 23 November 2022). However, there are also specific repositories for various types of models. Porubsky et al. [52] recommend BioModels (https://www.ebi.ac.uk/biomodels/, accessed 23 November 2022), a database for biological models that has a curation process which verifies whether uploaded models are reproducible.

Due to the difficulties in ensuring that future users have all requisite software installed in the correct versions, even when all code is included with a publication or deposited in an open repository it can prove difficult to run—especially when the code is not from the past couple of years. To ensure that model code can be run on as many computers as possible, Porubsky et al. [52], Sandve et al. [46], Waltemath & Wolkenhauer [41], and Rule et al. [49] suggest using either a virtual machine image or a web-based virtual machine such as Docker (https://www.docker.com, accessed 23 November 2022). Using a virtual machine, one can be sure that all necessary software, data, and model code will be available and able to run in any computing environment. We have instead opted for Binder (https://www.mybinder.org, accessed 23 November 2022) which allows users to run Jupyter notebooks in a web browser without needing to have the necessary software installed on their local machine.

The final suggestion we will discuss is making model code easy to understand. This can be achieved with thorough documentation, use of compartmentalization and functional programming, descriptive variable names, and using computational notebooks. However, for certain forms of model, there are standardized markup languages which are widely recommended in the literature, and which make model code easier for users to understand [33,41,48,50,52]. For systems biology, these include the systems biology markup language (SBML) and CellML. Unfortunately, as discussed in Medley et al. [38], these languages are limited in scope and do not support all forms of systems biology modeling. For instance, the models covered in this paper are currently unable to be adapted to SBML or CellML. However, we felt that it was important to mention these languages as they offer significant upsides for those models which they support.

VeVaPy followed biological modeling best practices as discussed above (Table 5), and this has made the five models reproduced for this research and our code, VeVaPy, very useful for many HPA axis modeling applications. To summarize, we have made the following efforts: the code for our VeVaPy package includes thorough documentation, including instructions for use; the demonstration models were implemented using Jupyter notebooks for improved readability and easier documentation; we have tracked the development process with the Git version control system and published it on GitHub, a freely accessible code repository; and we have provided instructions for accessing VeVaPy through Binder to facilitate its use on any Internet-connected computer.

We intend to follow up on the models used here for demonstration and to repurpose the model by Sriram et al. [23] to specifically match data that includes acute stressors (such as the TSST data used above). The most apparent modification to be made involves replacing the variable for stress input to a function for stress input that can change over the course of the simulation. This will allow for the introduction of an acute stressor (like a TSST) and then the cessation of the stressor afterwards. Another modification that we will test is the addition of delays between the release of ACTH and its action in the adrenal glands and between the production of cortisol and its feedback in the hypothalamus and pituitary. There are many possible routes that our modifications may take due to the many physiological processes not yet accounted for in the literature models (GRs in the hippocampus, fast vs. slow cortisol feedback, etc.). Using a model specifically designed to simulate stress tests will allow us to better understand the behavior of the HPA axis under acute stress—and the differences in this behavior between MDD, PTSD, and healthy control subjects.

Although we are hopeful that future published models will be more easily reproducible and include all data used by the authors for validation, we believe that the reproducibility problems discussed above can be eased to a degree by the development of robust tools for model V&V. We have begun the development of such tools in the field of HPA axis modeling. Although VeVaPy currently requires some programming knowledge to adapt beyond the five included models, we intend to further develop it into a graphical user interface for easily creating systems biology and systems pharmacology models. This will hopefully allow biologists with no experience in modeling to use our tools on many published mathematical models and improve the reach, and therefore the power to make an impact, of mathematical modeling in general. It is our firmly held belief that the development of tools to facilitate V&V of mathematical models, such as VeVaPy, will speed the pace of research and ensure that identification of valid models in the literature takes significantly less effort.

Supplementary Materials: The following supporting information can be downloaded at: https://www.mdpi.com/article/10.3390/e24121747/s1, Figure S1: Sriram Figure 3 Reproduction, Figure S2: Sriram Figure 4 Reproduction, Figure S3: Bangsgaard Figure 5 Reproduction, Figure S4: Somvanshi Figure 4 Reproduction, Figure S5: Malek Figure 3a Reproduction, Figure S6: Malek Figure 3b Reproduction, Figure S7: Bairagi Figure 2 Reproduction, Figure S8: Nelson TSST Data, Figure S9: Sriram et al. (2012) Nelson Data Comparisons, Figure S10: Bangsgaard & Ottesen (2017) Nelson Data Comparisons, Figure S11: Andersen, Vinther & Ottesen (2013) Nelson Data Comparisons, Figure

S12: Somvanshi et al. (2020) Nelson Data Comparisons, Figure S13: Malek et al. (2015) Nelson Data Comparisons, Figure S14: Bairagi et al. (2008) Nelson Data Comparisons, Table S1: Revised DSM-IV criteria for melancholic subtype [53], Table S2: Revised DSM-IV for atypical subtype [53], Table S3: Exclusion criteria [53], Table S4: PBK Model Reporting Template Completed for Model by Andersen et al. (2013), Table S5: PBK Model Reporting Template Completed for Model by Bairagi et al. (2008), Table S6: PBK Model Reporting Template Completed for Model by Bangsgaard & Ottesen (2017), Table S7: PBK Model Reporting Template Completed for Model by Malek et al. (2015), Table S8: PBK Model Reporting Template Completed for Model by Somvanshi et al. (2020), Table S9: PBK Model Reporting Template Completed for Model by Sriram et al. (2012), Table S10: TSST ACTH Data, Table S11: TSST CORT Data. References [22–27,29,53–59] are cited in the supplementary materials.

Author Contributions: Conceptualization: T.Z.; Data Curation: E.N.; Formal Analysis: C.P.; Funding Acquisition: T.Z. and E.N.; Investigation: C.P.; Methodology: C.P.; Project Administration: T.Z.; Resources: T.Z. and E.N.; Software: C.P.; Supervision: T.Z.; Validation: C.P.; Visualization: C.P.; Writing—Original Draft: C.P.; Writing—Review and Editing: C.P., T.Z. and E.N. All authors have read and agreed to the published version of the manuscript.

Funding: The work was funded by the United States Army grant number 1016183 ARMY W911NF-20-1-0192. and the National Institutes of Mental Health grant number NIMH K23MH067705. The APC was funded by the United States Army grant number 1016183 ARMY W911NF-20-1-0192.

Institutional Review Board Statement: The study was conducted in accordance with the Declaration of Helsinki, and approved by the Institutional Review Board of the University of Cincinnati (Protocol code #02-10-9-1, Approved 15 November 2006) and the Institutional Review Board of the Cincinnati Children's Hospital Medical Center (Protocol code #04-12-48, Approved 29 October 2007).

Informed Consent Statement: Informed consent was obtained from all subjects involved in the study. Written informed consent has been obtained from the patients to publish this paper.

Data Availability Statement: All data used in the study has been made available in the Supplementary Materials.

Conflicts of Interest: The authors declare no conflict of interest.

References

1. Thacker, B.H.; Doebling, S.W.; Hemez, F.M.; Anderson, M.C.; Pepin, J.E.; Rodriguez, E.A. *Concepts of Model Verification and Validation*; Schaller, C., Ed.; Report No.: LA-14167-MS; US Department of Energy, Los Alamos National Laboratory: Los Alamos, NM, USA, 2004.
2. IWGFR. *Specialists' Meeting on Predictions and Experience of Core Distortion Behaviour 1–4 October 1984*; International Atomic Energy Agency: Wilmslow, UK, 1984.
3. Nakagawa, M. Verification and Validation of Core Mechanical Performance Code ARKAS with IAEA Benchmark Problems, (II). *J. Nucl. Sci. Technol.* **1993**, *30*, 389–412. [CrossRef]
4. Hohne, T.; Kliem, S.; Bieder, U. IAEA CRP benchmark of ROCOM PTS test case for the use of CFD in reactor design using the CFD-Codes ANSYS CFX and TrioCFD. *Nucl. Eng. Des.* **2018**, *333*, 161–180. [CrossRef]
5. Schwer, L.E. An overview of the PTC 60/V&V 10: Guide for verification and validation in computational solid mechanics. *Eng. Comput.* **2007**, *23*, 245–252. [CrossRef]
6. Musuamba, F.T.; Bursi, R.; Manolis, E.; Karlsson, K.; Kulesza, A.; Courcelles, E.; Boissel, J.; Lesage, R.; Crozatier, C.; Voisin, E.M.; et al. Verifying and Validating Quantitative Systems Pharmacology and In Silico Models in Drug Development: Current Needs, Gaps, and Challenges. *CPT Pharmacomet. Syst. Pharmacol.* **2020**, *9*, 195–197. [CrossRef] [PubMed]
7. Musuamba, F.T.; Skottheim Rusten, I.; Lesage, R.; Russo, G.; Bursi, R.; Emili, L.; Wangorsch, G.; Manolis, E.; Karlsson, K.E.; Kulesza, A.; et al. Scientific and regulatory evaluation of mechanistic in silico drug and disease models in drug development: Building model credibility. *CPT Pharmacomet. Syst. Pharmacol.* **2021**, *10*, 804–825. [CrossRef] [PubMed]
8. Hicks, J.L.; Uchida, T.K.; Seth, A.; Rajagopal, A.; Delp, S.L. Is my model good enough? Best practices for verification and validation of musculoskeletal models and simulations of movement. *J. Biomech. Eng.* **2015**, *137*, 020905. [CrossRef] [PubMed]
9. Belmaker, R.H.; Agam, G. Major depressive disorder. *N. Engl. J. Med.* **2008**, *358*, 55–68. [CrossRef] [PubMed]
10. Gili, M.; Roca, M.; Armengol, S.; Asensio, D.; Garcia-Campayo, J.; Parker, G. Clinical patterns and treatment outcome in patients with melancholic, atypical and non-melancholic depressions. *PLoS ONE* **2012**, *7*, e48200. [CrossRef]
11. Fink, M.; Taylor, M.A. Resurrecting melancholia. *Acta Psychiatr. Scand. Suppl.* **2007**, *115*, 14–20. [CrossRef]
12. Smith, S.M.; Vale, W.W. The role of the hypothalamic-pituitary-adrenal axis in neuroendocrine responses to stress. *Dialogues Clin. Neurosci.* **2006**, *8*, 383–395. [CrossRef]

13. Hosseinichimeh, N.; Rahmandad, H.; Wittenborn, A.K. Modeling the hypothalamus-pituitary-adrenal axis: A review and extension. *Math. Biosci.* **2015**, *268*, 52–65. [CrossRef] [PubMed]
14. Oster, H.; Challet, E.; Ott, V.; Arvat, E.; de Kloet, E.R.; Dijk, D.J.; Lightman, S.; Vgontzas, A.; Van Cauter, E. The Functional and Clinical Significance of the 24-Hour Rhythm of Circulating Glucocorticoids. *Endocr. Rev.* **2017**, *38*, 3–45. [CrossRef] [PubMed]
15. Yehuda, R.; Teicher, M.H.; Trestman, R.L.; Levengood, R.A.; Siever, L.J. Cortisol regulation in posttraumatic stress disorder and major depression: A chronobiological analysis. *Biol. Psychiatry* **1996**, *40*, 79–88. [CrossRef] [PubMed]
16. Holsboer, F. The corticosteroid receptor hypothesis of depression. *Neuropsychopharmacology* **2000**, *23*, 477–501. [CrossRef] [PubMed]
17. Young, E.A.; Lopez, J.F.; Murphy-Weinberg, V.; Watson, S.J.; Akil, H. The role of mineralocorticoid receptors in hypothalamic-pituitary-adrenal axis regulation in humans. *J. Clin. Endocrinol. Metab.* **1998**, *83*, 3339–3345. [CrossRef]
18. Goodwin, B.C. Oscillatory behavior in enzymatic control processes. *Adv. Enzym. Regul.* **1965**, *3*, 425–438. [CrossRef]
19. Veldhuis, J.D.; Iranmanesh, A.; Lizarralde, G.; Johnson, M.L. Amplitude modulation of a burstlike mode of cortisol secretion subserves the circadian glucocorticoid rhythm. *Am. J. Physiol.* **1989**, *257 Pt 1*, E6–E14. [CrossRef]
20. Gonzalez-Heydrich, J.; Steingard, R.J.; Kohane, I. A computer simulation of the hypothalamic- pituitary-adrenal axis. *Proc. Annu. Symp. Comput. Appl. Med. Care* **1994**, 1010.
21. Liu, Y.W.; Hu, Z.H.; Peng, J.H.; Liu, B.Z. A Dynamical Model for the Pulsatile Secretion of the Hypothalamo-Pituitary-Adrenal Axis. *Math. Comput. Model.* **1998**, *29*, 103–110.
22. Bairagi, N.; Chatterjee, S.; Chattopadhyay, J. Variability in the secretion of corticotropin-releasing hormone, adrenocorticotropic hormone and cortisol and understandability of the hypothalamic-pituitary-adrenal axis dynamics—A mathematical study based on clinical evidence. *Math. Med. Biol.* **2008**, *25*, 37–63. [CrossRef]
23. Sriram, K.; Rodriguez-Fernandez, M.; Doyle, F.J. Modeling cortisol dynamics in the neuro-endocrine axis distinguishes normal, depression, and post-traumatic stress disorder (PTSD) in humans. *PLoS Comput. Biol.* **2012**, *8*, e1002379. [CrossRef] [PubMed]
24. Andersen, M.; Vinther, F.; Ottesen, J.T. Mathematical modeling of the hypothalamic-pituitary- adrenal gland (HPA) axis, including hippocampal mechanisms. *Math. Biosci.* **2013**, *246*, 122–138. [CrossRef] [PubMed]
25. Malek, H.; Ebadzadeh, M.M.; Safabakhsh, R.; Razavi, A.; Zaringhalam, J. Dynamics of the HPA axis and inflammatory cytokines: Insights from mathematical modeling. *Comput. Biol. Med.* **2015**, *67*, 1–12. [CrossRef] [PubMed]
26. Bangsgaard, E.O.; Ottesen, J.T. Patient specific modeling of the HPA axis related to clinical diagnosis of depression. *Math. Biosci.* **2017**, *287*, 24–35. [CrossRef] [PubMed]
27. Somvanshi, P.R.; Mellon, S.H.; Yehuda, R.; Flory, J.D.; Makotkine, I.; Bierer, L.; Marmar, C.; Jett, M.; Doyle, F.J. Role of enhanced glucocorticoid receptor sensitivity in inflammation in PTSD: Insights from computational model for circadian-neuroendocrine-immune interactions. *Am. J. Physiol. Endocrinol. Metab.* **2020**, *319*, E48–E66. [CrossRef]
28. Bremner, D.; Vermetten, E.; Kelley, M.E. Cortisol, dehydroepiandrosterone, and estradiol measured over 24 hours in women with childhood sexual abuse-related posttraumatic stress disorder. *J. Nerv. Ment. Dis.* **2007**, *195*, 919–927. [CrossRef]
29. Carroll, B.J.; Cassidy, F.; Naftolowitz, D.; Tatham, N.E.; Wilson, W.H.; Iranmanesh, A.; Liu, P.Y.; Veldhuis, J.D. Pathophysiology of hypercortisolism in depression. *Acta Psychiatry Scand. Suppl.* **2007**, *115*, 90–103. [CrossRef]
30. Golier, J.A.; Schmeidler, J.; Legge, J.; Yehuda, R. Twenty-four hour plasma cortisol and adrenocorticotropic hormone in Gulf War veterans: Relationships to posttraumatic stress disorder and health symptoms. *Biol. Psychiatry* **2007**, *62*, 1175–1178. [CrossRef]
31. Gupta, S.; Aslakson, E.; Gurbaxani, B.M.; Vernon, S.D. Inclusion of the glucocorticoid receptor in a hypothalamic pituitary adrenal axis model reveals bistability. *Biol. Med. Model.* **2007**, *4*, 8. [CrossRef] [PubMed]
32. OECD. *Guidance Document on the Characterisation, Validation and Reporting of Physiologically Based Kinetic (PBK) Models for Regulatory Purposes*; OECD: Paris, France, 2021.
33. Cucurull-Sanchez, L.; Chappell, M.J.; Chelliah, V.; Amy Cheung, S.Y.; Derks, G.; Penney, M.; Phipps, A.; Sheriff, R.S.M.; Timmis, J.; Tindall, M.J.; et al. Best Practices to Maximize the Use and Reuse of Quantitative and Systems Pharmacology Models: Recommendations From the United Kingdom Quantitative and Systems Pharmacology Network. *CPT Pharmacomet. Syst. Pharmacol.* **2019**, *8*, 259–272. [CrossRef]
34. de Pillis, L.G.; Radunskaya, A.E. Best practices in mathematical modeling. *Methods Mol. Biol.* **2012**, *929*, 51–74. [CrossRef] [PubMed]
35. Le Novère, N.; Finney, A.; Hucka, M.; Bhalla, U.S.; Campagne, F.; Collado-Vides, J.; Crampin, E.J.; Halstead, M.; Klipp, E.; Mendes, P.; et al. Minimum information requested in the annotation of biochemical models (MIRIAM). *Nat. Biotechnol.* **2005**, *23*, 1509–1515. [CrossRef] [PubMed]
36. Collins, F.S.; Tabak, L.A. Policy: NIH plans to enhance reproducibility. *Nature* **2014**, *505*, 612–613. [CrossRef] [PubMed]
37. Stodden, V.; McNutt, M.; Bailey, D.H.; Deelman, E.; Gil, Y.; Hanson, B.; Heroux, M.A.; Ioannidis, J.P.; Taufer, M. Enhancing reproducibility for computational methods. *Science* **2016**, *354*, 1240–1241. [CrossRef] [PubMed]
38. Medley, J.K.; Goldberg, A.P.; Karr, J.R. Guidelines for Reproducibly Building and Simulating Systems Biology Models. *IEEE Trans. Biomed. Eng.* **2016**, *63*, 2015–2020. [CrossRef]
39. Stodden, V.; Seiler, J.; Ma, Z. An empirical analysis of journal policy effectiveness for computational reproducibility. *Proc. Natl. Acad. Sci. USA* **2018**, *115*, 2584–2589. [CrossRef]
40. Baker, M. 1500 scientists lift the lid on reproducibility. *Nature* **2016**, *533*, 452–454. [CrossRef]

41. Waltemath, D.; Wolkenhauer, O. How Modeling Standards, Software, and Initiatives Support Reproducibility in Systems Biology and Systems Medicine. *IEEE Trans. Biomed. Eng.* **2016**, *63*, 1999–2006. [CrossRef]
42. Kim, Y.M.; Poline, J.B.; Dumas, G. Experimenting with reproducibility: A case study of robustness in bioinformatics. *Gigascience* **2018**, *7*, giy077. [CrossRef]
43. Kirouac, D.C.; Cicali, B.; Schmidt, S. Reproducibility of Quantitative Systems Pharmacology Models: Current Challenges and Future Opportunities. *CPT Pharmacomet. Syst. Pharmacol.* **2019**, *8*, 205–210. [CrossRef] [PubMed]
44. Miłkowski, M.; Hensel, W.M.; Hohol, M. Replicability or reproducibility? On the replication crisis in computational neuroscience and sharing only relevant detail. *J. Comput. Neurosci.* **2018**, *45*, 163–172. [CrossRef] [PubMed]
45. Munafò, M.R.; Nosek, B.A.; Bishop, D.V.M.; Button, K.S.; Chambers, C.D.; du Sert, N.P.; Simonsohn, U.; Wagenmakers, E.-J.; Ware, J.J.; Ioannidis, J.P.A. A manifesto for reproducible science. *Nat. Hum. Behav.* **2017**, *1*, 0021. [CrossRef] [PubMed]
46. Sandve, G.K.; Nekrutenko, A.; Taylor, J.; Hovig, E. Ten simple rules for reproducible computational research. *PLoS Comput. Biol.* **2013**, *9*, e1003285. [CrossRef] [PubMed]
47. Piccolo, S.R.; Frampton, M.B. Tools and techniques for computational reproducibility. *Gigascience* **2016**, *5*, 30. [CrossRef] [PubMed]
48. Mendes, P. Reproducible Research Using Biomodels. *Bull. Math. Biol.* **2018**, *80*, 3081–3087. [CrossRef]
49. Rule, A.; Birmingham, A.; Zuniga, C.; Altintas, I.; Huang, S.C.; Knight, R.; Moshiri, N.; Nguyen, M.H.; Rosenthal, S.B.; Pérez, F.; et al. Ten simple rules for writing and sharing computational analyses in Jupyter Notebooks. *PLoS Comput. Biol.* **2019**, *15*, e1007007. [CrossRef] [PubMed]
50. Schnell, S. "Reproducible" Research in Mathematical Sciences Requires Changes in our Peer Review Culture and Modernization of our Current Publication Approach. *Bull. Math. Biol.* **2018**, *80*, 3095–3105. [CrossRef]
51. Tiwari, K.; Kananathan, S.; Roberts, M.G.; Meyer, J.P.; Sharif Shohan, M.U.; Xavier, A.; Maire, M.; Zyoud, A.; Men, J.; Ng, S.; et al. Reproducibility in systems biology modelling. *Mol. Syst. Biol.* **2021**, *17*, e9982. [CrossRef] [PubMed]
52. Porubsky, V.L.; Goldberg, A.P.; Rampadarath, A.K.; Nickerson, D.P.; Karr, J.R.; Sauro, H.M. Best Practices for Making Reproducible Biochemical Models. *Cell Syst.* **2020**, *11*, 109–120. [CrossRef] [PubMed]
53. Nelson, E.B.; Strakowski, S.M.; Geracioti, T.; Sah, R.; McNamara, R. *Hypothalamic-Pituitary-Adrenal Axis Dysregulation and Phenomenology of Major Depression*; University of Cincinnati: Cincinnati, OH, USA, 2008.
54. Jelic, S.; Cupic, Z.; Kolar-Anic, L. Mathematical modeling of the hypothalamic-pituitary-adrenal system activity. *Math. Biosci.* **2005**, *197*, 173–187.
55. Day, J.; Rubin, J.; Vodovotz, Y.; Chow, C.C.; Reynolds, A.; Clermont, G. A reduced mathematical model of the acute inflammatory response II. Capturing scenarios of repeated endotoxin administration. *J. Theor. Biol.* **2006**, *242*, 237–256. [CrossRef]
56. Oliver, J.C.; Bland, L.A.; Oettinger, C.W.; Arduino, M.J.; McAllister, S.K.; Aguero, S.M.; Favero, M.S. Cytokine kinetics in an in vitro whole blood model following an endotoxin challenge. *Lymphokine Cytokine Res.* **1993**, *12*, 115–120.
57. Weigert, C.; Düfer, M.; Simon, P.; Debre, E.; Runge, H.; Brodbeck, K.; Häring, H.U.; Schleicher, E.D. Upregulation of IL-6 mRNA by IL-6 in skeletal muscle cells: Role of IL-6 mRNA stabilization and Ca^{2+}-dependent mechanisms. *Am. J. Physiol.–Cell Physiol.* **2007**, *293*, C1139–C1147.
58. Bangsgaard, E.O.; Hjorth, P.G.; Olufsen, M.S.; Mehlsen, J.; Ottesen, J.T. Integrated Inflammatory Stress (ITIS) Model. *Bull. Math. Biol.* **2017**, *79*, 1487–1509.
59. Rao, R.; DuBois, D.; Almon, R.; Jusko, W.J.; Androulakis, I.P. Mathematical modeling of the circadian dynamics of the neuroendocrine- immune network in experimentally induced arthritis. *Am. J. Physiol. Endocrinol. Metab.* **2016**, *311*, E310–E324.

MDPI
St. Alban-Anlage 66
4052 Basel
Switzerland
www.mdpi.com

Entropy Editorial Office
E-mail: entropy@mdpi.com
www.mdpi.com/journal/entropy

www.ingramcontent.com/pod-product-compliance
Lightning Source LLC
Chambersburg PA
CBHW040841250926
51850CB00007BA/1371